AF803156

THERAPEUTIC LANDSCAPES

Geographies of Health

Series Editors

Allison Williams, Associate Professor, School of Geography and Earth Sciences, McMaster University, Canada
Susan Elliott, Dean of the Faculty of Social Sciences, McMaster University, Canada

There is growing interest in the geographies of health and a continued interest in what has more traditionally been labeled medical geography. The traditional focus of 'medical geography' on areas such as disease ecology, health service provision and disease mapping (all of which continue to reflect a mainly quantitative approach to inquiry) has evolved to a focus on a broader, theoretically informed epistemology of health geographies in an expanded international reach. As a result, we now find this subdiscipline characterized by a strongly theoretically-informed research agenda, embracing a range of methods (quantitative; qualitative and the integration of the two) of inquiry concerned with questions of: risk; representation and meaning; inequality and power; culture and difference, among others. Health mapping and modeling, has simultaneously been strengthened by the technical advances made in multilevel modeling, advanced spatial analytic methods and GIS, while further engaging in questions related to health inequalities, population health and environmental degradation.

This series publishes superior quality research monographs and edited collections representing contemporary applications in the field; this encompasses original research as well as advances in methods, techniques and theories. The *Geographies of Health* series will capture the interest of a broad body of scholars, within the social sciences, the health sciences and beyond.

Forthcoming titles in the series

Sense of Place: Relationships between Place and Health
Edited by John Eyles and Allison Williams
ISBN 978 0 7546 7332 3

Primary Health Care
People, Practice, Place
Edited by Valorie A. Crooks and Gavin J. Andrews
ISBN 978 0 7546 7247 0

Space, Risk and Mental Health
Sarah Curtis
ISBN 978 0 7546 7331 6

Therapeutic Landscapes

Edited by
ALLISON WILLIAMS
McMaster University, Canada

Routledge
Taylor & Francis Group
LONDON AND NEW YORK

First published 2007 by Ashgate Publishing

Published 2016 by Routledge
2 Park Square, Milton Park, Abingdon, Oxon OX14 4RN
605 Third Avenue, New York, NY 10017

Routledge is an imprint of the Taylor & Francis Group, an informa business

Copyright © Allison Williams 2007

Allison Williams has asserted her right under the Copyright, Designs and Patents Act, 1988, to be identified as the editor of this work.

All rights reserved. No part of this book may be reprinted or reproduced or utilised in any form or by any electronic, mechanical, or other means, now known or hereafter invented, including photocopying and recording, or in any information storage or retrieval system, without permission in writing from the publishers.

Notice:
Product or corporate names may be trademarks or registered trademarks, and are used only for identification and explanation without intent to infringe.

British Library Cataloguing in Publication Data
Therapeutic landscapes
1. Medical geography 2. Landscape - Therapeutic use
3. Landscape - Health aspects 4. Environmental health
I. Williams, Allison, 1965-
615.5

Library of Congress Cataloging-in-Publication Data
Therapeutic landscapes / edited by Allison Williams.
p. cm.
Includes bibliographical references and index.
ISBN-13: 978-0-7546-7099-5
ISBN-10: 0-7546-7099-6
1. Landscape--Therapeutic use. 2. Landscape--Health aspects. 3. Medical geography. 4. Environmental health. I. Williams, Allison, 1965-
[DNLM: 1. Geography. 2. Therapeutics. 3. Environment Design. 4. Holistic Health. WB 700 T398 2007]

RA792.T455 2007
613--dc22

2007020564

ISBN 13: 978-0-7546-7099-5 (hbk)

Contents

PART 5 TRANSCENDING GEOGRAPHY: APPLICATIONS IN THE ANTHROPOLOGY OF HEALTH

List of Figures

List of Tables

List of Contributors

Gavin Andrews, PhD, Professor and Chair Department of Health, Ageing, and Society, McMaster University, Canada

Gavin's research has two central strands of inquiry. Geographies in clinical practice provide insights into the nature of everyday health care education and work. Meanwhile, geographical gerontology offers a spatial perspective to aging and older people's care. Elsewhere, he has branched out and researched historical geographies of health and geographies of sport and fitness. To date, Gavin has published over 75 journal articles and contributions to edited collections. He has also co-edited two books *Ageing and Place: Perspectives, Policy, Practice* and *The Sociology of Ageing: A Reader*. Gavin is the North American editor of the Blackwell periodical *International Journal of Older People Nursing*.

Matthias Braubach, EMPH, MSc, Technical Officer, Housing and Health WHO European Centre for Environment and Health

Matthias Braubach is the head of the Housing and Health unit at the World Health Organization European Centre for Environment and Health. He earned an MSc in environmental geography and urban planning at the University of Bonn, and a European Master of Public Health at the University of Bielefeld. With the WHO housing and health programme focusing on a number of housing issues, his main areas of work are the health impact of the urban environment, the quality of homes and indoor spaces, and the challenge of healthy housing for aging populations. He has also written several publications on housing and health issues and served on the scientific committee of various international conferences.

Damian Collins, PhD, School of Geography, Geology and Environmental Science, University of Auckland

Damian Collins is a Postdoctoral Fellow in the School of Geography, Geology and Environmental Science at the University of Auckland, New Zealand. His research interests centre on the relationship between health and place, and on the geographical dimensions of social and political conflict. Children and young people are a focus for both types of inquiry. Damian is currently working on a project which explores the changing nature of property on the New Zealand coast, where modest holiday homes and campgrounds are increasingly giving way to sumptuous, privatized residential landscapes.

Peter Collins, PhD, Senior Lecturer in Anthropology, Department of Anthropology, University of Durham, UK

Peter Collins teaches in the Department of Anthropology at the University of Durham. His main interests are in aesthetics, space and place, religion and qualitative methodology. He has recently co-edited *Locating the Field: Space Place and Context in Anthropology* (Berg, 2006), with Simon Coleman; *Reading Religion in Text and Context: Reflections of Faith and Practice in Religious Materials* (Ashgate, 2006), with Elisabeth Arweck; and *Religion, Identity and Change: Perspectives on Global Transformations* (Ashgate, 2004), with Simon Coleman. His *Continuity and Change: Bolton Quakers 1650-1990* (Edwin Mellen) is in press, as is *Keeping an Open 'I': Memory and Experience as Resources in Ethnography* (co-edited with Anselma Gallinat). He is currently working on the *Concise Oxford Dictionary of Anthropology* (Oxford University Press).

David Conradson, PhD, School of Geography, University of Southampton, UK

David Conradson is Senior Lecturer in Human Geography at the University of Southampton, UK. His research has explored the intersections between organizational space, subjectivity and wellbeing, with recent projects on the dynamics of both community drop-in centres and respite care environments. His work on these issues has been published in Environment and Planning A, Social and Cultural Geography, Health and Place and the volume *Emotional Geographies* (Ashgate, 2005). He has edited a special issue of *Social and Cultural Geography* on *Geographies of Care* (2003, 4:4) and is co-editor of *Landscapes of Voluntarism: New Spaces of Health, Welfare and Governance* (Policy Press, 2006).

Valorie Crooks, PhD, Assistant Professor, Department of Geography, Simon Fraser University, Canada

Valorie Crooks is an Assistant Professor in the Department of Geography at Simon Fraser University (British Columbia, Canada) specializing in health geography. Her areas of specialty include critically examining spaces of care provision, doctor-patient interactions, and illness experiences. She also conducts research regarding health-related social policies.

Sarah Curtis, DPhil, Professor of Health and Risk, University of Durham, UK

Sarah Curtis is a Professor of Health and Risk at the University of Durham, UK. She has extensive research experience in geography of health and health care. Her research focuses especially on why and how places are important for inequalities of health and wellbeing and access to health care. Her work also includes needs assessment in primary and community health care, and health impact assessment of regeneration schemes. Recently published research has been funded by Research Councils, major charities and health services agencies on topics including: therapeutic aspects of design of mental health facilities; the relationship of social and physical environments to variation in health and well-being of adolescents and variation in use of mental health services. Her books include *Health and Societies* (Arnold, 1996), co-authored with Ann Taket and *Health and Inequality: Geographical Perspectives* (Sage, 2004). Other experience includes: research, consultancy for UK Health Authorities, the Health Development Agency; the World Health Organization and she has served as non-executive director of an NHS Community and Mental Health Care Trust. She is Senior Editor, Medical Geography for the international journal *Social Science and Medicine*.

Malcolm P. Cutchin, PhD, Associate Professor, Occupational Science, University of North Carolina at Chapel Hill, USA

Malcolm Cutchin received his PhD in geography and holds an appointment as Associate Professor in the Division of Occupational Science, School of Medicine at UNC-Chapel Hill. He is also a Senior Research Scientist at the UNC Institute on Aging. His research interests include: social gerontology, especially processes of aging-in-place; health geography; and the philosophy of John Dewey. He has been funded by the National Institute on Aging and the National Cancer Institute for health geography projects, and he has published in journals such as *Social Science and Medicine*, *Health and Place*, *Journal of Gerontology: Social Sciences*, *Progress in Human Geography*, and the *Journal of Occupational Science*.

Joyce Davidson, PhD, Assistant Professor, Department of Geography, Queen's University, Canada

Joyce Davidson is Assistant Professor of Geography, cross-appointed with Women's Studies, at Queen's University, Ontario. Following the publication of *Phobic Geographies* (Ashgate, 2003), she has developed a research and teaching program focused around health, embodiment and emotion; current research examines virtual reality therapies for autism spectrum and anxiety disorders. Organizer of the First and Second (with Laura Cameron) Interdisciplinary Conference on Emotional Geographies (Lancaster, 2002, Queen's, 2006), she has co-edited special issues on this subject for *Gender, Place and Culture* (with Liz Bondi, 2004, 11:3) and *Social and Cultural Geography* (with Christine Milligan, 2004, 5:4). She has also co-edited the *Emotional Geographies* book (with Liz Bondi and Mick Smith, Ashgate, 2005). Davidson has published in sociology and philosophy as well as geography journals, and is co-author of *Subjectivities, Knowledges and Feminist Geographies* (Rowman and Littlefield, 2002).

Jody F. Decker, PhD, Geography and Environmental Studies, Wilfrid Laurier University, Canada

Jody Decker combined a career as a Registered Nurse with a doctoral degree (York University) and postgraduate fellowship (Queen's University) in geography, to work in the sub-disciplines of medical and cultural/historical geography. Her work has focused on the demographic impact of Indigenous groups in North America from diseases, contemporary impacts from infectious diseases on selected population groups, health impact assessment studies in cross-cultural groups, and gender and landscape studies. She is presently in the Department of Geography and Environmental Studies at Wilfrid Laurier University in Waterloo, Ontario, Canada.

Geoffrey DeVerteuil, PhD, School of Geography, University of Southampton, UK
Dr DeVerteuil is a Lecturer in the School of Geography at the University of Southampton. His academic background includes a BA in geography from McGill University (1991), an MA in Community and Regional Planning at the University of British Columbia (1993), and a PhD in Geography from the University of Southern California (2001). His research interests have consistently focused on vulnerable populations, such as the homeless, as well as the role of the state in managing the consequences of extreme inequality, as manifested through welfare state restructuring. More recently, he has extended these interests into the realm of health geography, including examining the shifting geographies of mental health care in the post-asylum period, and the challenges in sustaining therapeutic landscapes within marginalized spaces of the city.

Rhonda Donovan, MA, School of Geography and Earth Sciences, McMaster University, Canada
Rhonda Donovan is a graduate student at McMaster University. Her research interests are based in health and social geographies and include rural and women's health issues. Her graduate research involves examining the impact of home-based palliative care on family caregiver health and the home environment, with the aim of informing policy and planning specific to end-of-life care.

Dawn Einwalter, PhD, Planner and Project Manager, EDAW
Dawn Einwalter is a planner and project manager with EDAW, an international urban design firm. Her work focuses on urban revitalization, and has included the *Modesto Redevelopment Master Plan*, the *North Watt Corridor Plan*, and the *Reno Open Space and Greenways Master Plan*. Einwalter has been a resident of Nevada for over 10 years, where she focused on projects in rural Nevada as a planner with Lumos Engineering. She has carried out research in urban China and the western United States, where her investigations have emphasized the negotiation of behavior and meaning in public space.

Joshua Evans, PhD Candidate, School of Geography and Earth Sciences, McMaster University, Canada
Joshua Evans is a graduate student at McMaster University and a Doctoral Fellow in the Canadian Institute of Health Research Strategic Training Program in Health Care, Technology, and Place. His work addresses geographies of social welfare, disability and mental health. He is currently completing his doctoral thesis, which focuses on the everyday experiences of the shelter and social services system in Hamilton, Ontario.

Wilbert Gesler, PhD, Queen Mary College, University of London, UK
Dr Gesler received his PhD in Geography in 1978 from the University of North Carolina in Chapel Hill. From 1978 to 1981 he was a member of the Department of

Geography, Rutgers University; from 1981 to 1982 he was at the University of Sierra Leone, Freetown on a Fulbright scholarship; and from 1982 until retirement in 2003, he was in the Department of Geography at the University of North Carolina at Chapel Hill. Currently he is Research Professor in his Chapel Hill department and Honorary Visiting Professor at Queen Mary, University of London. Dr Gesler's teaching and research focus on health care delivery, spatial analysis, cultural geography, both quantitative and qualitative methods, and Africa. Beginning in the early 1990s, he developed the concept of therapeutic landscapes. He first applied the idea to three places with lasting reputations for healing (Epidauros in Greece, Bath in England, and Lourdes in France), and currently is using the idea to examine hospital design. He is the author, co-author, or co-editor of 11 books, 20 book chapters and over 60 refereed articles in more than two dozen different journals.

Bonnie Hallman, PhD, Associate Professor, Department of Environment and Geography, University of Manitoba, Canada

Bonnie C. Hallman is an Associate Professor in the Department of Environment and Geography, Research Affiliate with the Centre on Aging and Affiliate with the Arthur V. Mauro Centre for Peace and Justice at the University of Manitoba in Winnipeg, Canada. She joined the Environment and Geography Department at the University of Manitoba in 2001, having previously taught at California State University, Chico (1997–2001). Hallman's principal teaching and research interests are in the broad areas of social/cultural, gender, health and rural geographies. Early work focused on the geographies of family care-giving, particularly the resources and coping strategies of caregivers to elderly relatives. Three current research projects continue her longstanding interest in the everyday geographies of families and in the critical analysis of the spaces and places that structure and influence the resources and behaviors of families. These include: 1) exploring the potential for zoos as environments that can promote healthy, active living, particularly amongst older adults; 2) exploring the zoo as a landscape where the nature/family connection is inscribed into the zoo experience; and, 3) critiquing the historical creation and use of department store restaurants as a reflection of the rise of consumer capitalism and prevailing gender norms in order to understand current users of department store restaurants (with a particular focus on older adults). Last but far from least, Hallman is an active participant in the everyday, mundane geography of 'hockey goalie mom' and 'hockey referee spouse'.

Brian A. Hoey, PhD, Associate Professor, Department of Sociology and Anthropology, Marshall University, USA

Brian A. Hoey received his BA in human ecology from the College of the Atlantic (1990) and PhD in anthropology from the University of Michigan (2002). Recently a postdoctoral fellow at the Center for the Ethnography of Everyday Life in the Institute for Social Research, he is now an associate professor at Marshall University. As a Fulbright Scholar to Indonesia, Hoey studied the contested nature of constructing personally and culturally meaningful space within the process creating imagined and intentional community. He worked in far-flung agrarian settlements built from the ground up as part of a government-sponsored migration program. His dissertation research in Northwest

Lower Michigan explored non-economic migration where downsized and downshifting corporate workers relocate as a means of starting over in geographic places they believe provide necessary refuge to rethink work, family and personal obligations. A continuing interest in career change, personal identity and the moral meanings of work informs his ongoing research on *New Work*, unconventional arrangements of work, family and community life explored by so-called free-agents of the post-industrial economy. Hoey has a longstanding interest in the anthropology of place and, in particular, therapeutic ideals attached to particular places – including the case of purposive communities such as the moral treatment asylum. Together, his projects address a number of themes including personhood and place, migration, narrative identity and life-transition, community building, and negotiations between work, family, and self in different social, historical and environmental contexts. He has published on these and other subjects in the *American Ethnologist*, the *Journal of Anthropological Research*, the *Journal for Contemporary Ethnography* and *Ethnology*.

Dave Holmes, RN, PhD, Associate Professor, School of Nursing, University of Ottawa, Canada

Dave Holmes is Associate Professor at the University of Ottawa, Faculty of Health Sciences, and School of Nursing and at the Faculty of Medicine, Department of Psychiatry. He is also affiliated researcher at the University of Ottawa, Institute of Mental Health Research.

Anne-Cécile Hoyez, PhD, Associate Researcher, Département de Géographie, Université de Rouen, France

Anne-Cécile Hoyez' thesis, entitled 'The space-world of yoga: A social and cultural geography of globalisation of therapeutic landscapes', examines the process of globalization of social and cultural practices through the example of modern yoga, an Indian practice transplanted outside of India, in new places and spaces. She continues her research and teachings in France, in the geography department of the University of Rouen and Rennes. She is also a research associate in the Department of Social Sciences of the French Institute of Pondicherry (India). Her ongoing projects in cultural and social geography deal with the globalization of Indian medicine and the emergence of complementary medicine.

Robin Kearns, PhD, School of Geography, Geology and Environmental Science, University of Auckland, New Zealand

Robin Kearns, PhD (McMaster) is Professor of Geography at the University of Auckland. He has published *Putting Health into Place: Landscape, Identity and Well-being* (Syracuse University Press) and *Culture/Place/Health* (Routledge) (both with Wilbert Gesler), as well as numerous refereed articles and book chapters. His research interests include investigating the relationships between health and place, the cultural politics of place; and the social relations of research. Robin is an editor of three journals: the *New Zealand Geographer*, *Health and Place* and *Health and Social Care in the Community*.

Shaun Klassen, MA Candidate, Department of City Planning, University of Manitoba, Canada

Shaun is currently a graduate student in the Department of City Planning at the University of Manitoba. His academic background includes an undergraduate degree in Architecture from the University of Manitoba (2002). He is currently examining the social and material environments associated with formal treatment programs for substance abuse in his home city of Winnipeg.

Setha M. Low, PhD, Environmental Psychology, City University of New York, USA

Setha Low received her PhD in Cultural Anthropology from the University of California, Berkeley. She started her career as an Assistant and Associate Professor of Landscape Architecture and Regional Planning, City and Regional Planning, and Anthropology at the University of Pennsylvania. Dr Low is currently Professor of Environmental Psychology, Anthropology, and Women's Studies, and Director of the Public Space Research Group at The Graduate Center, City University of New York where she teach courses and trains PhD students in the anthropology of space and place, urban anthropology, culture and environment, and cultural values in historic preservation. She has been awarded a Getty Fellowship, a NEH fellowship, and a Guggenheim for her ethnographic research on public space in Latin America and the United States. She is widely published and lectures internationally on these issues. Her most recent books include: *Politics of Public Space* (2006 Routledge with Neil Smith, *Rethinking Urban Parks: Public Space and Cultural Diversity* (2005, University of Texas Press with S. Scheld and D. Taplin), *Behind the Gates: Life, Security and the Pursuit of Happiness in Fortress America* (2003, Routledge), *The Anthropology of Space and Place: Locating Culture* (2003, Blackwell with D. Lawrence-Zuniga), *On the Plaza: The Politics of Public Space and Culture* (2000, University of Texas), *Theorizing the City: The New Urban Anthropology Reader* (1999, Rutgers University Press), *Place Attachment* (1992, Plenum with I. Altman). Dr Low is currently President-elect of the American Anthropological Association and will take office November 2007. Her current research is on the impact of private governance on New York City coop residents, and she is writing a book on *An Anthropological Theory of Space and Place*.

Athena McLean, PhD, Senior Research Officer, Irish Centre for Social Gerontology, National University of Ireland and Professor of Anthropology, Department of Sociology, Anthropology and Social Work, Central Michigan University, USA

Athena McLean is a cultural anthropologist who has focused on the social production and political economy of medical knowledge and practices, mainly within psychiatry and dementia care. Dr McLean is also interested in social movements and epistemological concerns about the production of anthropological knowledge. She has studied the family therapy movement in the United States and psychiatric consumer movements both of primary consumers and their families, and has written several articles on these movements. Having spent 18 months conducting ethnographic research of two dementia care units in a major geriatric research and residential center in Philadelphia, she has written numerous articles and chapters about dementia caregiving. Her book, *The Person in Dementia: A Study of Nursing Home Care in the US*, was published in 2007 by Broadview Press. She co-edited a volume, *The Shadow Side of Research: Theorizing the Blurred Borders Between Anthropology and Life*, published by Blackwell Press in 2007. Dr McLean is currently a Senior Research Officer at the Irish Centre for Social Gerontology, National University of Ireland, on leave from her position as Professor of Anthropology in the Department of Sociology, Anthropology and Social Work at Central Michigan University.

Christine Milligan, PhD, Senior Lecturer, Institute for Health Research, Lancaster University, UK
Christine Milligan is Senior Lecturer in the Institute for Health Research at Lancaster University. She has published widely in international refereed journals and edited collections and has authored a book on *Geographies of Care: Space, Place and the Voluntary Sector* (Ashgate). She has co-edited two further volumes on *Landscapes of Health Welfare and Governance* (Policy Press) and *Celtic Geographies* (Routledge). She has been involved in a wide range of research projects funded by various agencies including research councils, the public, voluntary sectors and the European Union. Of particular interest to this volume, she has undertaken research and published refereed journal articles around older people and the therapeutic effect of gardening and social activity; and the impact of woodland on the mental well-being of young adults. Her current research interests focus on therapeutic landscapes, voluntary activism, informal care, transitions and older people.

Hester Parr, PhD, Department of Human Geography, University of Dundee, UK
Hester Parr is a Reader in Human Geography at the University of Dundee, Scotland, UK. She works on geographies of mental health, disability and illness and co-edited Mind and Body Spaces (1999, Routledge). Her work is published in a range of places including *Transactions of the Institute of British Geographers*, *Environment and Planning D: Society and Space*, *Cultural Geographies*, *Geoforum*, *Area* and *Health and Place*. She has a book forthcoming with Blackwell called *Mental Health and Social Space: Towards Inclusionary Geographies?*

Julie Pranikoff, MA, Environmental Psychology, The Graduate Center, City University of New York, USA
Julie Pranikoff is a PhD candidate in the Environmental Psychology program, with a concentration in Health Psychology, at The Graduate Center of the City University of New York. She received her BS from Cornell University in Human Development and Family Studies and her MA in Psychology from Hunter College. Julie's research examines how restorative environments, specifically gardens located on hospital grounds, affect patients' coping behaviors and their ability to adapt to illness.

Furthermore, she is interested in illness narrative creation and how personal meaning of illness is influenced by the environments in which narratives are formed.

Julie Sperling, MA, Department of Geography, University of Waterloo, Canada
Julie Sperling has an MA in Geography, a BA in Spanish and Latin American Studies and a BES in Environment and Resource Studies from the University of Waterloo. Her Master's research focused on cross-cultural notions of therapeutic landscapes using a case study of the Kaqchikel of San Lucas Tolimán, Guatemala.

Allison Williams, PhD, Associate Professor, School of Geography and Earth Sciences, McMaster University, Canada
Allison Williams completed her PhD at York University (Toronto, Canada) and, after holding permanent positions at both Brock University and the University of Saskatchewan, is now an Associate Professor in McMaster University's School of Geography and Earth Sciences. Her background in social and health geography continues to contribute to interdisciplinary research teams, three of which are currently examining urban quality of life, informal palliative caregiving, and rural health service delivery.

Robert Wilton, PhD, Associate Professor, School of Geography and Earth Sciences, McMaster University, Canada
Robert Wilton is an Associate Professor in the School of Geography and Earth Sciences at McMaster University. He earned a BA at the University of Hull (UK) and an MA and PhD from the University of Southern California in Los Angeles. His research is broadly concerned with the social geographies of exclusion. He is interested in the ways that 'problem' groups within society – such as persons with disabilities or people recovering from addiction – are excluded from valued environments, the ways in which this exclusion reproduces the marginal or problem status of such groups, and the political capacity of such groups to contest their exclusion.

Preface

It has been eight years since the publication of an 11-chapter collection of various research applications of the therapeutic landscape concept (*Therapeutic Landscapes: The Dynamic between Place and Wellness*, University Press of America, 1999). This first collection provided an introduction to the wide-ranging applicability of the therapeutic landscape concept – which was then less than a decade old, introducing how it could be used in health maintenance and well-being. Certainly the term has evolved extensively and, as this second collection illustrates, has a much wider horizon. My view is that this second collection marks a turning point in the concept's evolution, as it has been given focused attention by a group of scholars outside of social and health geography, those being self-identified health anthropologists.

Gratitude is extended to all contributors, who worked diligently to submit their work in a timely manner and who incorporated my editorial comments, large and small, with great generosity. I feel greatly rewarded to have worked with all of you in my role as editor, and thank you for the many insights you have shared in your work.

A very special thanks to Christine Heidebrecht and Peter Summers, whose research assistance has immeasurably eased the job of editing this collection. They have both worked diligently in the preparation of this collection, carrying the vast majority of responsibility for correspondence with the publisher and all contributors, and giving the detailed attention required to formatting and style requirements. Their commitment to this project is very much appreciated and will not be forgotten.

Appreciation is also extended to McMaster University for providing the leave time to work on this project; I cannot imagine working on a more enjoyable and worthwhile project during my leave.

Heartfelt gratitude to all who have allowed me the mental space and time to devote to this project, and who have ensured the balance needed for me to work effectively. This primarily includes my good friend Paula, my extended family, my partner Saeid, and my children Sarah and Cameron.

Allison Williams
Hamilton, Ontario
2007

Acknowledgments

Figure 2.1 *Listener* front cover, Summer 2005. (Reprinted with permission).

Figure 2.2 Advertisement for locums placed in university alumni magazines, 2005–2006 (Reprinted with permission).

Figure 9.1 A popular culture reflection of the significance of zoos and photographs in family life. © Baby Blues Partnership. King Features Syndicate.

Figure 9.2 Place identity at work. Photography by Mark Alexander. Created by Animal Instinct Advertising for Newport Aquarium.

Dedicated to Sarah and Cameron

Chapter 1

Introduction: The Continuing Maturation of the Therapeutic Landscape Concept

Allison M. Williams

Wil Gesler first developed the concept of therapeutic landscapes in the early 1990s, approximately 15 years ago. Since then, the concept has developed into a central component of health geographical thinking, being recognized at the 9th International Medical Geography Symposium (Montreal, 2001) as one of a few recent, internationally recognized contributions that health geographers have made to the larger discipline. Several books have been devoted to the subject and the number of peer-reviewed research papers appears to grow exponentially as each year passes. The centrality of the therapeutic landscape concept within the sub-discipline of health geography is confirmed with its inclusion in general health geography texts, the most recent being *Health and Inequality* (Curtis, 2004). While employed in a growing range of subject areas within health and social geography more broadly, the concept has been increasingly used in disciplines outside health geography, illustrating its flexibility and wide-ranging potential. This collection provides evidence of both the expanded use of the concept within the discipline of health geography – with 16 chapters illustrating a wide range of subject areas, as well as the innovative use of the concept in the anthropology of health, with four chapters illustrating various applications. This introductory chapter is broken up into three sections; the first section briefly traces the evolution of the concept, noting the central critiques. Second, a topical overview of the collection provides the foundation for the third and final section, which outlines ongoing trajectories and future directions.

The Evolution of the Therapeutic Landscape Concept

This section begins by first reviewing the concept's early application to traditional healing sites and carries through to contemporary applications in geography. Following, a discussion of the critiques of the concept, the current wide-ranging and ever-expanding use of the concept both within and outside geography is then reviewed. Due to the brevity of this review, not all pertinent work is discussed; only a few pieces of work have been noted to highlight the trajectory of the concept thus far.

The therapeutic landscape concept has its theoretical origins in cultural ecology, structuralism, and humanism, and provides a framework for analysis of natural and built, social, and symbolic environments as they contribute to healing and well-being in places – broadly termed landscapes (Gesler, 2003). Gesler's pioneering work on the Asclepian Sanctuary at Epidauros, Greece (1993), Marian Shrine at Lourdes, France (1996) and Roman Baths at Bath, England (1998) provided the template from which further applications emerged. Certainly, the ongoing examination of traditional sites continues to inform our understanding of the variant attributes or characteristics that should be given attention when creating a healing or healthful environment.

In addition to my own work in holistic health applications, which first suggested the idea of 'landscapes of the mind' (Williams, 1998), the first edited collection, entitled *Therapeutic Landscapes: The Dynamic between Place and Wellness* (Williams, 1999), illustrated the growing acceptance of the therapeutic landscapes concept within a socio-ecological model of health, with attention to a population health approach. In addition to covering traditional landscapes, the collection examined applications for marginalized populations, and applications in health care sites. These original three sub-sections, as used in the first collection – traditional landscapes, applications for marginalized and special populations, and health care applications – operate as categories for the work that followed, and still apply to the organization of the work contained within this current volume.

The use of the concept in examining the spaces and places for marginal and special populations have been many. Among them are numerous health promoting sites, including children's health camps (Kearns and Collins, 2000), gardens for the elderly (Milligan et al., 2004), community-based alcohol recovery programs (Wilton and DeVerteuil, 2006), respite centres (Conradson, 2005), and home for home-based caregivers (Williams, 2002). Related to this is the literary analysis of fiction that has used the therapeutic landscape as an interpretative framework (Baer and Gesler, 2004; Tonnellier and Curtis, 2005). Further, the application of the concept to specific cultural groups outside of the developed West illustrated the wide potential use of the concept (Wilson, 2003; Williams and Guilmette, 2001; Madge, 1998). As expected, a number of applications have been made to better-known health care settings. Included among these are those highlighted in Gesler's *Healing Places* (2003); hospital environments (Gesler et al., 2004; Curtis et al., 2007) and long-term care facilities (Andrews and Peter, 2006).

The evolution of the therapeutic landscapes concept has taken place, in part, due to numerous critiques (Gastaldo et al., 2004; Wakefield and McMullan, 2005; Conradson, 2005). One critique has been that healing places are not limited to places celebrated for their reputed healing qualities. Rather healing can take place in everyday, ordinary places, whether a residential backyard, a hospital room, or an imagined landscape. This critique has immeasurably extended the application of the concept beyond its earliest understanding, where reputed sites such as Lourdes, and Epidauros, made up the sites from which the concept was first developed. Another critique contests the nature of sites considered to be healing or therapeutic, arguing that such places can be simultaneously healthful and hurtful. A related critique has been made specific to the varied potential of any particular site for healing and/or health; what may be healing for one individual or group may not be for another. Finally, perception of whether a health-related landscape is therapeutic or not is context-dependent; that is, it is affected by local social and economic conditions and their associated changes. These latter critiques have certainly forced us to look at landscapes and places deemed to have therapeutic qualities more carefully, being sure to explore as many viewpoints as possible.

The concept has, not surprisingly, expanded in its reach beyond the discipline of geography. Certainly the idea of landscapes as therapeutic has been used extensively in landscape architecture, with the study of landscape design with respect to health and healing being well established before the concept of therapeutic landscapes was introduced in geography. Similarly, building design specific to health care facilities has been ongoing for approximately 30 years. The term therapeutic landscape is now being used formally in these literatures (Cooper Marcus, 2001). Environmental psychologists have a long history of research specific to therapeutic environments, specifically as they relate to natural environments; this is reviewed in the introduction to Part 5 in the present volume: *Transcending Geography: Applications in the Anthropology of Health*. Like geographers, anthropologists have borrowed from the psychology literature but, surprisingly, have only recently applied the therapeutic landscape concept to their work, as discussed in the introduction to Part 5. The four chapters making up this section thereby provides the initial employment of the term in anthropological research.

Further to anthropology, a number of other disciplines have been formally employing the therapeutic landscape concept, as we have come to understand it in geography; kinesiology, sport sociology, midwifery and nursing are among these disciplines. The meaning of summer camp, as therapeutic landscapes for youths with disabilities specific to identity development, has been researched by kinesiologists in Canada (Goodwin and Staples, 2005). The health experiences of members of a running club for sexual minorities highlight the social aspect of the therapeutic landscapes concept in research on the sociology of sport (van Ingen, 2004). The health professions of nursing and midwifery have recognized the importance of place in healing, noting how the therapeutic landscape concept incorporates social relations as well as the characteristics of the physical environment (Kennedy et al., 2004; Andrews, 2002). The growing use of the therapeutic landscapes concept outside of geography, and particularly in medical/health anthropology (this volume), illustrates its flexibility and wide applicability.

Common Threads and Notable Contributions

This collection is organized into five parts, based on the substantive content of the chapters contained within them. It becomes obvious when reading from the chapters that many of them could have been inserted in one of a number of the parts; each has been located in what has been determined to be the best fit, recognizing the need to keep parts as even as possible. Although each chapter is discussed in detail in the next section, they are briefly described here to provide the backdrop for the discussion of common threads and notable contributions. The first section, entitled *Traditional Therapeutic Landscapes: Natural and Built Environments*, examines the health-related qualities of specific environments, including: beaches, monastery retreats, yoga centres and mountains. This subject area continues to view places as therapeutic, hence the term 'traditional', albeit in a more critical manner than was the case ten years ago. *Therapeutic Geographies for Special Populations* makes up the second section, which explores the significance of therapeutic characteristics in the worlds of unique populations, including: addicts, support group users, urban neighbourhood residents and families. Part 3, *Applications in Health Care Sites*, provides examples of how health care sites and services are, to various degrees, therapeutic, as based on design, décor, signage, and meaning; sites include hospitals, assisted living residences and the home. The fourth part, *Contesting Landscapes as Therapeutic: Contemporary Advances*, further develops the critiques of the therapeutic landscape concept, as discussed in the previous section; Toronto gay bathhouses, an agrarian Guatemalan municipality, wooded areas in the UK, and the Soviet Gulag are the places explored. The fifth and final part, *Transcending Geography: Applications in the Anthropology of Health*, is devoted to the application of the therapeutic landscape concept in health anthropology, where ethnographic methods are used to explore: purposive community in suburban development, dementia care in long-term care facilities, art-making in public spaces, and a hospital as an aesthetic-therapeutic place.

What becomes evident when looking at the collection as a whole is a number of common threads that are woven throughout, particularly in substantive areas across the two disciplines represented. One such obvious thread is the interest in hospital environments, with three chapters devoted to examining quite different characteristics of these critical places of care (Chapters 10, 11 and 21). A second health care site of interest is residential care facilities for the aged, with two chapters being given focused attention on different types of facilities in the United States (Chapters 12 and 19). A related thread is the focus on the act and experience of caregiving, although one examines unpaid, informal care and the other paid, formal care provision (Chapters 13 and 19). The common interest in various art forms presents itself as a common theme addressed by two of the contributing anthropologists (Chapters 20 and 21). One of the most wide-ranging commonalities across the collection as a whole is the further development of the therapeutic landscape concept itself, which seems to work simultaneously with the use of innovative research methods. With this in mind, specific contributions are now highlighted, beginning with the geographical contributions.

Although the concept is further developed in various degrees in all geographical chapters, I would like to highlight four particularly notable contributions which, in my estimation, require particular mention. What is interesting to note in all four cases is the conceptual and/or methodological innovation employed. These four chapters will be briefly highlighted here, in consecutive order. Hoyez (Chapter 4) introduces the idea of the globalizing therapeutic landscape when reviewing the expansion of yoga centres across the globe, bringing together literatures which would otherwise stand in stark opposition to one another. In Chapter 7, Davidson and Parr provide evidence for the therapeutic potential of support groups, via modalities – the telephone, newsletter and the internet, which would otherwise be deemed banal in the sense of only providing information/resources. Sperling and Decker (Chapter 15) provide a stunning culturally specific examination of therapeutic landscapes through employing a novel methodological approach – photovoice – and, in so doing, outline both cultural and gender specificity. Finally, DeVerteuil and Andrews (Chapter 17) use documentary analysis of personal memoirs to illustrate how the Stalinist Gulag can be both hurtful and healing, and simultaneously physical and imagined.

All the anthropological contributions are original and new, as can be expected given the emergence of the TL concept in the discipline. Hoey (Chapter 18) extends the understanding of asylums, as based on moral treatment, to purposive communities. Einwalter (Chapter 20) introduces the idea of creating numerous art forms in public space as therapeutic. McLean (Chapter 19) discusses a moral landscape of dementia care, arguing for the need for improved quality of caregiving in nursing homes and, in so doing, coherently highlights the importance of the social aspect of therapeutic environments within the context of the psycho-social health of dementia patients. Finally Collins (Chapter 21) identifies and describes the therapeutic-aesthetic within the context of hospital restructuring in the UK, illustrating the extraordinary emphasis given to the aesthetic in the midst of growing public outcry regarding service availability and quality.

Topical Overview

Part 1: Traditional Therapeutic Landscapes: Natural and Built Environments

Collins and Kearns (Chapter 2) provide a well contextualized presentation of New Zealand beaches as landscapes with several possible meanings – therapeutic in some regards and risky in others. Current public health dialogue is well integrated and provides an ironic contrast to what beaches have traditionally meant, with regards to their therapeutic potential. Although suggested by others, the authors have gone to great lengths to provide the evidence needed to succinctly argue that not all therapeutic landscapes are, in fact, healthful. Whilst New Zealand beaches are highlighted, the argument can be applied to any number of nations who practise sun bathing.

Conradson (Chapter 3) explores the experience of stillness, a sought-after and increasingly required experience given the busy, technologically driven lifestyle characteristic of the Western world. The socio-geographical distribution of stillness is highlighted, with three 'spheres of the life-world' discussed – workplace, homespace and elsewhere. The latter sphere is then explored via a case study of two monastery retreat settings. A number of developments specific to the therapeutic landscapes concept are suggested, including the importance of: situating sites within their larger societal context, and contemplating the temporal dimensions of experience.

Hoyez (Chapter 4) explores a fascinating application of the therapeutic landscape concept – that being the globalization of yogic practice, from the continent of India across the world. Highlighted sights include Yogaville, located in Virginia (United States), as well as numerous international destinations. By first discussing the therapeutic qualities of yogic practice, as defined in ancient Indian texts, the healthful outcomes of yogic practice are outlined within a Western context. The place characteristics of yoga centres are described and applied to the therapeutic landscape concept, followed by a discussion of the globalization phenomenon.

Chapter 5 (Williams) provides an interpretation of the children's classic *Heidi*, written by Johanna Spyri in the 1800s. This chapter builds on previous analysis which interprets literary works using the therapeutic landscape as an analytical approach. Rural–urban contrasts are highlighted, where the hazardous world of city life in Frankfurt (where physical, emotional and spiritual illness appears to be rampant) is contrasted with a magical alpine world – where the physical, emotional and spiritual aspects of health and wellness are enhanced. The novel thereby presents a 'utopia' of sorts, symbolizing a traditional, yet everyday, therapeutic landscape. Spyri's *Heidi* reflects its times, as the novel was influenced by the nature-embracing Romantic Movement in Europe.

Part 2: Therapeutic Geographies for Special Populations

DeVerteuil, Wilton and Klassen (Chapter 6) extend the idea of the settings specific to substance abuse treatment and recovery as therapeutic landscapes, whilst also contributing to the geographies of marginalized groups and addiction treatment. The examination of a case study made up of three different facilities, located in variant urban environments (inner core, outer core and suburb) throughout Winnipeg (Manitoba, Canada) provide insight into the importance of place in substance abuse treatment.

Davidson and Parr (Chapter 7) shed new light on therapeutic geographies, identifying the 'emotional benefits of co-identification and support' via a non-physical therapeutic experience for clients of the United Kingdom National Phobics Society. They provide an excellent example of how the therapeutic landscape concept can be merged with the new and evolving area of emotional geographies.

Braubach (Chapter 8) provides a unique contribution to the literature as his chapter uses therapeutic landscape theory to frame his quantitative analysis of perceptual data collected in eight European cities. Using data from a WHO survey on housing and neighbourhood conditions and health, this chapter provides empirical data to support the preventive application of the therapeutic landscape concept in urban health.

Hallman (Chapter 9) argues that the zoological park offers a therapeutic landscape for families. She suggests that it is mainly parents of young children who accept that public places, like zoos, offer ideal opportunities to experience 'family life' through place identification. This is a novel contribution as little or no work has been done specific to the geographies of family life as it pertains to the therapeutic landscape concept.

Part 3: Applications in Health Care Sites

Gesler and Curtis (Chapter 10) provide a rich contribution to the collection, not only applying the therapeutic landscape concept to hospital design as it applies to the Private Funding Initiative for the National Health Service in the UK, but also providing insight into how hospital design can be improved if, indeed, all core aspects of the therapeutic landscape concept are accounted for holistically. A case study of perceptions of an acute mental health unit are discussed as informing physical, social and symbolic landscapes, and illustrate how design features either contribute or take away from the health and well-being of both patients and staff.

Crooks and Evans (Chapter 11) provide a unique contribution by examining the micro-space of the hospital waiting room and by specifically analysing the signage within. Conducting field work in waiting room spaces in three hospitals throughout the City of Hamilton (Ontario, Canada) included the collection of formal observations, field notes and photographs. Results suggest that messages contradicting the therapeutic are often communicated in these hospital spaces; these include messages of control, regulation and discipline.

Cutchin (Chapter 12) provides two contributions to the therapeutic landscape literature in his chapter; the first is engagement with gerontological knowledge and, the second, a critical lens of a site considered to be therapeutic – residential care environments for older people. Using research on assisted living residences in the United States, related layers of meanings are presented to provide a critical perspective on these residential sites as therapeutic landscapes. Three unhealthy aspects of these sites are discussed: commodified landscapes, liminal places and ambiguous places.

Donovan and Williams (Chapter 13) build on Williams' earlier work by examining how providing palliative and end-of-life care in the home affects caregivers' perceptions of home as therapeutic. Using three case studies, the findings point to the home as both a therapeutic (health-promoting) and non-therapeutic (health-detracting) environment, as demonstrated through a discussion of home along a number of themes. Policy and programme implications are then provided.

Part 4: Contesting Landscapes as Therapeutic: Contemporary Advances

Andrews and Holmes (Chapter 14) provide a novel contribution which strives to make a critical point that spaces/places considered to be therapeutic can, in fact, be the opposite. Using the example of gay bathhouses in Toronto (Ontario, Canada), the place of the bathhouse as therapeutic is turned on its head, as risky sexual behaviour pervades activities within this place.

Sperling and Decker (Chapter 15) begin to address the gap specific to both a non-Western and gendered perspective in therapeutic landscape research. Using a qualitative case study of an agrarian Guatemalan municipality, this chapter examines how Indigenous Mayan women and men differ in their construction and use of the therapeutic landscapes that surround them in their daily lives, the latter which are permeated by poverty and violence. This chapter demonstrates how therapeutic landscapes can be culturally defined, gender-specific and contingent on social, political and economic contexts.

Milligan (Chapter 16) discusses the 'duality' of the 'therapeutic' engagement with landscape using natural environments and, specifically, woodlands as a case in point. Qualitative data are used to provide further insight into the 'restorative and risk/fear-inducing' effects of the natural environment. This chapter makes an important contribution that not only ties a number of relevant different literatures together, but also asserts that negative reactions are subjectively experienced in landscapes otherwise considered therapeutic.

DeVerteuil and Andrews (Chapter 17) contribute to the evolution of the concept in a number of ways, arguing that places can be both healing and hurtful, are inclusive of ordinary/everyday geographies, and can be both physical and imagined. By first critically discussing the concept, the authors move on to illustrate, using personal memoirs, how the therapeutic landscape concept can, in fact, be extended using a reconstruction of the (un)healthy spaces of the Stalinist Gulag.

Part 5: Transcending Geography: Applications in the Anthropology of Health

Given the disciplinary focus of this section, a discipline-specific introduction is provided by Julie Pranikoff and Setha Low, the latter who is an established anthropologist with expertise in the study of space and place.

Hoey (Chapter 18) investigates the transition of an asylum landscape into a residential development based on the new-urbanism movement. Based on fieldwork carried out in Michigan's Grand Traverse region (United States), the idea of an 'intentionally constructed, purposive community' is examined as having therapeutic potential. In contrast, it is also examined as potentially reflecting class and racial homogeneity while possibly being socially isolating. Although asylums have been discussed extensively within health geography, the extension into the new-urbanism movement is new and captivating.

McLean (Chapter 19) discusses the ideal elements required to ensure nursing homes operate as therapeutic. While describing the physical changes required to obtain the ideal environment, as defined via the Culture Change movement, she argues that the more important change needed is in the quality of the caregiving provided. This main argument supports the importance of social (and indirectly, spiritual) relations in achieving a therapeutic landscape of long-term care for dementia patients.

Einwalter (Chapter 20) provides an interesting contribution to the understanding of landscapes as therapeutic, based on the importance of 'flexibility' in the subjective creation of roadside art/memorials. The therapeutic value of the roadside art/memorials to the creator is highlighted, outlining the regulations specific to creating and locating such art forms in public roadways.

Collins (Chapter 21) explores what has thus far often been assumed to exist specific to our understanding of the therapeutic landscape concept, that being the aesthetic as therapeutic. Using a case study approach, numerous examples of the aesthetic, including sculpture, architectural design, and abstract artwork located within a hospital environment created under the auspices of the Private Funding Initiative for the National Health Service in the United Kingdom, provide for the reader an opportunity to examine the relationship between the aesthetic and the therapeutic.

Ongoing Trajectories and Future Directions

Informed by the contents of this volume, this final section provides a few selected continuing lines of research, as well as future research directions, both of which are discussed in light of the continued maturation of the concept both within and outside the bounds of health geography. With the advent of anthropologists' involvement in conducting research on the therapeutic landscape concept, the possibility of filling the recognized gap in cultural understandings of place/experience and health (care) becomes more real. We know little about therapeutic landscapes, traditional and otherwise, outside of the Western world (see Chapter 15 for an exception); many Asian cultures, for example, have numerous traditional sites of therapeutic importance. If this challenge is undertaken, a better understanding of the therapeutic elements of health and healing traditions across the globe will extend to applications in health delivery systems and services in an increasingly multi-cultural and globalizing world.

The application of the therapeutic landscape concept to health care sites such as the hospital, the home, and facilities of care for the aged, have undoubtedly the greatest certainty of continued inquiry given the trajectory set by past research and most currently presented in this collection (Chapters 10–13, 19 and 21). Other care sites of continuing interest will likely include: respite facilities, complementary health clinics, and the numerous settings and services associated with relational and emotional geographies (Chapters 3 and 7), the latter which encompass both emotional and mental health concerns. What has been learned about health care sites through the application of the therapeutic landscape concept thus far has been criticized as being anecdotal. In order to reach firm conclusions and make sound recommendations, the effects of therapeutic landscape design on health and well-being for all involved will require systematic study.

As suggested above, emotional and relational geographies are a growing area of interest for many social and health geographers and provide a direction for therapeutic landscape research. In addition to counselling and various related therapies which address psychological and emotional issues (see Chapter 7), relational geographies – such as therapeutic touch, massage, and other such modalities – provide additional research possibilities. Research in emotional and relational geographies has the possibility to address the issues of patient and practitioners communication; how these relations are best achieved may be understood within the therapeutic landscape framework, where attention to the social and symbolic will become paramount, as suggested in Chapter 19. This type of research will continue to grow, as will geographies of the mind, where imagined geographies – such as accomplished via meditation – appear to becoming increasingly accepted as modalities for stress-reduction and enhanced focused concentration (see Chapter 4).

Continuing the research trajectory that began with the first collection (Williams, 1999), numerous types of population-specific health promoting sites will continue to be examined. In addition to the study of a number of recognized sites (such as health camps for various populations), recreational clubs (such as running clubs for sexual minorities), and substance abuse recovery programmes, there is a growing interest in studying sites that provide respite from the quick-paced, stressful lifestyle characteristics of the Western world, as suggested in Chapter 3 (and referred to in Chapter 4). Related to this, further interest in studying intentional spaces of purposive community or intentional communities (Chapter 18) will provide a more comprehensive understanding of everyday geographies as therapeutic.

Certainly the therapeutic landscape concept has moved beyond the celebration of traditional landscapes reputed for health and healing, as illustrated throughout this collection (particularly in Part 4). Although providing a critical stage in the development of the therapeutic landscape concept, the examination of traditional sites – which are still the focus of research and literary analysis (see Chapter 5), have been, for the most part, replaced by a much more discriminating understanding. One research direction which has yet to be fully realized is the examination of therapeutic landscapes reputed for their spiritual well-being (although see Chapter 3 for an example), as little is known about the relationship between such places and a broadly defined understanding of health. There is growing evidence of the relationship between being spiritually active and good health/well-being (as discussed in Chapter 4), reflected in the reduced onset of physical and mental illness and improved health status (George et al., 2000). Examining spiritually reputed sites would mean continuing the trajectory set for researching traditional sites, albeit in a more discerning manner; possible directions for ongoing research include the further examination of traditional sites (e.g. Mecca, Jerusalem), and reputed places of worship (e.g. shrines, sacred places). In addition, examining the effects of spiritual elements and spiritual activities within non-reputed places and spaces is needed, including in the geographies of health care, health promotion, and everyday and imagined geographies. The challenges of conducting research on spirituality are similar to those for therapeutic landscapes in general, and encompass the fact that spirituality is highly subjective, personal and individualistic. Further, there are limited approaches for measuring spirituality, due to variation in beliefs, practices and meanings, making the effect on health (care) difficult to measure. Tackling these challenges may be best achieved through interdisciplinary research approaches – certainly commonplace in today's academic milieu.

Finally, geographers, anthropologists and the other disciplines engaged in research on therapeutic landscapes will continue to provide further critical reflection of the concept; this will continue its maturation while contributing to the progress being made to real-world applications. The further examination of the above noted areas will take into account everyday geographies, being sure to pay heed to the larger socio-economic context, and discern both the healthful and hurtful aspects of experience, as dependent upon the individual or group concerned. Additional methodological approaches will likely be employed, further extending the perspective of and use of the concept.

Looking back on how far the concept has developed since its inception in the early 1990s provides a great stimulus for the future development and the ongoing application of the therapeutic landscapes concept in the study of healing, health and illness in place, space and experience.

Acknowledgments

Thanks to Wil Gesler for providing feedback on an earlier version of the first section of this introductory chapter.

References

Andrews, G.J. (2002), 'Towards a more place sensitive nursing research: an invitation to medical and health geography,' *Nursing Inquiry* 9 (4): 221–38.

Andrews, G.J. and Peter, E. (2006), 'Moral Geographies of Restraint in Nursing Homes: Worldviews on Evidence-Based Nursing', *First Quarter* 2006: 2–7.

Baer, L.D. and Gesler, W.M. (2004), 'Reconsidering the concept of therapeutic landscapes in J.D. Salinger's "The Catcher in the Rye"', *Area* 36: 404–13.

Conradson, D. (2005), 'Landscape, care and the relational self: Therapeutic encounters in rural England', *Health and Place* 11: 337–48.

Cooper Marcus, C. (2001), 'Hospital oasis – Through a participatory design process, a failed Tommy Church garden in San Francisco is reconfigured as an exemplary therapeutic landscape', *Landscape Architecture* 91: 36.

Curtis, S.E. (2004), *Health and Inequality: Geographical Perspectives* (London: Sage).

Curtis, S.E., Gesler, W.M., Fabian, K., Francis, S. and Priebe, S. (2007), 'Therapeutic landscapes in hospital design: A qualitative assessment by staff and service users of the design of a new mental hospital inpatient unit', *Environment and Planning C* (in press).

Gastaldo, D., Andrews, G.J., and Khanlou, N. (2004), 'Therapeutic landscapes of the mind: Theorizing some intersections between health geography, health promotion and immigration studies', *Critical Public Health* 14 (2): 157–76.

George, L.K., Larson, D.B., Koenig, H.G. and McCullough, M.E. (2000), 'Spirituality and health: What we know, what we need to know', *Journal of Social and Clinical Psychology* 19 (1): 102–16.

Gesler, W.M. (1993), 'Therapeutic Landscapes: Theory and a Case Study of Epidauros, Greece', *Environment and Planning D* 11: 171–89.

Gesler, W.M. (1996), 'Lourdes: Healing in a Place of Pilgrimage', *Health and Place* 2: 95–105.

Gesler, W.M. (1998), 'Bath's Reputation as a Healing Place', in R.A. Kearns and W.M. Gesler (eds) *Putting Health into Place* (Syracuse: Syracuse University Press) pp. 17–35.

Gesler, W.M. (2003), *Healing Places* (Lanham, MD: Rowman & Littlefield).

Gesler, W., Bell, M., Curtis, S., Hubbard, P. and Francis, S. (2004), 'Therapy by design: Evaluating the UK hospital building program', *Health and Place* 10: 117–28.

Goodwin, D. and Staples, K. (2005), 'The Meaning of Summer Camp Experiences to Youths with Disabilities', *Adapted Physical Activity Quarterly* 22: 160–78.

Kearns, R.A. and Collins, D.C. (2000), 'New Zealand children's health camps: Therapeutic landscapes and the contract state', *Social Science and Medicine* 51: 1047–59.

Kennedy, H., Shannon, M., Chauhorm, U. and Kravetz, M. (2004), 'The landscape of caring for women: A narrative study of midwifery practice', *Journal of Midwifery and Women's Health* 49: 14–23.

Madge, C. (1998), 'Therapeutic landscapes of the Jola, The Gambia, West Africa', *Health and Place* 4 (4): 293–311.

Milligan, C., Gatrell, A. and Bingley, A. (2004), 'Cultivating Health: Therapeutic landscapes and older people in northern England', *Social Science and Medicine* (58): 1781–93.

Tonnellier, F. and Curtis, S. (2005), 'Medicine, landscapes, symbols: "The Country Doctor" by Honore de Balzac', *Health and Place*, 11: 313–21.

Van Ingen, C. (2004), 'Therapeutic Landscapes and the Regulated Body in the Toronto Front Runners', *Sociology of Sport Journal* 21: 253–269.

Wakefield, S. and McMullan, C. (2005), 'Healing in places of decline: (re)imagining everyday landscapes in Hamilton, Ontario', *Health and Place* 11: 299–312.

Williams, A. (1998), 'Therapeutic landscapes in holistic medicine', *Social Science and Medicine* 46 (9): 1193–1203.

Williams, A. (ed.) (1999), *Therapeutic Landscapes: The Dynamic between Place and Wellness* (Lanham, MD: University Press of America).

Williams, A. and Guilmette, A. (2001), 'A Place for Healing: Achieving Health for Aboriginal Women in an Urban Context', *Canadian Journal of Native Studies* 21 (1):1–25.

Williams, A. (2002), 'Changing geographies of care: employing the concept of therapeutic landscapes as a framework in examining home space', *Social Science and Medicine* 55 (1): 141–54.

Wilson, K. (2003), 'Therapeutic landscapes and First Nations peoples: An exploration of culture, health and place', *Health and Place* 9: 83–93.

Wilton, R. and DeVerteuil, G. (2006), 'Spaces of sobriety/sites of power: Examining social model alcohol recovery programs as therapeutic landscape', *Social Science and Medicine* 63: 649–61.

PART 1
Traditional Therapeutic Landscapes: Natural and Built Environments

Chapter 2

Ambiguous Landscapes: Sun, Risk and Recreation on New Zealand Beaches

Damian Collins and Robin Kearns[1]

Contemporary public health promotion encompasses a broad range of messages. Prominent themes include the need for physical activity – much of which takes place outdoors – and the imperative of transforming individual behaviours and social landscapes so as to minimize exposure to risk. In New Zealand, beaches have long been thought to promote physical and psychological well-being by providing opportunities for relaxation and active recreation. While these benefits may still be found on the coast, awareness of attendant risks has increased and made beaches more ambiguous environments from the perspective of health promotion. This chapter reflects on this ambiguity, and what it means for understandings of beaches as therapeutic landscapes.

Beaches figure prominently in many New Zealanders' understandings of themselves and their nation, and sun exposure – whether incidental or intentional – has been a typical part of the New Zealand beach experience. As elsewhere, the suntanned body has been seen as a signifier of good health (McDermott et al., 2003; Chapman et al., 1992). However, with ozone depletion and increasing rates of melanoma and other skin cancers, social marketing campaigns have, since the mid-1980s, targeted beachgoers with warnings about the dangers of ultra violet radiation (UVR). These campaigns have typically employed threat appeals and other techniques intended to promote behavioural change by increasing wariness and anxiety vis-à-vis exposure of the skin to the sun. Thus, while the public health community understands health as more than the absence of disease, in this area it has adopted promotional strategies that clearly centre on disease prevention; moreover it seeks to create psychological discomforts which may disrupt their intended targets' sense of well-being.

In this chapter we consider what these promotional strategies mean for popular understandings of beaches, more than two decades after the first sun awareness/avoidance campaigns were launched in New Zealand. Specifically, we ask whether public health warnings about the risks of sun exposure are fundamentally changing perceptions of New Zealand beaches as therapeutic landscapes. To address this question, we draw on a range of primary and secondary materials, including a survey of beachgoers' attitudes towards tanning, sun exposure, and responsibility for sun protection. We conclude our chapter by reflecting on the significance of the new public

1 The authors acknowledge the assistance of Lorissa Edwards who was funded by a University of Auckland Faculty of Science Summer studentship.

health's central focus on risk for understanding sites of health promotion. If risk is variable, but essentially ubiquitous, can any landscape be considered therapeutic?

New Zealand Beaches as Therapeutic Landscapes

Beaches have a central place in the national imaginary in New Zealand, and are key sites for everyday recreation and domestic tourism, especially in summer (Perkins and Thorns, 2001). This prominence is determined in part by the accessibility of beaches to most New Zealanders: four of the country's five major urban regions are located on the coast – making trips to city beaches relatively easy propositions, especially for those who live in coastal suburbs. In addition, many beaches further from major population centres are well-established holiday destinations for domestic (and some international) tourists. Temporary residence at these sites has historically been made possible by affordable commercial campgrounds and modest holiday homes (Kearns and Collins, 2006).

In various ways, New Zealand beaches are 'obvious' candidates to be considered therapeutic landscapes. Following Conradson (2005), we observe that, almost irrespective of their location (for example, in sheltered harbours or on 'wild' ocean coasts), they possess four important qualities. First, they provide some physical and psychological distance from everyday routines and domestic demands. As Waddell (2003, 40) has observed in the Australian context, beaches promise 'anonymity, equality and stress deferral' and an escape from 'work identities'. It is useful to note that this was not always the case, and that prior to the advent of a leisure-oriented beach culture in the mid-to-late nineteenth century, the New Zealand coast was primarily a site of manual labour (Matthewman, 2004).

Second, beaches offer opportunities to be close to the natural environment – opportunities which may otherwise be rare for urbanites. While a proportion of the New Zealand coastline is suburbanized, commercialized and highly modified (for example, through provision of seawalls, piers and promenades – see Pawson, 2002), and much of the rest has been extensively modified by pastoral farming (Collins and Kearns, 2007), most beaches offer opportunities to encounter elements of the natural environment. These include not only the sometimes unpredictable waters, but also dune systems, remnant natural forest, various species of birds, fish and marine mammals, and – of course – exposure to the elements.

Third, beaches offer opportunities for either solitude or social interaction, depending on the preferences of the visitor. Thus, relatively deserted beaches may be found even during the peak summer season, but many – perhaps most – beach visits are social events, undertaken in the company of friends and/or family members. For these groups, the beach is likely to be a site not only of relaxation, but also one of social and physical activity oriented around pursuits such as picnicking, walking, swimming and game-playing.

Fourth, routine engagements with coastal places may shape individual and ultimately collective identity. By way of example, some New Zealanders define themselves partly in terms of their participation in particular activities at the coast (for example, as surfers, fishing enthusiasts, 'boaties'), and some families form deep place attachments to the coastal sites they routinely visit for shared holidays. More broadly, acts of visiting and staying at the coast connect individuals and families with *the* central site in New Zealand's leisure culture (Matthewman, 2004).

Unlike most visits to international beach resort locations (for example, Cancun, Tahiti, Phuket) – which tend to be planned, episodic and expensive – leisure-oriented trips to New Zealand beaches can be undertaken frequently and reasonably spontaneously, and traditionally at little cost. Taking seriously the health-promoting qualities of these sites is, then, consistent with recent calls for research into therapeutic landscapes to be attentive to ordinary places (Wakefield and McMullan, 2005). Moreover, visits to the New Zealand coastline have tended to be active in orientation. While passive sunbathing forms a *part* of many summer beach experiences, it is often accompanied by forms of recreation ranging from casual swimming and sports to more serious athletic activity.

Perkins and Thorns (2001, 197) suggest that '[t]hese pursuits, combined with visiting friends and relatives [while on holiday]' give New Zealand domestic tourism a unique flavour, which cannot be captured by the notion of a passive tourist gaze. The prominent place of physical activity at New Zealand beaches also underscores the point that the therapeutic elements of a particular landscape may depend less on its innate health-promoting elements, than on the qualities of the embodied engagements which occur there. For this reason, Conradson (2005, 346) suggests attentiveness to the 'therapeutic landscape experience', and not simply the landscape in and of itself.

The prominence of the coast as a site of activity also suggests a fifth way in which beaches fit the mould of a therapeutic landscape: they are places of exercise and routine physical activity for many New Zealanders. At the coast, individuals and families may act, intentionally or otherwise, in a manner consistent with public health exhortations to engage in exercise and active play. The landscape lends itself to a range of activities, both on land and in water, depending on the weather and the inclinations of visitors – activities which may be constructed as 'healthy, adventurous and morally virtuous' (Perkins and Thorns, 2001, 198).

While the therapeutic qualities of visits to the sea coast – particularly in terms of the availability of fresh air and clean water – were known in Europe from the late eighteenth century (Andrews and Kearns, 2005; Matthewman, 2004), as late as the mid-nineteenth century in New Zealand, 'bathing and swimming [were] fringe activities, indulged in by health fanatics, but regarded by many as unsafe and unseemly' (Daley, 2003, 119). From the 1890s, however, swimming came to be reconceptualized as a healthy activity and a marker of both personal responsibility and good citizenship. By 1910, swimming at the beach, 'rather than just dipping one's toes in the waves', was a popular activity (Daley, 2003, 127). This said, there was awareness from the outset of the risks that accompanied such exercise: not only the 'moral dangers' associated with gender-mixed bathing, but also the risk of drowning, which was partly ameliorated by the advent of surf lifesaving clubs (Daley, 2003). Today, New Zealand surf lifesavers are expected not only to fulfil their traditional role of promoting water safety, but also to role model sun-protective behaviour at beaches.

In New Zealand, time spent at the coast is strongly associated with both deliberate and unintentional exposure to (UVR). As elsewhere, the desire for a suntan is a relatively recent phenomenon: it was not until the 1930s that it was associated with good health (as opposed to manual labour), and tanning was recommended by health promoters (McDermott et al., 2003). As with swimming, this shift in perception occurred quickly, and 'sunbaking' rapidly became a 'national obsession', in keeping with an international shift towards recognizing the therapeutic qualities of sunlight (Daley, 2003, 131). Interestingly, in light of contemporary public health warnings, UVR exposure in childhood was deemed particularly valuable: children attending health camps in New Zealand were required to undertake an hour of naked sunbathing per day (Kearns and Collins, 2000), while the Plunket Society – a child health organization – was by the late 1920s instructing mothers to expose even infants to the sun throughout summer (Daley, 2003). One result of the increasing popularity of tanning was a marked shift in the clothing worn at the beach, with formal styles giving way to swimwear which allowed for increased exposure (and display) of bodies.

By mid-century, then, swimming and tanning were firmly ingrained as therapeutic activities, and both were focused on the coast. At the same time, the increasing availability of basic holiday homes along the New Zealand coast made the possibility of staying at the beach over summer holidays and select long-weekends a possibility for many families, including those on modest incomes (Keen and Hall, 2004). Consequently, beaches claimed a central place in the New Zealand imaginary – not necessarily because they were spectacular sites (although some were), but rather because they were accessible landscapes, integrated into annual routines, offering relatively safe and affordable opportunities for rest, recreation and respite.

Sun, Risk and Public Health Campaigns

Contemporary public health knowledge recognizes UVR exposure as something which has both beneficial and detrimental effects on human health. On the one hand, UVR enables the body to synthesize Vitamin D (which is essential for bone, joint and muscle health, and neurological function); on the other, excessive exposure is estimated to be responsible for at least 90 per cent of skin cancer cases (Cancer Society of New Zealand 2005; Armstrong, 2004). The photocarcinogenic effects of sunlight are of considerable importance in Australia and New Zealand, as the countries with the highest and second-highest incidences of skin cancer in the world, respectively (New Zealand Health Information Service, 2004). This prevalence has been connected to a variety of factors, including proximity to the equator, mild-to-warm climates, high sunshine hours, ozone depletion, outdoor lifestyles, and a high proportion of pale skin types.

In New Zealand, over 50,000 cases of skin cancer are reported each year (in contrast to approximately 16,000 cases of all other cancers combined), and on average 1700 of these are the most dangerous form, malignant melanoma (O'Dea, 2000). Skin cancer causes some 300 deaths per year, including 250 from melanoma. Unsurprisingly, in light of these figures, reducing New Zealanders' exposure to UVR is a public health priority – something which has been reflected in health promotion messages for over two decades.

The first sun protection campaign in New Zealand centred around the internationally known 'Slip! Slop! Slap!' message, first developed in Victoria,

Australia in 1980. In 1993, the Cancer Society of New Zealand and the Health Sponsorship Council (a government-funded health promotion agency) adopted a second Victorian initiative – 'SunSmart', a comprehensive 'sun-related behaviour message and brand' (SunSmart, 2006). It did not abandon the 'Slip! Slop! Slap!' message (and indeed added a fourth imperative – 'Wrap!' – to encourage sunglass-wearing), but incorporated it into a broader programme. This encompasses conventional advertising campaigns, as well as sponsorship (of surf lifesavers), public education (such as that around the UV Index), policy and resource development (including UVR protection in health and safety guidelines for outdoor workers), and research (for example, into the epidemiology of skin cancer) (Montague et al., 2001). As with other social marketing programmes, it seeks to effect long-term changes in personal and institutional attitudes, and human behaviour, to achieve a socially desirable outcome: in this case, to reduce the incidence of UVR-related morbidity and mortality.

By increasing public awareness of a significant health risk, and promoting changes in behaviours and environments so as to minimize this risk, SunSmart potentially alters experiences and perceptions of the everyday landscapes in which bodies are exposed to UVR. Indeed, the far-reaching consequences of the programme are highlighted in a recent review of its effectiveness in Australia:

> A significant proportion of sport and leisure organisations, workplaces … have adopted sun protection policies .... Weather forecasting, the fashion industry, solariums, building design, industrial awards, occupational health and safety practices, sunscreen production and pricing, hat, swimwear and shade manufacture have all been influenced by the program.
>
> At the beach, hats and protective swimwear are the norm for children and are frequently worn by adults. Personal shade structures are commonplace, and lifesavers demonstrate and promote SunSmart behaviour … (Montague et al., 2001, 3).

SunSmart, then, is part of a broader new public health discourse which works, in large part, by problematicizing the spaces in which exposure to risk occurs (see Brown and Duncan, 2002). In the case of UVR-related risk, potentially any outdoor space is problematic, although beaches as sites of intentional exposure have been particular targets for attention – as the above quotation suggests. While sun protection messages continue to focus on the beach context, recent initiatives have also been attentive to places in which there is an institutional responsibility to manage risk (for example, workplaces, schools). In New Zealand, school environments are rendered particularly problematic by virtue of their traditional orientation around spacious, unshaded fields – in which pupils are required to play and exercise. Recent research has highlighted significant shifts in attitudes towards sun protection at the primary school level (Collins et al., 2006), although high schools have proved more reluctant to change (Jopson and Reeder, 2004), possibly reflecting a view that high school students are old enough to manage their own risk in this area.

Sun protection is primarily envisioned as an individual responsibility and choice, at least for adults. Accordingly, a range of approaches is apparent at New Zealand beaches over summer, with some individuals and groups adopting a spectrum of sun-protective behaviours (including bringing their own shade structures, and wearing extensive clothing on the beach and UVR-protective shirts in the water) while others nearby continue to tan (and occasionally burn). The popular association of tanning with good health and beauty has proven difficult to break, in spite of its relatively recent origins (McDermott et al., 2003),

New Zealanders have a 'long-standing cultural belief in the healthiness of the outdoors' (Collins et al., 2006, 438) and, in broad terms, tanned skin continues to be associated with a healthy, active lifestyle – even though, in some cases, it is acquired indoors, through application of sprays and lotions, or use of a sunbed. Similar trends have been observed elsewhere (see Garvin and Wilson, 1999), including in Australia, where a survey of young people in the mid-1990s found that the majority remained 'positive about acquiring a light to medium tan' (Lupton and Gaffney, 1996, 147). This situation prevailed in spite of over 15 years of social marketing promoting an unambiguous message of sun protection and avoidance, and associated efforts to reduce social acceptance of the tanned body (Garvin and Eyles, 2001).

Threat Appeals and their Implications

Montague et al. (2001, 7) observe that in the late 1990s, SunSmart messages became increasingly 'hard-hitting … designed to shock people into action'. While threat appeals, which seek to elicit types of emotional arousal (for example, fear, anxiety) conducive to a desired behavioural change, had always formed part of the social marketing of sun protection in New Zealand and Australia, they became more forceful in order to increase risk perception. This change in emphasis was driven by research suggesting that the community was becoming 'complacent' about UVR risk, and that 'adolescents in particular needed graphic prompts to spur them on to do what they know they should do' (Montague et al., 2001, 3). As with the use of threat appeals in the marketing of tobacco control, shocking visual and verbal warnings about the targeted behaviours (entailing intentional or incidental sun exposure) were accompanied by messages about how the audience could undertake change, in order to reduce health risks.

The use of threat appeals in health promotion has been credited with considerable success in the area of smoking cessation (Hill et al., 1998; Biener and Taylor, 2002), but has also been criticized for its often negative tone, and narrow focus on disease prevention (Hastings and MacFadyen, 2002). While graphic and explicit health warnings facilitate emotional engagement, demand public attention, and increase individuals' motivation to change (Biener and Taylor, 2002), over time they may also overwhelm their audience:

> [Health promoters] fragment health into different topics and engage in an odd kind of consequence proliferation, where smoking, driving, having sex, and a dozen other behaviours vie with each other to present more dramatic risks. In the process, we end up telling people that if the cigarettes don't get them, a drunk driver, sexually transmitted disease or an over-abundance of butter surely will. The net effect is to turn health promotion from a tremendous opportunity for people to enhance their enjoyment of life, into a clutch of disparate and capricious threats (Hastings and MacFadyen, 2002, 74).

Figure 2.1 ***Listener*** **front cover, Summer 2005 (reprinted with permission)**

In the case of sun protection, social marketing has involved graphic portrayals of severely burnt bodies (often portrayed at beach locations), and bodies deformed as a result of surgery to remove skin cancers. This imagery is not employed to the complete exclusion of other techniques (for example, one current SunSmart campaign in New Zealand features an animated tiger prawn light-heartedly advising on the best ways to cover up in the sun), but is nonetheless dramatic. We suggest that this imagery and associated messages have significant potential to change public perceptions of outdoor environments.

Warnings around UVR risk may transform beaches, in particular, by inserting fear and disease-related anxiety into landscapes understood by many to be sites of relaxation and healthy practices. Clearly, they explicitly target one element of the 'therapeutic experience' enjoyed at many New Zealand beaches – that of deliberate sunbathing and tanning. The public health imperative of avoiding this behaviour is underpinned by the mantra 'tanned skin is damaged skin'. Moreover, a second SunSmart message – sun avoidance during the 'peak UV hours' of 11 am to 4 pm – imposes significant limits on many other leisure-related uses of New Zealand beaches, which are generally undertaken when air and water are warmest. Strict sun avoidance is difficult to achieve on the unshaded dunes and beaches which characterize much of the New Zealand coast, and all but impossible when in the water. Tellingly, a recent cover story in a popular national magazine asked 'is *any* sun safe?' (*New Zealand Listener*, 2005), and portrayed a surfer on a beach – a young man who had been diagnosed with skin cancer at age 20, and whose surgical scars were hidden by a wetsuit (see Figure 2.1).

In considering the potential of anxiety-invoking public health messages around UVR exposure and skin cancer to alter popular understandings of beaches and beach behaviours as therapeutic, we do not suggest threat appeals are necessarily inappropriate. Nor do we wish to imply that concern for health risks more generally should be cast aside in favour of the unfettered pursuit of pleasure – a position which is obviously open to abuse by commercial marketers (Monbiot, 2006). What we do contend is that these messages, in bringing about their goal of behavioural change, have the ability to transform understandings of the relationship between health and place. This transformative potential is not necessarily restricted to the case of sun protection and beaches. One could also consider, for example, the apparent success of social marketing in portraying smoke-filled bars and restaurants as places which pose serious risks to human health (as opposed to places of relaxation or excitement).

It is also important not to over-state the influence of social marketing, which is frequently inhibited by competing messages from commercial marketing. While the ability of commercial marketers to resist public health messages around tobacco is increasingly proscribed, in other areas this is not the case: commercial marketing budgets for motor vehicle manufacturers, for example, overwhelm the budgets available to those promoting alternative forms of transport. In the case of sun protection, the competition is often indirect – commercial marketers do not so much sell UVR (which is, of course, free in its natural form), but sell locations and experiences in which exposure to the sun is a central attraction. As the CEO of the Health Sponsorship Council observed, when asked about the effectiveness of New Zealand's SunSmart campaign:

> Every holiday is about lying on a beach in the sun, under a palm tree. Plus we have the sunbed industry, and the cultural construction of lovely summer days – 'get out there and enjoy the sun.' Whereas [cigarettes are] unacceptable, there is competition around sun protection messages (Iain Potter, interview).

Figure 2.2 Advertisement for locums placed in university alumni magazines, 2005–2006 (reprinted with permission)

Thus, while sun protection campaigns have been operative in New Zealand for over 20 years, their central message is by no means uncontested – something which may go some way to explaining the ongoing allure of tanning. Indeed, tanning and sun exposure remain common, seemingly unproblematicized tropes in much commercial marketing. By way of example, one recent advertisement, prominently placed on the back cover of university alumni magazines in New Zealand, portrayed five tanned young people wearing minimal attire at a sun-drenched beach, and instructed readers to 'live the lifestyle' (see Figure 2.2). What made this otherwise routine combination of image and imperative noteworthy was that the advertisement was promoting an employment service for locum doctors in New Zealand and Australia. In other words, it was appealing to health care professionals in countries where skin cancer is exceptionally common, and represents a significant cause of premature morbidity. The ambiguity is clear: while sun-soaked beaches appear in public health discourses as sites of risk, they have retained powerful reputations for being places of leisure, pleasure and lifestyle. With this in mind, we turn to our survey of beachgoers in Auckland, New Zealand.

Perceptions of Sun Exposure on Auckland Beaches

In order to gain insight into how New Zealanders perceive sun-related risk in the coastal environment, we conducted a questionnaire survey of 100 adult beachgoers in Auckland during the summer of 2005/06. It was carried out during the warmest hours of the day (specifically, between 12 and 4 pm), when large numbers of potential participants were present at two selected beaches – Long Bay, a white sand beach on the city's east coast, and Piha, a black sand beach on the west coast. As UVR is at its peak during these times, and both beaches lack natural forms of shade, our sample was biased towards those willing to accept some exposure to the summer sun (for example, it was unlikely to include strict sun avoiders). The survey included a broad range of questions regarding attitudes towards tanning and sun protection, and behaviours in the sun, especially at the beach. Basic demographic information was also collected, and respondents were asked to identify their skin type on a standardized chart – from Type I (lightest and most prone to sunburn) to Type VI (darkest and least sensitive to the sun). The questionnaire, and associated procedures, were approved by the University of Auckland Human Participants Ethics Committee.

Our sample included 57 women and 43 men, with an average age of 34. Most respondents (71) were New Zealanders, who we predicted would have at least basic knowledge of SunSmart messages. All skin types were included: the most common was Type III (average Caucasian skin: burns moderately, tans gradually), reported by 44 per cent of respondents, followed by Type II (fair Caucasian skin: burns easily, tans minimally) (26 per cent), and Type IV (olive skin: burns minimally, always tans well) (21 per cent). Most of our respondents were, then, more-or-less prone to burning – a significant risk factor for skin cancer – and only two respondents reported the darkest skin type. Moreover, respondents reported being on the beach for an average of 80 minutes prior to completing the questionnaire, confirming that it was a potentially significant site of UVR exposure.

Respondents' attitudes towards tanning suggest resistance to efforts to reduce the social desirability of tanned skin. When asked whether tanned skin was more attractive than pale skin, a majority (62 per cent) said yes, and an additional 10 per cent were in partial agreement. In addition, most respondents (69 per cent) continued to believe that there was a safe way to tan. Interestingly, many of these appeared to have picked up on certain aspects of the SunSmart message, and equated them not with the avoidance of tanning, but with safe, albeit limited, exposure to the sun. Thus, 20 respondents suggested use of sunscreen allowed for safe tanning ('Yes, use sun protection, for example SPF 20 to slowly build up a tan'; 'Yes, with sunscreen – applying it all the time'), and 11 thought a safe tan could be acquired by exposing one's skin in short bursts and/or by avoiding exposure at peak times ('Yes, 20 minutes at a time is enough'; 'Yes, certain times of exposure, for example, not in the high noon sun'). Both approaches are contrary to public health warnings: for example, that 'a tan ... is a sign of skin damage' and that '[s]unscreen should not be used to extend time in the sun or to achieve a suntan' (Cancer Council of Australia, 2003, 1). This said, 28 respondents made less problematic observations to the effect that a safe tan could be acquired 'out of a bottle' or in the form of sprays. Three believed that the use of sunbeds was safe (whereas in fact they emit UVR), while 30 stated that no tan was safe.

Although many beachgoers' attitudes to tanning were favourable, this did not correspond with a majority actively seeking a tan. Indeed, only 45 per cent of respondents reported that they were trying to tan at the time of the interview. This group reported attitudes strongly reflecting twentieth-century cultural constructions of tanning: 14 were trying to tan because it looked healthier ('to get a healthy glow'; 'I look healthier and feel better about myself. I don't look sick and pasty'), 12 offered other reasons relating to appearance ('because I look slimmer'; 'It saves on make-up – I don't have to wear foundation and my clothes look better'), and nine tanned for enjoyment ('because I like it!'). Perhaps the most startling response to this question came from a 20-year-old woman who reported: 'I think it looks better even though I know it's going to give me cancer.' Thus, even exaggerated awareness of skin cancer risk may not, in and of itself, bring about a shift in attitudes and behaviours: the social marketing of sun protection is by no means a straightforward task.

This said, a majority of respondents (54 per cent) stated they were not tanning, and many gave reasons relating to the health dangers of sun exposure. For 18 respondents, this was a foremost concern, with eight specifically mentioning skin cancer ('Not now, I'm scared of skin cancer') Indeed, one respondent perceived tanning as radically anti-therapeutic: 'I know how dangerous it is and how vicious the sun is.' An additional 10 non-tanners focused less on health risk *per se*, than on the propensity of their skin to burn: ('It's hard work, and I burn too easily'; 'I don't tend to tan. I tend to burn'). Another 12 respondents framed their decision not to tan in personal terms, offering explanations such as 'I can't be bothered'.

Respondents aged 30 and under (exactly half the sample) were more likely to be tanning, and to view tanned skin as attractive, than those aged over 30 (see Table 2.1). At one level this is consistent with research pointing to the enduring appeal of tanning among younger people (Lupton and Gaffney, 1996; Garvin and Wilson, 1999). However, the observed differences in behaviour and perception across the two age groups were not found to be statistically significant using chi-square tests. While it might be expected that younger people have a greater tendency to resist health-related threat appeals (Montague et al., 2001), it bears emphasizing that sun protection has been marketed in New Zealand since the mid-1980s: younger people have been exposed to warnings about UVR and tanning for most, if not all, of their lives.

Table 2.1 Attitudes towards tanning by age, Auckland beach survey

	Respondents ≤30 years (*n*=50)		Respondents >30 years (*n*=50)	
Do you find tanned	Yes	No	Yes	No
skin attractive?	70%	20%	54%	36%
Are you currently	Yes	No	Yes	No
trying to tan?	66%	34%	24%	74%

Not all percentages add to 100, as table includes definitive 'yes' and 'no' answers only, and omits data for more ambiguous responses, as well as for those who declined to respond.

While dividing the sample according to whether or not respondents were tanning at the time of interview offers some insight into sun-related attitudes and behaviours, it risks creating the impression of a polarized population – that is, of two mutually exclusive groups, one of which accepts the messages of social marketing vis-à-vis sun protection, and another which does not. On closer inspection, however, the majority of tanners were undertaking certain measures to protect themselves from the sun, such as wearing sunscreen (36/44), sunglasses (31/44), and/or protective clothing (24/44). In addition, three were making use of umbrellas to shade their bodies, at least partially. The proportions of non-tanners making use of sunscreen and sunglasses were little different, although they were considerably more likely to wear protective clothing and make use of shade. Respondents were not questioned about the SPF value of their sunscreen, although we note that most sunscreens currently offered for sale in New Zealand advertise a protection factor of at least 15. Notably, only three tanners reported taking no precautions against excessive exposure (with one candidly stating 'I put on the oil and fry!'), as did two non-tanners (both of whom claimed they never burned). Any suggestion of stark differences between tanners and non-tanners in terms of their UVR-related knowledge was further dispelled by the ability of all respondents to identify at least one health risk associated with sun exposure, with 92 per cent mentioning skin cancer.

Our survey also included a number of questions intended to encourage reflection on sun exposure *at the beach*, and not merely tanning or UVR risk in general. First, we asked respondents whether their attitudes to sun protection were different at the beach than at other outdoor locations. Most responded that they were not (64 per cent). Of those who answered in the affirmative, many observed that they took extra precautions in the beach environment: 'I am not nearly as worried [in other places] so am less inclined to wear sunblock away from the beach'; 'I am probably more careful at the beach than I am anywhere else. There is not much shade at the beach so you have to be a bit more careful'; 'I am more aware of the sun when I am at the beach and come prepared for a day in the sun'. While these comments point to awareness of the risks of sun exposure in a beach environment, several participants noted that increased preparedness was necessary in part because of the tendency to wear less clothing at the beach.

Second, we informed respondents that many primary schools in Auckland were building shade structures to protect pupils (see Collins et al., 2006), and then asked whether similar protection should be provided at beaches where there were few trees (such as those at which we were conducting our survey). In total, 74 per cent of respondents supported this idea, with many suggesting that umbrellas and tents should be available for hire (as is the case at some international resorts). Those opposed to this suggestion were generally of the view that the provision of shade was an individual responsibility ('No, it's an individual thing to bring an umbrella'; 'No, people should take responsibility for themselves'), while some were also concerned for aesthetic values ('No, beaches are better natural'; 'No, it would be hideous').

Third, we presented respondents with a scenario in which a young child was in direct sunlight at a beach, seemingly unprotected – that is, without a hat or other protective clothing – and asked whether they would step in and say anything to the child or caregivers. Just under one-quarter of respondents thought that they would intervene under these circumstances, whereas 76 per cent said they would not. For a small number, a reluctance to intervene was due to the (not unreasonable) notion that the child could be assumed to be wearing adequate sunscreen, at least in the absence of obvious burning ('No, not unless they were extremely burned, I'd imagine they would have suncream on'). A more common response, however, was that intervention would involve crossing the boundaries of privacy which prevail, even on public beaches. Specifically, respondents suggested that interfering in the parenting practices of others was deeply inappropriate, and that parents essentially had the right to be left alone ('No, it's not really my place – it's the parents' responsibility'; 'No, I don't involve myself with other people's children'). This held true even among those who thought they would be genuinely concerned for the well-being of the child in question: 'No, I would feel like it because I know the dangers of getting burned, the child would have pain. But I don't know that I'd tell them ...'; 'No, I would probably just bitch about it but don't want to tell other parents how to do their job'. For other respondents, respect for privacy and parental autonomy involved a degree of self-preservation: eight expressed concern that any intervention would be met with a hostile, even violent response ('No, I would get hit over the head by the parent!').

Thus, while the social marketing of UVR protection appears to have influenced some beach practices, particularly in terms of widespread use of sunscreen, it does not seem to have heightened public awareness of risk to the point that beachgoers are generally willing to intervene in the practices of others. Social expectations around privacy remain intact: sun protection is envisioned as an individual choice at the beach (although not at all schools or workplaces, where institutional authority can insist on protective practices). Accordingly, adults are deemed responsible for themselves, and parents are deemed responsible for young children. This allows for freedom to disregard sun protection messages, or to interpret them in ways consistent with desires to tan.

The final question in our survey asked whether there were any factors influencing personal behaviour in the sun which had not been discussed. A number of respondents remarked that they took precautions at the beach primarily because their skin, or that of their children, was fair and prone to freckling – and they knew this put them at risk. While these responses could be attributed in part to the success of social marketing messages, other participants reported more *unmediated* knowledge of the health risks associated with sun exposure. Specifically, three had personally contracted some form of skin cancer during their lives, four had friends who had been diagnosed with skin cancer, and nine reported a family history of the disease ('My mum, dad and grand-dad have had skin cancers cut off so I am acutely aware of what the sun can do'). Given the remarkably high incidence of skin cancer in New Zealand, knowledge and attitudes vis-à-vis sun exposure can be shaped by myriad factors beyond official social marketing campaigns.

Conclusion

In this chapter, we have argued that a concern for the health dangers associated with UVR exposure has complicated the reputation of New Zealand beaches as therapeutic landscapes. They continue to be understood as places of leisure and pleasure, where certain elements of a healthy lifestyle may be pursued, but they are also sites of risk. Indeed, beaches are *central* to UVR-related risk in New Zealand, as largely unshaded sites of both deliberate and unintentional sun exposure. Ozone depletion in the Southern Hemisphere has heightened the risks associated with such exposure, contributing to a situation whereby a key ingredient of the proverbial 'good day at the beach' – the sun – is now also envisioned by experts, and some members of the public, as a significant threat to human health. This ambiguity is, arguably, the outcome of a broader trend, whereby the new public health emphasizes the risks associated with activities often conceived of as pleasurable (for example, eating, drinking alcohol, having sex, speeding) (cf. Brown and Duncan, 2002).

Our case study lends support to claims that social marketing campaigns such as SunSmart have brought about at least partial changes in individual attitudes and behaviours (see Montague et al., 2001). In particular, there has been widespread acceptance of the need to apply sunscreen to reduce sun damage to the skin – although this behaviour is often incorporated within discourses of 'safe tanning' and 'healthy and attractive tans', as opposed to the practice of sun avoidance promoted by many public health messages. We also found evidence to suggest that, given the high incidence of skin cancer in New Zealand, people's knowledge of, and attitudes towards, sun exposure may be shaped by a range of factors beyond formal public health campaigns.

This said, threat appeals appear to have been taken seriously by a significant minority, even among those willing to go to the beach and risk some level of exposure. When beachgoers offer comments such as 'the sun is very dangerous', we see evidence of a double-coding: relatively direct sunlight both facilitates and problematicizes many enjoyable beach experiences. Moreover, as some beaches offer more opportunities for shade than others, and as UVR levels vary significantly both seasonally and diurnally, the nature of this risk is not unchanging – and the therapeutic qualities of beach experiences may vary accordingly.

Moving beyond the specificity of beaches, we can suggest that the increasingly omnipresent discourse of risk (Beck, 1992) complicates the notion of therapeutic landscapes. While particular places continue to offer opportunities for activities and experiences which are psychologically and/or physically therapeutic, promoters of the new public health also encourage us to recognize the risks present in these milieux, and to modify our behaviours accordingly. In some instances, these behavioural changes may be dictated by powerful institutions. In others, individuals are encouraged to adopt the right (risk-averse) attitudes and behaviours through social marketing campaigns which work in part by instilling feelings of fear, anxiety and guilt.

The imperative of recognizing risk, and responding accordingly, is particularly acute in those outdoor environments which offer opportunities for physical activity, encounters with 'nature', and escape from quotidian concerns – precisely those environments often considered conducive to therapeutic experiences. Yet, in New Zealand as elsewhere, these sites remain popular sites for recreation and relaxation. Exposure to risk appears to be an *inescapable* element of many embodied engagements with the landscape. All places are to some extent 'risky', and therefore candidates for public health interventions.

In concluding, we suggest three important implications flow from our research. First, even in landscapes commonly understood as therapeutic, it may not be possible to evade the necessity of 'being responsible' by internalizing public health messages. Second, public health has a seemingly unbounded mandate for scrutinizing and judging the risks inherent in the geographies of everyday life. Finally, public health messages work, in part, by seeking to transform popular understandings of the relationship between health and place.

References

Andrews, G.J. and Kearns, R.A. (2005), 'Everyday health histories and the making of place: the case of an English coastal town', *Social Science and Medicine* 60: 2697– 713.

Armstrong, B.K. (2004), 'How sun exposure causes skin cancer', in D. Hill, J.M. Elwood and D.R. English (eds) *Prevention of Skin Cancer* (Dordrecht: Kluwer Academic Publishers).

Beck, U. (1992), *Risk Society* (Newbury Park, CA: Sage).

Biener, L. and Taylor, T.M. (2002), 'The continuing importance of emotion in tobacco control media campaigns: A response to Hastings and MacFadyen', *Tobacco Control* 11: 75–7.

Brown, T. and Duncan, C. (2002), 'Placing geographies of public health', *Area* 33: 361–69.

Cancer Council of Australia (2003), 'Tanning, Ageing and Beauty' (published online January 2003) <http://www.cancer.org.au/content.cfm?randid=299825>

Cancer Society of New Zealand (2005), 'Position Statement: The Risks and Benefits of Sun Exposure in New Zealand' (published online March 2005) <http://www.sunsmart.org.nz/skin-cancer-control-in-nz/position-statements/risks-and-benefits-of-sun-exposure.aspx>

Chapman, S., Marks, R. and King, M. (1992), 'Trends in tans and skin protection in Australian fashion magazines, 1982 through 1991', *American Journal of Public Health* 82 (12): 1677–80.

Collins, D. and Kearns, R. (2008), Uninterrupted Views: Real Estate Advertising and Changing Perspectives on Coastal Property in New Zealand, *Environment and Planning A* (forthcoming).

Collins, D., Kearns, R. and Mitchell, H. (2006), "An Integral Part of the Children's Education": Placing Sun Protection in Auckland Primary Schools', *Health and Place* 12 (4): 436–48.

Conradson, D. (2005), 'Landscape, care and the relational self: Therapeutic encounters in rural England', *Health and Place* 11: 337–48.

Daley, C. (2003), *Leisure and Pleasure: Reshaping and Revealing the New Zealand Body 1900–1960* (Auckland: University of Auckland Press).

Garvin, T. and Eyles, T. (2001), 'Public health responses for skin cancer prevention: the policy framing of sun safety in Australia, Canada and England', *Social Science and Medicine* 53: 1175–89.

Garvin, T. and Wilson, K. (1999), 'The use of storytelling for understanding women's desires to tan: lessons from the "Field"', *The Professional Geographer* 51 (2): 296–306.

Hastings, G. and MacFadyen, L. (2002), 'The limitation of fear messages', *Tobacco Control* 11: 73–5.

Hill, D., Chapman, S. and Donovan, R. (1998), 'The return of scare tactics', *Tobacco Control* 7: 5–8.

Jopson, J. and Reeder, A. (2004), *Sun protection in New Zealand secondary schools: obstacles and opportunities* (Dunedin: Department of Preventative and Social Medicine, University of Otago).

Kearns, R. and Collins, D. (2000), 'New Zealand Children's Health Camps: Therapeutic Landscapes Meet the Contract State', *Social Science and Medicine* 51 (7): 1047–1059.

Kearns, R. and Collins, D. (2006), "On the Rocks?' New Zealand's Coastal Beach Landscape and the Case of Rangitoto Island', *New Zealand Geographer* 62 (3): 228–36.

Keen, D. and Hall C.M. (2004), 'Second Homes in New Zealand', in C.M. Hall and D.K. Muller (eds) *Tourism, Mobility and Second Homes: Between Elite Landscapes and Common Ground* (Clevedon: Channel View Publications) pp. 174–95.

Lupton, D. and Gaffney, D. (1996), 'Discourses and practices linked to suntanning and solar protection among young Australians', *Health Education and Research: Theory and Practice* 11 (2): 147–59.

Matthewman, S. (2004), 'More than sand: theorising the beach', in C. Bell and S. Matthewman (eds) *Cultural Studies in Aotearoa New Zealand: Identity, Space and Place* (Melbourne: Oxford University Press) pp. 36–53.

McDermott, L., Emmison, M. and Lowe, J. (2003), 'Changing Discourses and Popular Attitudes to Suntanning', in J. Skinner, K. Gilbert and A. Edwards (eds) *Some Like it Hot: The Beach as a Cultural Dimension* (Oxford: Meyer and Meyer Sport) pp. 91–109.

Monbiot, G. (2006), 'Just follow the money', *Guardian Weekly*, 17 February.

Montague, M., Borland, R. and Sinclair, C. (2001), *SunSmart twenty years on: What can we learn from this successful health promotion campaign?* (Victoria: Anti-Cancer Council of Australia).

New Zealand Health Information Service (2004), *Cancer, New Registrations and Deaths* (Wellington: New Zealand Health Information Service).

O'Dea, D. (2000), *The Costs of Skin Cancer to New Zealand* (Wellington: Cancer Society of New Zealand).

Pawson, E. (2002), 'On the Edge: Making Urban Places', in E. Pawson and T. Brooking (eds) *Environmental Histories of New Zealand* (Melbourne: Oxford University Press) pp. 200–13.

Perkins, H.C. and Thorns, D.C. (2001), 'Gazing or Performing? Reflections on Urry's Tourist Gaze in the Context of Contemporary Experience in the Antipodes', *International Sociology* 16: 185–204.

SunSmart (2006), 'Skin Cancer Control in NZ', <http://www.sunsmart.org.nz/> (home page), accessed 21 October 2006.

Waddell, T. (2003), 'The Not So Secret Life of Us at Home and Away: Cashing in on Beach Culture', in J. Skinner, K. Gilbert and A. Edwards (eds) *Some Like it Hot: The Beach as a Cultural Dimension* (Oxford: Meyer and Meyer Sport) pp. 40–57.

Wakefield, S. and McMullan, C. (2005), 'Healing in places of decline: (re)imagining everyday landscapes in Hamilton, Ontario', *Health and Place* 11: 299–312.

Chapter 3

The Experiential Economy of Stillness: Places of Retreat in Contemporary Britain

David Conradson

A fragmented and rushed temporality is typical of a growing majority of the population in the rich countries.
Eriksen (2001, 148)

Introduction

The subjective experience of stillness has often been integral to human encounters with therapeutic landscapes. In areas of natural beauty, such as alpine or coastal settings, people may thus find themselves lost for words. This arrest of speech, where human volubility is disrupted, typically reflects both a sensory or cognitive overwhelming and a heightened consciousness of the present moment. The passage of time appears to be slowed. At sites of spiritual pilgrimage, we may observe practices of reflection and contemplation that foster a similar shift in perception. In connecting with the Divine or reconsidering one's circumstances in view of a metaphysical plane of reference, new forms of awareness may emerge. While the capacity of particular socio-environmental settings to induce stillness is a complex matter, a slower rhythm of activity and the distance afforded from everyday routines are often important. Cultural understandings of landscape may also be significant, with the legacy of post-Enlightenment Romanticism of particular salience in Western countries (Glacken, 1976).

But what exactly is stillness? In this chapter stillness is understood in psychosocial terms, as an internal state of calm in which a person becomes more aware of their immediate, embodied experience of the world and less concerned with events occurring 'out there'. The experience of stillness may encompass an enhanced clarity of thought, or act as a portal to new modalities of feeling. Importantly, this affective state is not dependent upon physical inactivity or geographical stasis. Nor is it necessarily coincident with auditory silence. An individual may thus subjectively experience a train or plane journey, when there is noise, movement and various forms of activity, as a space of relative stillness. Such a perception may emerge in part because travelling provides some insulation from external demands, or at least lowers expectations regarding a quick or immediate response to them. The psychosocial states which follow may emerge in part because of this diminution of relational intersections.

Responding to these ideas, this chapter considers how the potentially therapeutic environments of two Benedictine monasteries can be related to a broader experiential economy of stillness. It is argued that stillness has been subject to growing valorisation in Western countries in recent years. With pressures on many configurations of work–life balance, the scarcity of stillness has contributed to heightened public interest in places of retreat (see, for example, The Retreat Association 2007; Whiteaker, 2004). Many of those who visit such places of retreat – including Christian monasteries, Buddhist meditation centres and more generically designated wellbeing centres – appear to be looking for environments that facilitate rest and renewal. In contrast to historic practices of religious pilgrimage, however, these visitors may not express any personal affiliation to the belief systems that infuse a particular place of retreat. Their engagement may be prompted by more loosely formulated notions of spirituality (Bruce, 2002; Davie, 1994; McLeod and Usdorf, 2003), or it may simply reflect an appreciation of stillness in more secular terms.

Theoretically, an appreciation of socio-economic context is thus critical to understanding how and why particular landscapes come to be designated or experienced as therapeutic. Perhaps because some of the cases examined have been ancient (Gesler, 1993) or coincident with relatively undeveloped natural environments (Palka, 1999), the relation between particular therapeutic landscapes and their wider context has not always received sustained attention in the literature. Wider societal trends and social policy matters have certainly been explored, as evident in studies on topics such as home care (Williams, 2002), children's summer camp environments (Thurber and Malinowski, 1999; Kearns and Collins, 2000) and collective gardening programmes (Milligan et al. 2004). But there are arguably benefits to be gained from giving greater attention to the economic dimensions of context. In particular, by considering how the experiential properties associated with a particular setting are subject to variable evaluation and distribution within wider society, we can gain useful insights into why a place is regarded as therapeutic. As I have argued elsewhere, 'in order to understand a particular therapeutic landscape experience, it is useful to give attention to the broader relational configurations within which it occurs' (Conradson, 2005a, 346)

In developing these points, the chapter has two main sections. In the first, I reflect upon the variable evaluation and distribution of stillness within contemporary British life. Particular attention is given to the cultural assumption that stillness is most easily attained in locations beyond the everyday sites of work and home, in places that are in some sense 'elsewhere'. Although focused on the UK, many of the points made are relevant to other Western nations, including the United States, Canada, Australia, New Zealand and the countries of Western Europe. In the second part of the chapter, I examine two examples of places of retreat, each rooted in the Benedictine tradition of Christian spirituality (de Dreuille, 2002; Knowles, 2004). Drawing on field visits and interviews with staff, I consider how the practices and temporal rhythms of these communities contribute to an experiential environment facilitative of internal stillness for visitors. At the same time, the inclusion of such places in published retreat directories has seen them drawn into a broader experiential economy which valorises speed, connectivity and regular electronic communications. As these monastic communities interface with society, the ethos inherent to their work may thus be subjected to various forms of pressure. In conclusion, I outline some more general points regarding therapeutic landscape research.

The Variable Evaluation and Distribution of Stillness

> The tyranny of the present is not boredom or the lack of difference and colour and excitement in our lives but the opposite: we are over-excited, bombarded by stimulation, information, possibilities, connections and access. It is claimed that our lives are too busy; that we are trying to do too much too often; electronic communications speed up our ability to do things and the pace and extent of our transactions. We are bereft of time to commit to things we consider important like long-term relationships, our children, careful planning and leisure (Franklin, 2003, 13).

Stillness, it would seem, is a much desired experiential state in Western countries these days. Reflecting in part the pressures of intensified work and home commitments (MacInnes, 2005), significant numbers of people appear to be seeking ways of cultivating and preserving internal tranquillity. A diverse range of social practices have arisen to meet these needs. We can observe the popularity of yoga, tai chi and other forms of therapeutic mind–body practice, for instance (Thrift, 2000a). We might also note the growth of slow tourism and holidays that emphasise a relative absence of activities, all as part of deliberate efforts to facilitate personal wellbeing within a culture that by and large valorises speed (Eriksen, 2001; Franklin, 2003). While each of these social phenomena is multi-faceted, an important dimension of their attractiveness would seem to be the promise and possibility of more restful psychosocial states.

If the valorisation of stillness in Western societies reflects work–life balance pressures, it is also connected to the development and dynamics of a broader experiential economy (Pine and Gilmore, 1999). As O'Dell (2005) has argued, experiential economies focus on the commodification of particular modes of feeling and being, whether relatively ephemeral or more enduring. The experiences offered by many contemporary night-clubs or dance spaces are part of such an economy, for what is being sold typically centres on a particular form of embodied rhythm and engagement rather than any discrete product. Adventure tourism activities, such as bungy jumping and ski-diving, similarly involve the commercialisation of heightened sensory experience (Cloke and Perkins, 2002). In each case, these developments reflect the pervasive tendency of capitalism to colonise new areas of social and psychic life, in an effort to extract a profit (Thrift, 2000b).

Alongside the market for intensive, energetic activities of this type, we can also observe a growing interest in what might be described as 'slow experiences' (Franklin, 2003). Here the emphasis is on the diminution rather than amplification of sensory stimuli, and upon cultivating less intense rhythms of activity than those which characterise many Western lifestyles. A range of commercial and non-commercial responses to the demand for slow experiences can now be identified, including holiday cottages marketed on the basis of isolation and solitude, yoga-based breaks in mountain locations, and various forms of Buddhist retreat. Part of what is being offered here is the promise of fewer external intrusions, with an accompanying narrative which suggests that such an environment will foster personal renewal and restoration.

Thinking about therapeutic landscapes in relation to experiential economies alerts us to a number of things. Firstly, the notion of economy asks that we consider how particular experiential states are *differentially valued*. The value ascribed to a psychosocial state such as relaxation, happiness or excitement may vary significantly within and between social groups, for instance. Such variation is often linked to the relative accessibility of the experience for a particular person or group. In the UK, for instance, studies of work–life balance have indicated that many people feel that their lives have become more stressful in recent years (MacInnes, 2005). Working hours are amongst the longest in Western Europe, a significant number of the so-called 'professional classes' declare that they suffer from stress, and levels of absenteeism through work-related illness undoubtedly constrain workforce productivity. In this context, efforts to combat work–life balance issues typically seek to either reduce overall hours or to enable more flexible patterns of working (so that employees experience less conflict with childcare responsibilities, for example). An equally important, albeit less easily measurable, dimension of work–life balance might be described as the capacity to achieve or attain stillness. It is here that we can connect with the growing interest in various forms of slow experience and therapeutic mind–body practices.

Secondly, the notion of economy invites us to notice that access to particular experiential states is *differentially distributed*. The conditions which facilitate stillness, for example, are variably distributed in both social and geographical terms. Not everyone has an equal ability to control their work schedule, for instance, or similar resources to address key tasks. We can also think of places where stillness is more likely to be enabled than others. In a work environment such as the trading floor of a stock exchange, where profitability is closely linked to the speed of deals being made, the reflective dimensions of stillness would generally be seen as antithetical to productivity. The same might be said for many labour-intensive manufacturing settings. By contrast, an artist or writer might actively seek to cultivate internal stillness in their work setting – by virtue of quietness and the absence of external interruption – because this is understood as somehow conducive to quality results. The variable distribution of stillness is thus closely linked to its valuation by key actors in each setting.

If the accessibility of a particular experiential state varies across space, then a third point is that geographic mobility is one route towards new modes of thinking and feeling. By changing place, humans are able to reposition themselves in relation to the social and economic relations that shape the emotional dynamics of particular settings. Leaving the office so as to walk in a park is a well-practised form of relaxation, for example. Similarly, spending a weekend away from home and work commitments may be experienced as rejuvenating. As strategies for thinking and feeling differently, these forms of geographic mobility recognise the variable accessibility of particular experiential states across space (Conradson, 2005b).

Table 3.1 The variable evaluation and distribution of stillness in contemporary Britain

ACTIVITY SPHERE	WORKSPACE	HOMESPACE	'ELSEWHERE SETTINGS'
Key influences upon the experiential texture of spaces within these activity spheres	• Employers and management • Regulatory authorities • Legislation (e.g. labour law, health and safety directives)	• Household members • Those financially responsible for property, such as tenant(s) or owner(s) • Government initiatives (e.g. work–life balance policies)	• Private companies (e.g. slow tourism and wilderness holiday firms; gyms offering yoga classes) • Government bodies (e.g. National Park committees) • Community, religious and spirituality groups (e.g. walking collectives, monasteries, retreat centres)
Dominant evaluation of stillness as a psychosocial state	• 'UNPRODUCTIVE INACTIVITY'	• TYPICALLY VALUED BUT UNDER PRESSURE	• 'PRODUCTIVE NON-ACTIVITY'
Social practices that facilitate or marginalise stillness	• Techniques to minimise stillness may be evident in highly structured work environments, such as time-sheeting and clocking in/out procedures	• Some evidence of efforts to reduce activity, such as the 'slow food' movement and personal meditation. But such practices are highly variable by income and class status	• Techniques to cultivate stillness include: engagement with nature, such as walking in scenic settings; lowering levels of activity; and therapeutic mind–body techniques (e.g. yoga and pilates)
Distribution of stillness	• Stillness is a relatively scarce psychosocial state within many contemporary workplaces	• Stillness is unevenly present in contemporary home-spaces, with significant variation by household composition, education, income, and position in life course	• Stillness is often perceived to be more easily achieved in 'elsewhere settings' than in work or home spaces

Drawing these arguments together, Table 3.1 schematically depicts the variable evaluation and distribution of stillness in contemporary British society. It does so with reference to three overlapping spheres of the life-world: workspace, homespace and 'elsewhere' settings. The first refers to the domain of the formal economy, where labour is undertaken; the second to domestic environments, including the living and sleeping arrangements of a household, however configured; and the third term is intended to denote other important locations, such as leisure and community environments.[1] A selection of the key influences upon the experiential texture of these spheres is noted. Within many businesses, supervisors and managers exercise significant control over the working environment, for example. In contrast, domestic settings are often more able to be shaped by their residents, although landlords may also be involved. Reflecting their diversity, 'elsewhere' environments are subject to a wide range of influences, depending on whether they are commercial (e.g. wilderness cottages and spa complexes, public sector (e.g. national parks) or voluntary sector developments (e.g. religious-based retreat centres).

An indication of the dominant evaluation of stillness in each of these spheres is then noted. In many contemporary workplaces, perhaps particularly in business but also the public sector, it is suggested that stillness tends to be framed as *unproductive inactivity*. With the exception of work in which reflection or contemplation are explicitly foregrounded, stillness is often conflated with stasis and thus seen as down-time and inimical to profit generation. In contrast, the inhabitants of most home settings typically place significant value upon stillness and the conditions which facilitate it, such as the ability to control one's own schedule and to deflect or defer external demands. Despite this general cultural endorsement of stillness in the domestic sphere, it is nevertheless under pressure by virtue of the increasing complexity of managing work, friendships, family and other social commitments (Urry, 2004).

In the 'elsewhere' settings of wider social life, a growing interest in slow experiences and therapeutic mind–body practices is evident. Such activities suggest an alternative evaluation of stillness as *productive non-activity.* This assessment is arguably connected to the less positive framings evident in some workplaces, and to the difficulty of achieving internal stillness in many domestic settings. Such difficulties in part reflect the increasingly porous nature of the boundary between home and work. We can also observe the domestic intrusions of cold-calling (electronic systems seeking to sell goods and services) and increased internet connectivity (Eriksen, 2001). Given the relative elusiveness of the social conditions conducive to internal stillness in many home settings, it is unsurprising that 'elsewhere' has

1 It is important to stress that this conceptualisation is schematic. One can immediately think of situations which trouble its assumed boundaries, such as the self-employed woman who works at 'home', the unemployed man who knows no geographical workplace, or the homeless youth whose home-space is considerably more dispersed than that of a fixed residence. While acknowledging the significance of these sets of circumstances, the table is intentionally constructed with reference to the narrower group of people that (a) are in paid employment of some kind beyond the home and (b) possess a fixed residential environment. The rationale for this choice is that it reflects the predominant profile of those who visit retreat centres in contemporary Britain.

become a conceptual and geographical category into which all manner of hopes are invested. In such circumstances, the possibility for renewal through temporary relocation is understandably attractive.

Places of Retreat: Alton Abbey and Elmore Abbey

> Stilling the tongue gradually leads to a stillness of mind – but not immediately. At first, we become very aware of a babble of voices within, a never-ending stream of thoughts and images and emotions (Tyers, 2007, 12).

Stillness has thus become increasingly noticed and valued as an experiential state in contemporary Britain, both within and beyond the commercial economy. This valorisation can be seen in the frequent references to tranquillity and renewal in many holiday advertisements, as well as in the popularity of various forms of meditation and yoga. A further consequence of this valorisation is that sites already characterised by conditions conducive to stillness have become the subject of growing public interest. Monasteries are one focus of such attention, although the stillness they afford typically emerges more as a by-product of worship and reflection than as its primary aim. For those who seek refreshment, however, the historic openness of monasteries to guests means they become potential places of retreat.

The research reported upon here examined two such places in southern Britain: Alton Abbey and Elmore Abbey. Both are active monastic communities in the Benedictine tradition, with a group of men committed to collective life and spiritual development. In examining these places as potentially therapeutic environments, three related issues will be discussed: (1) the nature and rhythms of their everyday practices; (2) the ways in which guests experienced their visits, drawing on initial research and relevant secondary materials; and (3) how these communities are being drawn into a wider experiential economy of stillness, in a manner that at times sits uneasily with their ethos of contemplative reflection.

Methodologically, the examination of these issues draws on participant observation visits to each community (undertaken as part of a personal day retreat); interviews with Guestmasters (the monk responsible for welcoming guests); unstructured and informal discussions with fellow day visitors who were present at the time; and examination of web-based and print materials that describe the monasteries and their work. Taken together, these techniques and materials enabled a broad understanding of each setting.

Rhythms of Practice: Worship, Work and Welcome

Founded in 1895, Alton Abbey is situated in a small, rural village in the southern English county of Hampshire. Although subject to fluctuation, at the time of writing the monastic community consisted of six permanent members. Alongside this core group, members of the public are able to engage with the collective life of the Abbey in a number of ways. The most intensive of these is Oblature, a form of spiritual affiliation in which an individual commits to privately following the spirit of St Benedict's rule – a text described in more detail below – without a residential

Figure 3.1 The grounds of Alton Abbey, Hampshire, England

Source: Author's photograph

commitment to monastic life. Then there is Companionship, which involves regular participation in worship and occasional retreats. Less formally, a person may come to the Abbey as an occasional residential guest or simply as a day visitor. Around 400 individuals undertake residential stays each year of various forms. Finally, it is possible to become a 'Friend of Alton Abbey', which is essentially a financial commitment by previous guests to support the work of the Abbey in the future.

Located some distance to the west, in the county of Berkshire, Elmore Abbey is also an Anglican Benedictine monastery. The community here is slightly smaller, with five brethren, a number of Oblates, and around 200–250 residential guests per year. In addition, there is a significant number of day visitors. Both visitors and guests come to a functioning, residential community, such that the Abbey is not dependent upon their presence to create a particular ethos or sense of atmosphere.

The setting of each monastery is quiet and relatively secluded, with attractive gardens surrounding the main buildings, and rolling English farmland beyond (Figure 3.1, Figure 3.2). For those who visit, the experiential environment is closely shaped by the Rule of St Benedict, a particular exposition of faith and practice for Christian monastic life. Benedict was born in Italy, at Nursia, around 480 AD, and became committed to a monastic existence in his early twenties. His Rule, which is thought to have been written around 530 AD, sought to chart a *via media* or middle way between strict asceticism on the one hand, which was the dominant form of monastic practice at the time, and forms of communal living that would have been relatively indistinguishable from the wider society in which they existed. Importantly, Benedict argued that a base-line level of physical comfort was not necessarily in

Figure 3.2 Elmore Abbey, Berkshire, England

Source: www.elmoreabbey.org

conflict with a life dedicated to worship and individual holiness. His rule has been highly influential in shaping Christian monasticism in subsequent centuries, both within Europe and, more recently, in North America and Australasia (Dean, 1989).

The Rule was essentially intended as a guide for Christians who had determined to live, pray, work and eat together under the authority of an Abbott. In 73 short chapters, the text offers instruction on rhythms of prayer and worship; on the importance of humility, silence and deference in community relations; and on the value of practical work, both to maintain an Abbey and its surroundings, and to provide an opportunity for physical exercise. The emphases of the Benedictine monastic tradition are thus sometimes summarised by the phrase 'work and pray'. Although initially written for men, it is important to note that the Benedictine rule has also been adopted and embodied by communities of women.

The daily routine of each Abbey is structured around the Benedictine practice of regular prayer and sung worship at intervals throughout the day. At Elmore, this begins at 0530 with Matins and then progresses via Terce (1000), Sext (1200), None (1600), Evening Prayer (1800) and the Compline (2030), or 'completion of the work of God'.[2] A similar pattern is followed at Alton Abbey. Interspersed between these times of worship are shared meals, opportunities for private and collective study, and physical work (for

2 The names of some of these worship sessions reflect the Latin for the third, sixth and ninth hours of the day respectively.

example, gardening). Guests are invited to share in prayer and worship as they wish, but there is no compulsion or requirement to do so. Outside of the scheduled services, silence is observed wherever possible. This is to allow those present to engage in the Benedictine practice of *recollection*, a discipline which entails a deliberate cultivation of a reflective stance towards one's own life and its relation to God or the Divine.

In the 53rd chapter of the Rule, which specifically addresses the issue of welcoming visitors from outside the community, it is noted that guests are to be received 'as Christ himself'. Such hospitality has in fact been a distinctive and enduring feature of Benedictine monasticism. Although the guest is afforded courtesy and care during their stay, the Rule specifies that the monastic community should not otherwise engage or interact with such persons except by special permission from the Abbott. The primary focus for the monastic community is worship and work, with hospitality an important but nonetheless ancillary dimension of their activities. In this sense, becoming a guest at a monastery is quite different from attending a dedicated retreat centre, especially if attendance at the latter facility is undertaken as some kind of paying customer.

The Experiences of Guests: Space-Times of Stillness

By virtue of the call to 'restraint of speech' (chapter 6 of the Rule) and structured worship times, the experiential texture of each Abbey stands in contrast to the home and work-spaces of most guests. A sense of this alterity is often encountered prior to visiting, for if one wishes to enquire about a potential retreat, the preferred method is to write a letter or telephone during specified, non-worship hours. The process is thus rather different than the customer-focused interactions of the hospitality industry. Elmore Abbey has recently included an email address on its website, but there remains a sense that external requests are managed so as to preserve a particular rhythm of monastic life. In contrast to the widespread expectation in 'fast societies' of quick responses to electronic communications (Eriksen, 2001), the intention is that worship and reflection continue to be centre-stage.

As one travels down the countryside lanes that lead to each monastery, this transition from a faster to a slower world is also discernible. The landscaped gardens and trees visible upon approach seem to signal tranquillity, if only by virtue of their association with bucolic notions of rurality. Having arrived, the hospitality extended to guests has a number of elements: an initial welcome from the Guestmaster, with guidance to the accommodation; an assurance of confidentiality with regard to any personal information volunteered or disclosed; an implicit invitation to take off social masks, in terms of encouraging an attitude of authenticity and being oneself; an emphasis on being and becoming rather than doing; and an encouragement to engage with the scenic natural environment in which each Abbey is set. There are parallels here with the therapeutic environments which counsellors and psychotherapists seek to foster, in that an effort is made to fashion conditions in which an individual's thoughts and feelings are both taken seriously and contained (Hazler and Barwick, 2001). Indeed, one of the Guestmasters made explicit reference to his understanding of the Abbey as an environment in which one might be 'held'. Certainly, there is some resonance between the attitudes demonstrated by the monks towards guests and the Rogerian formulation of unconditional positive regard, as articulated with the person-centred or humanistic traditions of counselling and psychotherapy (Mearns and Thorne, 1999; Nelson-Jones, 2001). This is one reason why it is perhaps as appropriate to think of these monasteries as potentially therapeutic environments as much as therapeutic landscapes (for a discussion of the distinction, see Conradson, 2003).

During their stay, guests dine with the community and are served their food by the monks in a manner which affirms their value, independent of any wider calculation of professional or personal status. Indeed, the meal operates only obliquely to the monetary economy which normally accompanies being served food by strangers, as in a restaurant. More generally, there are significant opportunities for guests to tell their stories and be listened to. Both Abbeys offer one-to-one discussions with a member of the monastic community, either informally or within a more structured practice of spiritual direction (where an individual reviews their personal spiritual journey). There are also opportunities to engage with the physical landscape in which each monastery is set, by wandering in the grounds that surround the buildings, taking time to sit beneath the trees or to observe the local wildlife (principally birds, but sometimes small mammals, such as squirrels, and occasionally deer in the surrounding fields).

The rhythmic schedule of each day, along with a more general commitment towards silence, fosters an environment conducive to internal stillness. Although this stillness is in some senses a by-product of the worshipping community, it is also evidently attractive to guests. Each monastery enables people to achieve some degree of physical and mental distance from the demands of their everyday lives. As one resident mentioned when he heard the phone ringing at Alton Abbey, 'I could hear it ringing and the wonderful thing was that at the same time realising that it wasn't for me, that I didn't need to answer it'. Relocation, of course, never provides total respite from an individual's circumstances, if only because these are often linked to personality, which is less easily left behind. Nevertheless, in the context of reduced external demands, several guests spoke of becoming more closely attuned to their own thoughts and feelings, rather than considering how to respond to those of others.

Obtaining distance from everyday demands is central to many contemporary tourist settings, but the monastic emphasis on contemplative spirituality is arguably distinctive. For some guests, particularly those with an acknowledged or evident personal faith, visiting was clearly about making space and time for connection with God or some understanding of a transpersonal, metaphysical domain. Reflecting on this point, one of the monks thus spoke of the Abbey as a 'portal of grace', a sacred environment where the long-standing rhythms of prayer and worship somehow facilitated encounters with the Divine. It is not the place of this chapter to consider the veracity of such narratives, or the philosophical bases on which they might be examined (Wynn, 2005), but suffice to say that various formulations of the idea of a sacred setting were evident among a number of guests. There are links here to other practices of movement between nominally profane and sacred spaces, as explored in work on the geographies of pilgrimage and religion (Gesler, 1996; Kong, 2001; Wynn, 2006).

It was not always the case that the monasteries were experienced as calming environments. In each Abbey, reference was made to those who approached the stillness on offer somewhat instrumentally, as an experiential good to be acquired simply by virtue of locating oneself in the right place. Such individuals, the monks felt, did not always appreciate that personal stillness was a relational achievement which also required cultivation by the individual in search of it, as opposed to passive receipt. It was remarked that those who arrived with a sense of urgency did not always, in their urgency, find the relief they sought. It was as if internal tension made it more difficult to access the sense of stillness which permeated each Abbey environment. For yet other guests, stillness acted as a context in which difficult feelings began to emerge, such that their visit involved a measure of discomfort and perhaps even temporary distress. There was clear awareness within the monastic community of the potential for such experiences to arise during a retreat, however, and counselling and spiritual direction were made available to guests with this in mind.

A Meeting of Worlds: Contrasting Appropriations of Stillness

When visiting each monastery, there is a sense in which guests are moving between two rather different worlds: that of wider British society on the one hand, with its capitalist economy and particular cultural values, and the contemplative environment of the Benedictine tradition on the other. It is not the case, of course, that some impermeable boundary exists between these two spheres. Monastic communities inevitably buy in various goods and services, engage in external communications, and are influenced by wider church and legal directives. And yet their intersection with the capitalist economy is not always without dissonance.

Although the monasteries considered here unreservedly welcomed guests, a degree of distance was sought from organisations that sought to market similarly therapeutic environments in commercial terms. As one example of such commodification, we might consider the British company Center Parcs. In advertising holidays in scenic, forested surroundings, its website offers the following narrative:

> *Magical Moments*. Step into 400 acres of beautiful forest and discover another world. Space to relax and play. Time to spend with family and friends. A short break at Center Parcs will create a lifetime of memories (www.centerparcs.co.uk).

Amongst the promise of a scenic landscape and 'time to spend with family and friends', there is a clear invocation here of a more restful temporality and rhythm of being. Immediately below the text is a link to the booking pages, however, which swiftly reminds us that money will have to change hands to appropriate this particular opportunity. In contrast, neither Abbey considered here is a commercial venture. While both ask guests for a contribution to cover costs, they do not seek to profit through such transactions. To do so would be counter to the ethos the monks seek to cultivate in their community life. So while the stillness that monasteries offer is clearly valued by members of the public, its valuation is not primarily registered in monetary terms.

Having made this point, a more commercial valorisation of stillness does impinge upon monastic environments. *The Good Retreat Guide*, a directory of 'over

500 places to find peace and spiritual renewal in Britain, Ireland, France, Spain and Greece' (Whiteaker, 2004, cover page), is perhaps illustrative of the growing intersection between a wider therapeutic economy and monastic communities in Western Europe. Now in its fifth edition, the Guide's entries for Alton and Elmore Abbey read as follows:

> [Alton Abbey] is a place where you may find stillness and, hopefully, that reflection which may lead to worship and prayer with the community. On offer are Benedictine hospitality and a very wide range of retreats tailored to guest's needs (Whiteaker, 2004, 86).
>
> [Elmore Abbey] is a distinguished Anglican abbey with a new and beautiful church whose oak columns seem like trees in a woodland. The little cloister courtyard at the entrance could not be prettier or more charming. All the refurbishment, rebuilding and designing have been done with obvious care and considerable talent. Everywhere is elegant and peaceful and so too seems the community of monks that live here. There is a large oblate membership who come often, and they have first priority on the guest accommodation (Whiteaker, 2004, 85).

One of the monks interviewed expressed a measure of ambivalence about inclusion in this Guide, and one Abbey had in fact declined to provide the requested details. Its directory entry was compiled nevertheless. The perceived difficulty was that the monasteries would be rendered primarily as 'places of retreat', a category which they saw as eliding their prior and wider existence as communities for worship and spiritual growth. Moreover, the layout of the Guide is akin to a commercial accommodation guide, such as one might consult when searching for a bed and breakfast facility in an unfamiliar area. It was felt that one subject position thus invited of readers would be less that of a guest seeking retreat, with echoes of the medieval tradition of faith-based pilgrimage, and more of a Western consumer expecting a product in exchange for payment. The latter kind of instrumentality was seen as antithetical to Benedictine formulations of stillness and community.

The intersection of religious and commercial interests is not new, of course, as the presence of merchandise stalls in the heart of pilgrimage sites such as Lourdes makes clear (Gesler, 1996). In the Christian tradition, concerns over such juxtapositions may be traced back to New Testament accounts of Jesus of Nazareth's eviction of the moneylenders from a Jewish temple. In their efforts to hold open a certain space for worship, contemplation and personal renewal, contemporary monastic communities, such as Alton and Elmore Abbey, are forced to negotiate similar issues. In accord with the Benedictine tradition, they thus continue to welcome guests, whilst also seeking to maintain a certain distance from the more commercialised elements of the experience economy.

Conclusion

In considering two monastic communities in contemporary Britain – settings where stillness is not only valued but explicitly cultivated – a number of points about therapeutic landscapes emerge. Firstly, and as noted initially, in order to understand the therapeutic properties attributed to particular settings or landscapes, it is analytically productive to examine the broader societal context in which they exist. The growing attractiveness of places of retreat in Britain is rendered more intelligible when we consider the pressures on many configurations of work–life balance, for stillness has become somewhat scarce. It is also important to note the turn towards spirituality in the wake of declining adherence to traditional forms of religious belief and practice in many Western countries (McLeod and Usdorf, 2003). The examples considered demonstrate the complexity of such processes, for some elements of the popular interest in spirituality are here prompting an engagement with environments rooted in a Christian tradition.

A second point is that there is value in considering further the economic dimensions of therapeutic landscapes. Many studies to date have focused on their positive features, perhaps overlooking the labour relations that underlie the apparent tranquillity of spa retreats or the financial accessibility of scenic natural environments. So we might reflect further on how the commercial economy impinges upon therapeutic landscapes, affecting their operation and shaping the expectations of visitors more generally. In this study, it was noted that the broader societal commodification of stillness was perhaps complicit in producing individuals who approach monasteries more as deliberative consumers than contemporary pilgrims. At the same time, it is important to recognise that the worth ascribed to experiential states such as stillness is not solely reducible to their exchange value in the capitalist economy.

Finally, it is important to consider the temporality of therapeutic landscape experiences. What is the rhythm of activity offered in a particular setting, and how does this compare to that of wider society, or to the lifestyle of a particular individual or social group who are visiting? How should we think of therapeutic landscapes in relation to the various 'slow movements' that are emerging with respect to food and travel (Franklin, 2003)? These are questions for further research, but alongside their social and environmental features, it would seem that therapeutic landscapes of the future will increasingly be appreciated for their capacity to slow us down. The preservation of places that offer an alternative temporality is perhaps not assured in the face of growing attention from inhabitants of fast societies, but it is undoubtedly important. For productivity is not always coincident with speed and, in a turning world, still points have more than commercial value.

Acknowledgments

I am grateful to Allison Williams for editorial comments on an earlier version of this chapter, and to Nick Clarke for suggesting some of the references on fast and slow time.

References

Bruce, S. (2002), *God is Dead: Secularisation in the West* (Oxford: Blackwell).

Cloke, P. and Perkins, H.C. (2002), 'Commodification and Adventure in New Zealand Tourism', *Current Issues in Tourism* 5: 521–49.

Conradson, D. (2003), 'Spaces of Care in the City: The Place of Drop-In Centres', *Social and Cultural Geography* 4: 507–25.

Conradson, D. (2005a), 'Landscape, care and the relational self: therapeutic encounters in southern England', *Health and Place* 11: 337–48.

Conradson, D. (2005b), 'Freedom, Space and Perspective: Moving Encounters with Other Ecologies', in J. Davidson, L. Bondi and M. Smith (eds) *Emotional Geographies* (Aldershot: Ashgate) pp. 103–16.

Davie, G. (1994), *Religion in Britain since 1945: Believing without Belonging* (Oxford: Blackwell).

de Dreuille, M. (2002), *The Rule of St. Benedict: A Commentary in Light of World Ascetic Traditions* (New York: Paulist Press).

Dean, E. (1989), *Saint Benedict for the Laity* (Collegeville, Minnesota: The Liturgical Press).

Eriksen, T.H. (2001), *Tyranny of the Moment: Fast and Slow Time in the Information Age* (London: Pluto Press).

Franklin, A. (2003), *Tourism: An Introduction* (London: Sage).

Gesler, W.M. (1996), 'Lourdes: healing in a place of pilgrimage', *Health and Place* 11 (2): 173–85.

Gesler, W.M. (1993), 'Therapeutic landscapes: Theory and a case study of Epidauros, Greece', *Environment and Planning D* 11: 171–89.

Glacken, C.J. (1976), *Traces on the Rhodian Shore: Nature and Culture in Western Thought from Ancient Times to the End of the Eighteenth Century* (Berkeley: University of California Press).

Hazler, R. and Barwick, N. (2001), *The Therapeutic Environment: Core Conditions for Facilitating Therapy* (Buckingham: Open University Press).

Kearns, R. and Collins, D. (2000), 'New Zealand children's health camps: therapeutic landscapes meet the contract state', *Social Science and Medicine* 51: 1047–59.

Knowles, D. (2004), *The Monastic Orders in England* (Cambridge: Cambridge University Press).

Kong, L. (2001), 'Mapping "new" geographies of religion: politics and poetics in modernity', *Progress in Human Geography* 25: 211–33.

MacInnes, J. (2005), 'Work-life balance and the Demand for Reduction in Working Hours: Evidence from the British Social Attitudes Survey 2002', *British Journal of Industrial Relations* 43: 273–95.

Magical Moments <www.centerparcs.co.uk> accessed January 2007.

McLeod, H. and Usdorf, W. (eds) (2003), *The Decline of Christendom in Western Europe, 1750–2003* (Cambridge: Cambridge University Press).

Mearns, D. and Thorne, B. (1999), *Person-centred Counselling in Action* (London: Sage).

Milligan, C., Gatrell, A. Bingley, A. (2004), '"Cultivating health": therapeutic landscapes and older people in northern England', *Social Science and Medicine* 58 (9): 1781–93.

Nelson-Jones, R. (2001), *Theory and Practice of Counselling and Therapy* (London: Continuum).

O'Dell, T. (2005), 'Experiencescapes: blurring borders and testing connections', in T. O'Dell and P. Billing (eds) *Experiencescapes: Tourism, Culture and Economy* (Copenhagen: Copenhagen Business School Press) pp. 11–33.

Palka, E. (1999), 'Accessible wilderness as a therapeutic landscape: Experiencing the nature of Denali National Park, Alaska', in A. Williams (ed.) *Therapeutic Landscapes: The Dynamic between Place and Wellness* (Lanham: University Press of America) pp. 29–53.

Pine, J. and Gilmore, J.H. (1999), *The Experience Economy* (Boston, MA: Harvard Business School).

The Retreat Association (2007), *Retreats 2007: Annual Magazine of The Retreat Association* (London: The Retreat Association).

Thrift, N. (2000a), 'Still life in nearly present time: the object of nature', *Body and Society* 6: 34–57.

Thrift, N. (2000b), 'Performing cultures in the new economy', *Annals of the Association of American Geographers* 90: 674–92.

Thurber, C. and Malinowski, J. (1999), 'Summer camp as a therapeutic landscape', in A. Williams (ed.) *Therapeutic Landscapes: The Dynamic between Place and Wellness* (Lanham: University Press of America) pp. 53–71.

Tyers, J. (2007), 'The Gift of Silence', *Retreats 2007*, Annual Magazine of The Retreat Association, 256 Bermondsey Street, London SE1 3UJ, <www.retreats.org.uk>.

Urry, J. (2004), 'Small worlds and the new social physics', *Global Networks* 4: 109–30.

Whiteaker, S. (2004), *The Good Retreat Guide*, 5th edn (London: Random House).

Williams, A.M. (2002), 'Changing Geographies of Care: Employing the Concept of Therapeutic Landscapes as a Framework in Examining Home Space', *Social Science and Medicine* 55: 141–54.

Wynn, M. (2005), *Emotional Experience and Religious Understanding: Integrating Perception, Conception and Feeling* (Cambridge: Cambridge University Press).

Wynn, M. (2006), *God, pilgrimage, and acknowledgement of place*, unpublished manuscript, Department of Theology, University of Exeter, Exeter EX4 4QH, UK.

Chapter 4

From Rishikesh to Yogaville: The Globalization of Therapeutic Landscapes

Anne-Cécile Hoyez

Introduction

This chapter aims to analyse therapeutic landscapes according to the way that they are produced and reproduced in the globalization process. Therapeutic landscapes have been studied in the field of geography since 1992. Wil Gesler was the first to emphasize the importance of settings of spaces and places in the healing process (Gesler, 1992; 1993; 1996; 2003; 2005). Following this work, others further developed the concept by introducing social and cultural considerations (Kearns and Gesler, 1998; Williams, 1998; 1999). Today, research in the area of therapeutic landscapes considers the relationship between the physical structure of a place, social relations and cultural practices.

In this chapter, I consider the therapeutic landscape concept to incorporate different environmental, cultural and social elements that play a role in healing and well-being. I also view therapeutic landscapes as globalizing entities. This chapter expands on my earlier work on the globalization of yoga (Hoyez, 2005). Globally, the practice of yoga is increasing, particularly in Western countries and in yoga's 'source country', that being India. The globalization of yoga practices and yoga centres emerges from a complex process that has evolved quickly since the turn of the twentieth century.

Studying the globalization of yoga entails the consideration of a socio-cultural perspective as yoga is a worldwide phenomenon that influences cultural and social practices. (Strauss, 2005; Smith, 2007). Globalization of yoga reflects the increasing connectedness of places and people, as described by Giddens (1991), as well as the globalization of practices and ethnoscapes (Appadurai, 1996). The globalization of yoga can be modeled via the construction of social links, localizing and multiplying the types of practices. This idea is recurrent in the notion of the 'global village' (Friedman, 1994) or the 'global ecumene' (Hannerz, 1989; 1992; 1996).

This chapter will demonstrate that therapeutic landscapes are produced and reproduced worldwide. Furthermore, this chapter illustrates a geographical approach to the interaction between the macroscale (globalization, the world space), and the microscale (localization, the body in space and as a space). Space and place are always co-dependent. Space is seen as one of society's dimensions and products (Lefebvre, 1974). Space can be material, immaterial and/or ideal and can be perceived, represented and/or imagined. Place is, to a certain extent, related to a location, a space in which the concept of distance is invalidated (Lévy and Lussault, 2003). Place is one of the smallest geographical entities where we find social interactions, and where experience defines values and realities attached to the place (Tuan, 1977).

The data informing this chapter was collected during my doctoral fieldwork which I carried out in Indian and French yoga centres. The fieldwork was based on inquiries and interviews with people practising and teaching yoga, and on observations in yoga centres. At these centres, I attended yoga classes not as a 'yogi' but as a social scientist, observing the practices and perceptions of individuals using such places. In total, two and a half years were spent in India (the primary study settings were Pune, Chennai and Rishikesh), and two and half years in France (Rouen and Paris).

This chapter is organized into four sections. In the first section, I describe the links between health, healing and yoga, and how these relationships have developed. In the second section, I illustrate the manner in which therapy and yoga take place in space. Once the environmental, cultural and social elements characteristic of the practice of yoga have been described, two concrete examples of production and reproduction of yogic therapeutic landscapes are provided: the cases of Rishikesh and Yogaville. Lastly, I theorize the links that unite the two concepts of therapeutic landscapes and globalization.

Health and Healing in Yoga

Initially, I will provide an overview of the practice of yoga, and how the relationship between yoga and health and healing has been constructed through time. The linkages between health and yoga are vast, as will be demonstrated. Yoga can have influences on disease, healing, well-being, as well as offering cultural, political and sometimes spiritual rewards. This extensive scope is a result of reinterpreting yoga in modern societies. Indeed, through the process of its globalization, yoga has been incorporated into many different cultures and societies, and its knowledge has consequently been transformed and reinterpreted.

Defining Yoga: A Difficult Task

Essentially, yoga is a complex discipline that involves practices of the physical body, while incorporating philosophical, religious or cultural concerns (Eliade, 1975). Classically defined, the word 'yoga' refers to the etymological Sanskrit root 'yug' which means 'to unite, yoke, join'. There are many debates, however, regarding the definition of yoga, the first one of which is concerned with whether it should be defined as a method, a philosophy or a practice. Definitions of yoga vary according to different authors. Consensus is shared on one point: yoga is holistic, and relies on eight levels of practice, from the most practical to the most spiritual.[1]

---

1 The eight limbs are *yama* (self-restraint), *niyama* (religious observances), *asana* (postures), *pranayama* (regulation of breath), *pratyahara* (abtract of the senses), *dharana* (concentration), *dhyana* (meditation), *samadhi* (superconscious state).

Although Hindu ideology does not require the practice of yoga, yoga can be defined as a physical experience organized around diverse notions emanating from Hinduism (Flood 1996; 2004; Michaels, 2004; Stutley, 2003). Many scholars emphasize that yoga in itself can be a vector for social relations, and can carry ideological, religious and/or therapeutic values (Filliozat, 1987). Fields (2001) believes that the diversity inherent in yoga can be analysed depending on historical phases, each one marked by the rising of a new text, or a new religious reform movement. More generally, yoga is characterized by its total freedom of interpretation. This explains the large numbers of books published on yoga, written by scholars and lay practitioners alike. As a result, the question of defining yoga in contemporary society has become quite difficult. Therefore, it is more suitable to describe it as a rigorous practice that takes different forms depending on the way people look at it through time.

Modern yoga, as defined by De Michelis (2004, 187), is a 'technical term to refer to certain types of yoga that evolved mainly through the interaction of Western individuals interested in Indian religions and a number of more or less Westernised Indians over the past 150 years'. The term 'modern' is in reference to the historical context (the beginning of the modern age), and to the new geographical area covered by yoga. Indeed, yoga is entering into so-called modern societies, and, in the midst of these, in urban spaces. Overall, the foundation of modern yoga is based on postures ('modern postural yoga', according to De Michelis) and the effect of these postures on the body and mind.

Yoga: Ancient and Contemporary Healing

We can find traces of yoga in classical texts that are dated from the Middle Ages. At that time, however, these texts had relatively little influence and were not systematized (Filliozat, 1979). Moreover, they were used by gurus[2] as a basis for the oral transmission of yogic practice. Examples of such texts include: the *Yoga-Sutras of Patanjali* (dated between the second and eighth centuries),[3] the *Hatha-Yoga Pradipika* (dated from the fourteenth or fifteenth century), the *Siva Samhita* (eighteenth century), and the *Gheranda Samhita* (seventeenth century).

The *Yoga-Sutras of Patanjali* is likely the work that is the most often referred to today. According to Patanjali, yoga is a system of philosophy as well as a source of practical information, with related body and mind techniques and a contemplative method. Patanjali also created a 'memorandum' for ascetics who practise yoga. He chose some techniques verified by secular experiences, and then formulated, systematized and integrated them into a philosophy, explaining their pragmatic value and justifying their role in obtaining salvation.

The *Hatha-Yoga Pradipika* focuses on the details of the physical practice of yoga, and also provides insight into ways of achieving health and longevity. Indeed,

---

2 Literally, the word guru means 'the one who removes darkness'. A guru is considered as a guide or a teacher. A guru is supposed to initiate his disciples and to enlighten them in their spiritual or intellectual life.

3 According to some authors, the *Yoga-Sutras of Patanjali* can be dated between 200BCE – *Before Common Era* – and 200 CE – *Common Era*.

it introduces the power of yoga to prevent or cure physical or mental disorder, by demonstrating that the practice of yoga regulates the actions of the heart, lungs, and of blood circulation.

The *Siva Samhita* is a Sanskrit text about yoga. It consists of listings of yoga's concepts and cognitive principles. It is comprised of five chapters that concern the following aspects: (1) centrality of yoga and its methods (principles of *pranayama* – breathing techniques, and *asanas* – postures); (2) means of obtaining certain powers (*siddhis*) and of knowing the concepts of a philosophy of existence; (3) definition and role of spirit and *maya* (illusion); (4) microcosm (the incarnation of the macrocosm in the human being); and (5) functions of the body, *kundalini* (a latent force in the body). The *Siva Samhita* is often referred to by advanced students of yoga who intend to extend their practice to spirituality.

The *Gheranda Samhita* is a text that takes the form of a dialogue between Gheranda, the teacher, and Chandakapali, the pupil. The text focuses on *Ghatasthayoga* (*Ghata* refers to the body), which means 'yoga based on a body approach'. This text gives more information about the body than others, and provides more details regarding personal hygiene.

Yoga as a system of knowledge is said to be rooted in ancient times; the rhetoric of yoga relies on the principle of a renewed ancient practice. Today, the re-emergence of these texts introduces broader issues of concern to the medical and social sciences, those being reinterpretations of health, healing and well-being in different societies.

Healing and Yoga in the West

Yoga as a modern discipline has undergone many transformations in its practice and knowledge. At the turn of the twentieth century, when yoga was brought out of India by gurus as well as brought back from India into the West by travelers, it received attention from intellectuals and scholars. Many publications appeared in France, Germany and Great Britain. Among them, the publications which retained most of the attention were those underlining the benefits of yoga for health. As a result, yoga has been anchored in the field of health, therapy, healing and well-being (Hoyez, 2007a; 2007b).

The introduction of yoga in Europe, for example, in the mid-twentieth century was made through a series of public performances. The fascination provoked by the physical experiences of yogis and ascetics was then interpreted as demonstrating the efficacy of yoga on health. Since that time, the potential of controlling the body and emotions through yoga has been relayed by the medical field, through scientific investigation.[4] Having revealed the therapeutic properties of yoga through scientific

4 Dr Thérèse Brosse was one among the first to lead a research study in India in 1936 to measure, with medical instruments (for example, electrocardiograms and pneumographs), the respiratory and cardiac capacities of some ascetics engaged in a regular practice of yoga. This study, led under the guidance of the Cardiology laboratory of the Broussais Hospital, was supported by many specialists (including Jean Filliozat). Its aim was scientific: '*to study experimentally the mechanism of functional diseases that reveal, at the time, the primary importance of the subject's mental situation in the production and regression of certain cardio-*

observations, some scientists tried to 'pierce the secrets of yoga' by establishing medical research groups. Western scientists contributed by identifying some of the physiological effects of yoga: cardiac frequency, arterial pressure, chemical and biological constants, pressure measures (digestive, pulmonary), volumetric respiratory values with ventilation frequency, skin resistance, electroencephalogram, and so on.

Furthermore, yoga was incorporated into philosophical, medical as well as psychological approaches (Fields, 2001; Berthelet-Lorelle, 2003; Alter, 2004). Many authors, including famous psychoanalysts, began to integrate yoga into Freudian and Jungian conceptions of health (Jung, 1973).

During the 1950s, most psychologists considered psychological health as an essential element to: free one from mental conflict, assist with social integration, and/or reach emotional maturity. At that time, the link between yoga, psychology and health became quite evident and recognized by those interested in the body–mind connection. Since that time, yoga has been organized and defined between a mingling of classical texts and psychological technical language. For example, many publications on health and yoga available today insist on yoga's psychological effect on health, linking Sanskrit terms to psychoanalytic terminology:

> Each one with their methods, Patanjali and Freud teach us that it [the symptom] is the cause of suffering (kleshas). Taking roots in the unconscious (avidyâ), this position is made of pride (asmitâ), of desire (râga), of denial (dvesha) and of fear (abhinivasha). (Berthelet-Lorelle, 2003).

These types of interpretations rely on a renewal of metaphysical discourses.[5] Hence, yogic practice has evolved into a discourse on health, emphasizing the transcendence of the mind–body connection. The modern utilization of 'old words', which were merely details in some minor texts, are now part of these discourses.

There is also a medical field in which yoga has made its mark, that being complementary healthcare. Since 2000 the WHO guidelines for complementary medicine have included yoga (WHO, 2002). This final assertion by the WHO substantiates the incorporation of yoga into the health field. Moreover, there is a growing number of academic resources, including MEDLINE, the Cochrane Library, the Lancet Publishing Group, as well as medical reviews such as the *American Journal of Psychiatry*, and the *Journal of Clinical Epidemiology*, which consider yoga as a complementary medicine.[6]

---

vascular disorders' (Brosse, 1976). At the same time, during the 1920s, similar research was being conducted in an Indian institute, in Lonavla, near Pune, Maharashtra (Alter, 2004).

5 The term 'discourse' is used, in this chapter, not only to reflect oral interactions between individuals, but also in reference to all the devices contributing to the social construction of reality that aim to maintain, reinforce and/or subvert the society (Lévy and Lussault, 2003).

6 Between 1966 and 1996, more than 33,000 articles are listed in MEDLINE under headings related to 'complementary medicine', with an average of 30 per year between 1966 and 1971, 42 between 1972 and 1986, 150 after 1986 (Barnes et al., 1999). Among these articles, a large number deal with yoga as a complementary therapy.

We have seen that, today, health is the primary reason for adopting a yoga practice. In parallel, other motivations are superimposed with health concerns and are also found to ultimately be health-related (such as concepts of well-being, mental relaxation, and so on). We will now see that yoga practice and the outcomes that result take space, place and time, which brings into consideration the therapeutic landscapes concept.

Place, Health and Yoga

In this section I will demonstrate the importance of place to yoga. The knowledge of the subjective experience of yoga practice is vital to understanding the value and significance of space and place. In this section, I will illustrate the ways in which place is perceived and produced for health purposes.

Places for Yoga

The setting for yoga practice is meant to be an idyllic place, where humans and nature are united, and where social interactions are peaceful. Such a place is referred to in classical texts as an '*ashram*'.[7] Symbolically, the location of the *ashram*, which is remote from everyday life, is supposed to reflect the state of mind of the yogi. In ancient yoga practice, there are several elements necessary to visualize or experience a certain sense of place. Some socio-cultural practices characterize places of yoga. Moreover, as we have seen within the various interpretations and definitions of yoga practice, the perception of place in yoga reflects a duality between the past and present.

In ancient texts, we find repeated descriptions of how the setting for yoga practice should be arranged. In texts such as the *Hatha-Yoga Pradipika*, we can find elements that describe the ideality of the place. These elements of place are presented as basic to the practice of yoga. Some 'natural' elements are seminal, such as a location near a mountain, river or forest. These elements evoke a certain divine construction of the world. Moreover, they symbolize the notion of purity and well-being inherent to yoga. The place, as well as the individual, must be pure. For example, in the *Hatha-Yoga Pradipika*, the first chapter is dedicated to the spatial context in which the yogi should perform his *asanas* (the *asanas* being described in the following chapters of the book). The text instructs that the yogi should be alone, in an isolated structure that is not too large and is free from natural hazards. The place should have only one opening to the outside and should be well maintained. According to most modern commentators of the *Hatha-Yoga Pradipika*, however, these conditions are quite unrealistic. They should therefore be understood as spatial metaphors concerning how yogis should behave in their social and natural environment.

---

7 *Ashram*, in Hindu orthodoxy, is understood as a stage of a Brahmin's life. During his *ashram* period, the Brahmin is supposed to retire from his day-to-day life and to meditate (Biardeau, 1995). Little by little, the period of the *ashram* has been understood as the place of retirement.

As it is impossible to live in such isolation, modern understandings of what should constitute a yoga setting rely on the idea of a place that allows for temporary isolation from day-to-day life. The place where yoga is practised should maintain boundaries from the outside world for a set period of time; these include boundaries from noise, pollution, social disruptions, and so on.

The setting in which yoga is practised can also be viewed as a place where a community receives spiritual guidance from a guru. Some socio-cultural activities can take place in these settings. These activities can include: the gathering of so-called cultural images of India, such as icons and people (priests, gurus, ascetics); social activities such as services to the poor (education, food, health); or assemblies of greater numbers of people (lectures, conferences, musical or dance programmes).

Today, however, *ashrams* tend to be established as 'yoga centres'. The use of these centres dates back to the mid-twentieth century, when gurus opened schools specializing in yoga in cities in both India and the West. This movement towards yoga centres as the place for yoga practice is consistent with the urbanization of yoga. The number of yoga centres located in urban spaces is ever-increasing. There has been a renewed understanding of what is reflected in the ancient texts, and the production of spatial imaginary and symbolism is found in the design of the built environment, such as in accordance with the interpretation of the 'natural' environment. To adapt yoga centres, which are enclosed places, to meet people's needs, controlled heat, lighting and sound are needed. The yoga setting can be considered to be a place adjusted for the urban inhabitant's need for a small 'island' of peace and serenity in the midst of a hectic urban life.

Yogic Therapeutic Landscapes

Yogic therapeutic landscapes, within the traditional concept, consist of: (a) natural elements, (b) cultural elements, (c) emotions, and (d) social relations. All of these elements have a very strong link with space and place.

'Natural' elements are part of the natural physical world, with mountains, water, forests, and so on, nearby. Cultural elements refer to the knowledge of Hindu or Indian practices, as they are supposed to be in India. These cultural elements are, therefore, subject to political instrumentation by diverse groups, such as Hindu nationalists (McKean, 1996).[8] Emotions refer to the set of physical and mental sensations perceived by individuals through their practice of yoga. Lastly, social relations are present in two forms. Yoga involves social interactions between persons practising at individual yoga centres, as well as the construction of social relations on a global

8 During the past two decades, Hindu nationalist parties and organizations (RSS – Rashtrya Swayamsevak Sangham/National Volunteers' Union, VHP – Vishwa Hindu Parishad/World Hindu Council, BJP – Bharatya Janata Party/Party of Bharatya People) show a growing interest in the so-called Ancient Hindu knowledge (medicine, science, yoga, astrology, and so on). These organizations claim the paternity of a worldwide knowledge that has been long oppressed and suppressed by colonizers (British, Christian, missionaries and Muslim presence). Their goal is to maintain and develop this Hindu knowledge in the Indian territory, in schools and administrations, and to enhance Hindu identity through these practices.

scale. Indeed, there is a very strong sense of 'unity in diversity' in the worldwide community of yogis. The key elements that comprise these social links, within worldwide knowledge and practice, can be traced to India as a source country.

The worldwide phenomenon of yoga corresponds with the production and reproduction of spaces and places. Moreover, when it comes to the reproduction of practices and behaviours, it directly reflects the therapeutic landscape concept.

Globalized therapeutic landscapes of yoga are rooted in the idea and representation of a pristine environment. Typically, however, there is a gap between this ideal and reality. Some yoga centres are busy and polluted, while in others, the pristine environment is created via wall hanging of nature scenes. There is a permanent adjustment in space, place and time, as well as an adjustment in reality. These adjustments can be realized through simple modifications of the infrastructure (sound isolation of the room, heat control, cleanliness), which can help to maintain distance between the real world and the world of the yogi. In addition, adjustment is made through the body as a space and place of healing and well-being. The body itself thus becomes a place that the individual can use to create distance from the external world.

Therapeutic Landscapes: the Body/the World in Space and Place

Interpretations of spaces and places in yoga are incorporated in the production of therapeutic landscapes. Moreover, we will see how these therapeutic landscapes are globalized as they are reproduced in many countries.

Examples: from Rishikesh to Yogaville

Yoga is originally an Indian practice, although it is now widely practised in and outside of India, as shown in the following examples of Rishikesh and Yogaville.

Rishikesh is situated 220 kilometres north of New Delhi, at the foot of the Himalayas. Rishikesh is a place that has attained a certain reputation as it is mentioned in the *Puranas*.[9] The name 'Rishikesh' is derived from an ancient etymology: *Rishi* means wise man, and *kesh* means abode (Stutley, 2003). Rishikesh is considered to be an auspicious place for Hindu devotees, and it is the starting point for the main Himalayan pilgrimages. As such, Rishikesh is known as a spiritual place. Today, however, Rishikesh has also acquired the reputation of a healing place and a place meant for peace of mind and well-being, as some internationally known gurus have established *ashrams* and/or yoga centres there. The Lonely Planet tourist guide considers Rishikesh to be the 'world capital for yoga' (The Lonely Planet, 2000).

However, although Rishikesh is currently a well-known and frequented place, there are few sources of economic and demographic development in the city. The strong identity of Rishikesh is based on its reinterpretation as a mythical place that has attracted the wise for more than a millennium (Strauss, 2002a; 2002b; 2005). The development and reputation of Rishikesh are especially reliant on the 1960s period

9 The *Puranas* are Sanskrit texts written from the fourth century BC and about 1000 AD.

of counter-culture and hippies. Since the Beatles and other artistic personalities came to Rishikesh in the 1960s to meditate, create and experience an alternative way of life, many other hippies have followed in their steps (Saltzman, 2000). Today, the place where the Beatles stayed is closed and reclaimed by the jungle, but the density of hippies young and old is still increasing, as well as the number of activities linked to yoga.[10] Concurrently, Rishikesh is increasingly frequented by the Delhi middle class, who find there a quiet and relaxing place easily accessible from the capital. Additionally, since the 1980s, the Ministry of Tourism of Uttaranchal[11] has promoted Rishikesh as an ideal spiritual and tourist place, and the emphasis on yoga is exemplified through the organization of an annual international festival of yoga.

The city is full of places to practise yoga, and the city attracts saints (holy men, not necessarily turned toward ascetism), gurus and teachers, of both Indian and international origin. Rishikesh now has a complex identity, characterized by both the environment and the population, all of which is part of the broader landscape. The Ganges, the Himalayas, the jungle, Hindu temples, gurus, monks and ascetics are important markers of the representation of the yogic therapeutic landscape. Rishikesh is characterized by a strong sense of place. Sometimes, it is referred to as a spiritual place; spirituality is expressed in a discourse that incorporates notions ranging from beauty and purity to ideas of the environment. In other cases, Rishikesh can be seen as a place with a blend of individual identities and global knowledge.

The city is frequented and exploited for its beneficial atmosphere. As a therapeutic landscape, it gathers specific natural attributes meant to be good for the mind and the body, and linked to health and well-being. Rishikesh is a central place for yoga and health. However, not all yogis can visit Rishikesh, so yoga practitioners have created an alternative solution, that being the establishment of a yogic place.

This is what a student of Swami Sivananda (1887–1963), the founder of the Divine Life Society, one of the most frequented *ashrams* in Rishikesh, has done. This student, named Swami Satchidananda (1914–2002), is well known for his presentation of 'integral yoga' at the Woodstock festival in 1969. From there, he became popular in the United States and Europe. He was asked to stay in the United States by devotees who helped him to develop an ashram called 'Yogaville'.

Yogaville is located in Virginia, United States, and was built in 1972 according to the tenets of a yogic therapeutic landscape – near a river and a forest, in a lovely environment. Within Yogaville, yoga draws on symbols from Hindu ideology. The

---

10 Rishikesh attracts other therapeutic activities, such as ayurvedic clinics (ayurveda being one of the medicines of India) and medical shops, reiki (an Asian system of healing through body energy), Indian and/or Thai massage, and healing music. Consequently we find many therapeutic activities in Rishikesh that have re-emerged from Indian or Asian knowledge, located in a destination for international yogis.

11 Uttaranchal is one of the 35 Federal States and Union Territories of India. It was created in 2000, after partition from the Uttar Pradesh state. As all the other states, it has its own federal government and ministers, who apply the guidelines given by the central government and who are in charge of the development of local issues (tourism and transport mainly).

design of the main building, the LOTUS Temple[12] is shaped according to Hindu elements originating from India:

> An aerial view of the LOTUS reveals its unique *yantra*-like design. Just like a mantra is the sound one would hear in deep meditation, a *yantra* is a physical symbol one would have the vision of in deep meditation. [...]
>
> The Shrine is also built in the shape of a lotus flower. This sacred flower contains all the five elements: it grows out of the mire (earth) into the water and air. The pink petals represent the fire element. The element ether – the space the other elements fulfill – is always present. The geographic location of the LOTUS is also very special. The five elements are strong at this location: earth (the foothills of the Blue Ridge Mountains), water (the James River and the LOTUS Lake), fire (the hot summer temperatures and the homas or fire ceremonies conducted at the Shrine), air (the fresh air of this rural region), and ether. When these five elements are in balance, a strong healing energy is created. [...]
>
> When we get to the end of the reflecting pond, the two carved granite elephants that greet us are symbols of the earth element and strength and knowledge. The elephant is also an Eastern symbol of the removal of obstacles. [...] (Alan Deva, 2007).

The main building has the form of a lotus and the main gate recalls the temples of South India (*gopuram* and large pillars supporting a large arch). The colours chosen (pink, white and light blue) represent Indian rather than Western styles.

As the toponym indicates, Yogaville is meant for yoga. People can go there for a short stay, or for weekly programmes. Activities range from collective yoga and meditation classes to silent retreats and alternative therapies, including massage.

In Yogaville, the purpose is to 'live' yoga or to 'experience' yoga in a special landscape reminiscent of India and Indian healing. At least once a month there is a seminar on the adaptation of yoga *asanas* to healing and well-being. Yogaville is a reproduction of what should be an optimum environment for practising yoga and achieving its many benefits.

The LOTUS Temple is said to be a place of worship for every kind of faith. Within Yogaville, however, there is a place for Hindu worship called the Nataraja Shrine (a temple dedicated to the god Shiva). Many sculptures representing gods related to Shiva stand there. Yogaville's website declares that 'These welcoming Deities were created in India and sent here in 1993' (Satchidananda Ashram Yogaville, no date). Additionally, many religious rituals are performed there according to Hindu practices. The presence of this shrine in the site indicates links with India while reflecting the strong cultural identity that shapes the many global therapeutic landscapes of yoga.

We have seen that the practice of yoga is ideally characterized by a particular space and place. In the two previous examples, the therapeutic landscapes of yoga are clearly linked to those of Indian origin. Although the practice of yoga is often

12 LOTUS is written in capital letters because it is the acronym for Light Of Truth Universal Shrine. It should not be confused with the Lotus Temple of the Bahai Community in Delhi.

associated with India, the globalization of yoga centres illustrates that the Indian reference is not always present.

Beyond Rishikesh and Yogaville: Take a Yoga Tour of the World's Capitals!

In Western cities, there are many places where yoga can be practised. These urban places can be seen as operating as therapeutic landscapes as, when there, people experience healing while carrying out their yoga practice. These urban places have a distinctive sense of place, which nurtures yogis to seek calm, wellness and positive effects on their health.

This set of emotions is part of the globalization process as they can be experienced in any number of global cities. For example, reporters from the *Yoga Journal* suggest a 'yoga tour' in a metropolitan area. These 'yoga tour[s] of …' are destined to help or appeal to the traveling yogi. They usually draw a rapid portrait of the city concerned (often evoked under negative terms), and make a list of principal yoga centres to visit to find ambiance and leisure. On the *Yoga Journal* website,[13] we can find information about practising yoga in 33 international cities, among them New York, Los Angeles, Paris and Bangkok.[14] Paris is seen as a city of good taste and good wine, which is not incompatible with the practice of yoga (Rynn, 2001). The principal yoga centres are listed and described with reference to the main tourist destinations (Arc de Triomphe, Place de la Bastille, Palais Royal, Pompidou Centre), making them easy to locate and access.

New York, 'the city that never sleeps', has the reputation of being full of people who are stressed and ill-mannered; the exception is in yoga centres where it seems that the reputation of New Yorkers has the potential to be changed: 'Who knows? With so many Manhattanites becoming devoted yogis, the cliché of the rude, harried New Yorker may soon be a thing of the past' (Parch, 2000).

On the other side of the United States, Los Angeles is seen as a very attractive city for yoga centres. According to a journalist from the *Yoga Journal*, if Los Angeles has attracted so many gurus and yogis, it is 'thanks in part to the temperate climate, miles of beaches and parks galore, getting into shape in this town seems almost mandatory. Yoga has hit the City of Angels with the same fervor of the modern fitness movement that began here' (Dunn, 2001).

In the above description, it seems that the 'natural' environment determines the place for yoga. In contrast, Bangkok, Thailand, is depicted as repulsive because of pollution, noise and congestion (Barendsen, 2004). Although this may be the case, activities of yoga and meditation can enhance day-to-day life: 'The chaos can be overwhelming, but Bangkok's challenges make its rewards – meditating at street shrines or enjoying a 50-cent papaya salad – that much more potent. And both

---

13 http://www.yogajournal.com, accessed 7 March 2007.

14 http://www.yogajournal.com/travel/toc_1.cfm, accessed 7 March 2007. On this page, we find diverse articles written by the *Yoga Journal* reporters (Rynn, about Paris; Parch, about New York; Dunn, about Los Angeles; Barendsen, about Bangkok; Bray Smith about Hawaii).

Thais and Western expatriates have found that yoga is a great retreat from a hot and crowded city.' Here, yoga centres appear as places of escape from urban life.

Hawaii is seen as an idyllic place in and of itself, owing to its spectacular natural scenery. Among other types of complementary medicine (ayurvedic medicine and tai chi), yoga too finds a place in this setting. The *Yoga Journal* declares on its website: 'Practicing yoga against the tropical backdrop is a dream for most people. In Hawaii, it's a daily reality, an example of the profound connection between human breath and elemental energy that seems to be inherent in the islands themselves'.[15] As in Los Angeles, the scenery offered in the Hawaiian Islands fits perfectly with the practice of yoga (Bray Smith, 2002).

The practice of yoga is primarily a matter of achieving physical and mental well-being in space, which emanates from its practice. The increasing number of people practising yoga leads to the production of places dedicated to yoga practice across the world. This phenomenon also appeals to new definitions and practices of yoga, which compels the reproduction, in space and place, of emotions and sensations emanating from the mind and body experience.

Theorizing Global Therapeutic Landscapes

The globalization of yogic places as therapeutic landscapes is a complex phenomenon that implies the presence of different worlds, differentiated by economic, social, cultural, environmental and political characteristics. In addition, global yogic therapeutic landscapes bring together hybrid systems and include many different social networks (including those of yogis, gurus, teachers, writers and medical practitioners).

The production and reproduction of the therapeutic landscapes of yoga remind us of the notions of deterritorialization and decentralization evoked by Deleuze and Guattari (2004). Indeed, there is no central power, no centralized hierarchy that directs the production of yogic therapeutic landscapes. There are neither fixed boundaries nor barriers in the creation of a yogic therapeutic landscape. Yoga is deterritorialized and reterritorialized elsewhere, potentially changing the meaning of the practice.

The process and dynamics of the globalization of yogic therapeutic landscapes are presented in Figure 4.1. The first level of the diagram illustrates that yoga is primarily linked to therapy and well-being. This leads into the production or creation of therapeutic landscapes, combining settings of the place ('natural' environment), emotions and socio-cultural practices and behaviours.

We know that yoga is a globalized practice. Consequently, the therapeutic landscapes of yoga are globalized too. The globalization of yogic therapeutic landscapes is accompanied by the production and reproduction of spaces and places. In this process, environmental elements, emotions and socio-cultural practices and behaviours are reproduced, leading to the production of new places, reflecting global hybridity.

15 http://www.yogajournal.com/travel/668_1.cfm, accessed 7 March 2007.

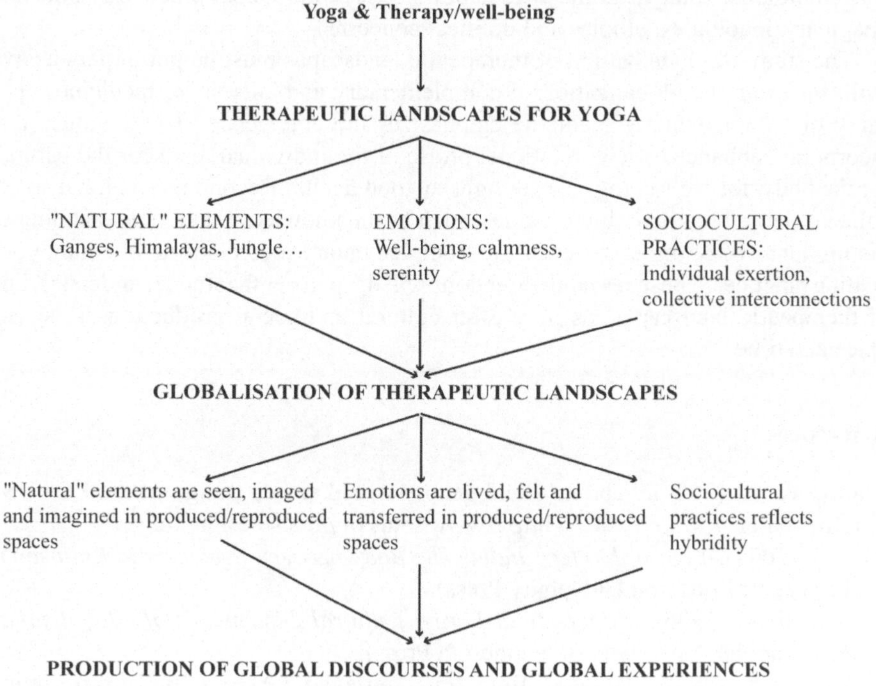

Figure 4.1 Globalization of therapeutic landscapes

Conclusion

The globalization of yoga seems to have played a large role in its 'rejuvenation' from ancient times. Yogic therapeutic landscapes are produced and reproduced for groups, by groups, with all types focused on a discourse of the individual and the body. There is consequently a permanent articulation between the macro scale and the micro scale. Obviously, as Latour writes: 'there is no zoom going from macrostructures to microinteractions … [since] both micro and macro are local effects of hooking up to circulating entities' (Latour, 1999). Behind the globalization of therapeutic landscapes, there is, therefore, a multiplicity of local actions that are practised, represented and generalized, and consequently act on other places and people, of varying degrees of proximity.

Rooted at the level of the body and consequently in the frame of 'techniques of the self', globalization of therapeutic landscapes appears to consider 'governmentality'. Governmentality concerns the 'processes proposed or prescribed to individuals to fix up their identity, to maintain it or to transform it thanks to a control on the Self or a knowledge of the self by the self' (Foucault, 1979; 1994). This consideration recalls charges that we must see globalization of therapeutic landscapes in a precise epistemological context: at the articulation between individual practices, collective imaginary, ideologies, utopias and political concerns.

The study of globalization of therapeutic landscapes must be put in perspective with, for example, globalization of complementary and alternative medicines. The study of complementary medicine emphasizes the emergence of new values that incorporate enhanced social relations: praise of the individual, care for the natural, for the body, for well-being, for the right to good health. Beyond these global social values emerge broader cultural values, such as the knowledge and understanding of the mixture of healing experiences, as complementary medicines invoke numerous healing practices. These research directions will deepen our theoretical understanding of therapeutic landscapes, as they offer cultural and social productions in space, place and time.

References

Alan Deva, B. (2007), 'The mystical geometry and sacred design of the LOTUS', retrieved on 7 March 2007 <http://www.lotus.org/docs/design.htm>.

Alter, J. (2004), *Yoga in Modern India. The Body between Science and Philosophy* (Princeton: Princeton University Press).

Appadurai, A. (1996), *Modernity at Large: Cultural Dimensions of Globalisation* (Minneapolis: University of Minnesota Press).

Barendsen, K. (2004), *Bangkok Yoga Tour*, retrieved 7 March 2007 from <http://www.yogajournal.com/travel/1368_1.cfm>.

Barnes, J., Abbot, N.C., Harkness, E.F. and Ernst E. (1999), 'Articles on Complementary Medicine in the Mainstream Medical Literature. An investigation of MEDLINE, 1966 through 1996', *Archives of Internal Medicine* 159: 1721–5.

Berthelet-Lorelle, C. (2003), *La sagesse du désir. Le yoga et la psychanalyse* (Paris: Seuil, Coll Couleurpsy).

Biardeau, M. (1995), *L'Hindouisme, Anthropologie d'une Civilisation* (Paris: Flammarion, Coll. Champs).

Bray Smith, T. (2002), *Hawaiian Islands Yoga Tour*, retrieved 7 March 2007 from <http://www.yogajournal.com/travel/668_1.cfm?ctsrc=sectnav>

Brosse, T. (1976), *Etudes Expérimentales des techniques du yoga. Expérimentation Psychosomatique* (Paris, EFEO).

De Michelis, E. (2004), *A History of Modern Yoga, Patanjali and Western Esoterism* (London: Continuum International Publishing Group).

Deleuze, G. and Guattari, F. (2004), *A Thousand Plateaus* (London: Continuum Publishing Group).

Dunn, S. (2001), *Los Angeles Yoga Tour*, retrieved 7 March 2007 from <http://www.yogajournal.com/travel/568_1.cfm>.

Eliade, M. (1975), *Techniques du Yoga* (Paris: Gallimard, Coll. FOLIO Essais).

Fields, G. (2001), *Religious Therapeutics. Body and Health in Yoga, Ayurveda and Tantra* (Albany: State University of New York Press).

Filliozat, J. (1979), *Yogasataka. Texte médical attribué à Nagarjuna. Textes sanskrit et tibétain* (Pondichéry: Institut Français d'Indologie).

Filliozat, J. (1987), *Les philosophies de l'Inde* (Paris: Presses Universitaires de France).

Flood, G. (1996), *An Introduction to Hinduism* (Cambridge: Cambridge University Press).

Flood, G. (2004), *The Ascetic Self: Subjectivity, Memory and Tradition* (Cambridge: Cambridge University Press).

Foucault, M. (1979), 'On governmentality', *Ideology and Consciousness* 6: 5–22.

Foucault, M. (1994), *Dits et Ecrits 1954–1988. Tome IV 1980–1988* (Paris: Gallimard, NRF).

Friedman, J. (1994), *Cultural Identity and Global Process* (London: Sage).

Gesler W.M. (1992), 'Therapeutic landscapes: medical issues in light of the new cultural geography', *Social Sciences and Medicine* 34: 735–46.

Gesler, W.M. (1993), 'Therapeutic landscapes: theory and a case study of Epidauros, Greece', *Environment and Planning D* 11: 95–105.

Gesler W.M. (1996), 'Lourdes: healing in a place of pilgrimage', *Health and Place* 2: 95–105.

Gesler, W.M. (2003), *Healing Places* (Lanham, MD: Rowman and Littlefield Publishers).

Gesler, W.M. (2005), 'Therapeutic landscapes: An evolving theme', *Health and Place* 11: 295–7.

Giddens, A. (1991), *The Consequences of Modernity* (Stanford: Stanford University Press).

Hannerz, U. (1989), 'Notes on the global ecumene', *Public Culture* 1: 66–75.

Hannerz, U. (1992), 'The Global Ecumene as a Network of Networks', in A. Kuper (ed.) *Conceptualizing Society* (New York and London: Routledge) pp. 34–56.

Hannerz, U. (1996), *Transnational Connections: Culture, People, Places* (London: Routledge).

Hoyez, A-C. (2005), *L'espace-monde du yoga. Une géographie sociale et culturelle de la mondialisation des paysages thérapeutiques* (PhD Thesis, University of Rouen, 492pp. Retrieved 7 March 2007 from Centre National de la Recherche Scientifique <http://tel.archives-ouvertes.fr/tel-00011237)>.

Hoyez, A.-C. (2007a), 'Globalisation and health in complementary healthcare. Yoga between therapy and culture', in A. Vaguet (ed.) *Health in India* (New Delhi: Manoharlal Banarsidas) (forthcoming).

Hoyez, A.-C. (2007b), 'The "world of yoga"; the production and reproduction of therapeutic landscapes', *Social Science and Medicine* (forthcoming).

Jung, C.G. (1973), *Ma vie. Souvenirs, rêves et pensées* (Paris: Gallimard, Coll. Folio).

Kearns, R. and Gesler, W.M. (1998), *Putting Health into Place. Landscape, Identity and Well-Being* (Syracuse: Syracuse University Press).

Latour, B. (1999), 'On recalling ANT', in J. Law and J. Hassard (eds) *Actor Network Theory and After* (Malden, Mass.: Blackwell) pp. 15–25.

Lefebvre, H. (1974), *La production de l'espace* (Paris: Anthropos).

Lévy, J. and Lussault, M. (eds) (2003), *Dictionnaire de la géographie et de l'espace des sociétés* (Paris: Belin).

The Lonely Planet (2000), 'India' (Lonely Planet, 1072 pp, Ed. 2000).

McKean, L. (1996), *Divine Entreprise: Gurus and the Hindu Nationalist Movement* (Chicago: The University of Chicago Press).

Michaels, A. (2004), *Hinduism: Past and Present* (London: Princeton University Press).

Parch, L. (2000), *New York Yoga Tour. Where to recharge in the city that never sleeps*, retrieved 7 March 2007 from <http://www.yogajournal.com/travel/249_1.cfm>

Rynn, M. (2001), *Paris Yoga Tour*, retrieved March 7th, 2007 from <http://www.yogajournal.com/travel/540_1.cfm>.

Saltzman, P. (2000), *The Beatles in Rishikesh* (New York: Viking Studio).

Satchidananda Ashram Yogaville, no date, retrieved on 7 March 2007 from <http://www.yogaville.org/Yogaville/Nataraja_Shrine/Nataraja_Shrine.html>.

Smith, B. (2007), 'Body, Mind and Spirit? Towards an Analysis of the Practice of Yoga', *Body and Society*, 13(2), 25–46.

Strauss, S. (2005), *Positioning Yoga. Balancing Acts Across Cultures* (Oxford: Berg).

Strauss, S. (2002a), '"Adapt, adjust, accommodate": the production of yoga in a transnational world', *History and Anthropology* 13: 231–51.

Strauss, S. (2002b), 'The master's narrative: Swami Sivananda and the transnational production of yoga', *Journal of Folklore Research* 39: 217–41.

Stutley, M. (ed.) (2003), *A Dictionary of Hinduism: Its Mythology, Folklore and Development 1500 BC–1500 AD* (New Delhi: Munshiram Manoharlal).

Tuan, Y.-F. (1977), *Space and Place: The Perspective of Experience* (Minneapolis: University of Minnesota Press).

Williams, A. (1998), 'Therapeutic landscapes in holistic medicine', *Social Science and Medicine* 46: 1193–1203.

Williams, A. (ed.) (1999), *Therapeutic Landscapes: The Dynamic between Place and Wellness* (New York: University Press of America).

World Health Organization (2002), *WHO Strategy for Traditional Medicine 2002–2005* (Geneva: WHO).

Chapter 5

Healing Landscapes in the Alps: *Heidi* by Johanna Spyri

Allison M. Williams

The expanded use of the therapeutic landscape concept in numerous areas of research within social and health geography has moved it well beyond its initial application, that being the understanding of the healing aspects of reputed places. One of the numerous applications that have met with interest by a few core proponents of the concept is its use in interpreting literary works. Lending from cultural geographers, who have embraced the use of literary works – whether as an illustration or, more commonly, to use as the central data source in geographical analysis (Preston and Simpson-Housely, 1994; Simpson-Housely and Norcliffe, 1992; Gilbert and Simpson-Housely, 1997) – health geographers are now too using this medium in their work. Of the few health geographers who have embarked on literary analysis (Gesler, 2000; Baer and Gesler, 2004; Tonnellier and Curtis, 2005) many, if not all, to my knowledge, have used the therapeutic landscape concept in approaching the interpretation of the text, albeit to varying degrees. Gesler (2000) first examined Mann's *The Magic Mountain,* tracing the main character's journey in attaining knowledge of both disease and health. Baer and Gesler (2004) then embarked on an analysis of Salinger's *The Catcher in the Rye* and, in so doing, extended the concept of therapeutic landscapes more thoroughly by highlighting the concept's shortcomings. This resulted in an expanded understanding of the concept, which encompassed everyday landscapes as ambivalent, nuanced spaces. Most recently, Tonnellier and Curtis (2005) examined Balzac's *The Country Doctor* for aspects of the social and physical environment, as perceived by the author, as having an impact on health and well-being. The interpretations of Mann's *The Magic Mountain* (Gesler, 2000), Salinger's *The Catcher in the Rye* (Baer and Gesler, 2004) and Balzac's *The Country Doctor* (Tonnellier and Curtis, 2005) illustrate the potential of the therapeutic landscape concept in interpreting literary works and in so doing, bringing attention to the socially constructed ideas about health (care).

Of the four types of environments associated with the therapeutic landscape concept – natural, built, social and symbolic (Gesler, 2003), it is the symbolic that are, as noted by Baer and Gesler (2004, 406) '... perhaps ... most amenable to analysis of fictive geographies', due to symbolic environments being the product of people's perceptions and meanings of health and illness. This chapter explores the symbolism of the Swiss Alps in Johanna Spyri's well-known children's classic, *Heidi* (Spyri, 1880/2000), illustrating how a popular work of fiction depicts the therapeutic landscape concept, as they affect physical, emotional and spiritual health. The main character and protagonist is a young girl named Heidi who lives modestly high in the Swiss Alps with her grandfather. The Swiss Alps are idealized in the novel as a therapeutic landscape, producing physical, emotional and spiritual health for those who experience them. While extending the idea of therapeutic landscapes as lived, everyday places, and providing a reflection of the socially constructed ideas about health (care) at the time, this analysis builds on the previous work which interprets literary works using the therapeutic landscape as an analytical approach.

Although some people perceive the mountains as claustrophobic, the general perception shared today is that mountainous environments are healthful. Alpine environments have and continue to symbolize many things, including beauty, greatness and good health. A stay in high altitudes has been a long-standing medical prescription, remaining valid today with more and more doctors recommending stays in high altitudes as good preventative medicine (Auer, 1982). Many spas, such as those found in the Swiss Alps (for example, St. Moritz), continue to offer the healing alpine waters and related activities promoting health and wellness. The idea of alpine environments as therapeutic is well illustrated in *Heidi.* Written in the 1800s, the novel was influenced by the Romantic movement in Europe, which embraced – rather than feared – nature, the latter which was the case a century or so earlier. Although more than a century has passed since *Heidi* was first published, the symbolism of the alpine geography in physical, emotional and spiritual well-being, as depicted in the novel, is still very much in the human psyche today.

After a brief description of the author, the essence of the story is reviewed. Following this, the contrasts between the two main place settings of the book – the city of Frankfurt and the Swiss Alps – are discussed, presenting stereotypical images of contrasting environments that represent ideals rather than reality. The urban realm of Frankfurt symbolizes illness and all things unhealthy, unlike the Swiss Alps which symbolize health and all things that contribute to wellness. The idealized healthful and healing aspects of the Alps, as depicted in the story, are then elaborated on, using succinct examples from the novel to illustrate the impacts on physical, emotional and spiritual health.

The Author

Johanna Spyri (1827–1901) drew upon the memories of her own happy childhood when writing *Heidi* and other stories of life in the mountains of Switzerland. She was born in Hirzel, near Zürich, Switzerland. Her home was on the high Swiss Plateau, overlooking Lake Zürich. Beyond it lay alpine pastures and towering mountain peaks. In this idyllic setting, Johanna spent the first 25 years of her life. In 1852 she married a young lawyer, and moved to Zürich. The Spyris had one son, who died in childhood.

When Johanna Spyri was 43 years old, she began to write stories about the people and scenes of her youth. The wholesome tales brought international fame to the Swiss homemaker. Although Spyri wrote in German, *Heidi* has been translated into many languages and has been the subject of several motion pictures. In addition to

Heidi – first published in two parts in 1880–81, her books include *The Little Alpine Musician*, *Uncle Titus*, *Gritli*, and *Veronica*.

The Story

Orphaned, Heidi's aunt leaves her with her grandfather, who lives quite isolated, high in the Swiss Alps. Although only a young school-age girl, she develops a strong sense of community and place identity in the agrarian environment; she befriends a young goat herder, Peter, and becomes acquainted with his grandmother, Grannie. Quite suddenly, Heidi's aunt returns and secretly brings her to the city of Frankfurt, where she is to be the playmate for Clara, an invalid daughter of a wealthy businessman named Mr Sesemann. Heidi fulfills her role well but, whilst living in Frankfurt, becomes unwell and yearns for her rustic mountain home. A doctor friend of the Sesemann family, Dr Classen, prescribes that she is sent home immediately, to the dismay of Clara. Within a short time in her beloved mountains, Heidi's health returns. The two friendly acquaintances she made whilst in Frankfurt – Clara and Dr Classen – visit her in the mountains. Similar to Heidi, both Dr Classen and Clara experience healing in the alpine environment. Heidi's grandfather, Uncle Alp, cares for and nurses both Dr Classen and Clara in their sojourn in the mountains. Although he is ridiculed as being a hermit by the townsfolk living below, Uncle Alp represents the ultimate example of a person who has learned to live in intelligent harmony with their environment.

Urban–Rural Contrasts

The story's contrasts are not easy to miss. The authenticity of Heidi's alpine home is not only real to her, but to the two city folk (Dr Classen and Clara) who visit her in the mountains, further contrasting the city of Frankfurt as an uncaring, unauthentic environment.

The total joy that Heidi experiences while living high up in the alpine mountains contrasts with the total despair that all but destroys her when she is kept away from it, in Frankfurt. Similarly, the healing and healthful experiences of Dr Classen and Clara, when visiting the mountains, contrast with the hardship experienced in their Frankfurt home.

These contrasting elements suggest that the mountains are symbolic of positive health in what is otherwise a less-than-healthful world. A positive sense of place is strongly associated with the mountains, as elucidated in the protagonist, who believes that the mountains symbolize good health for those who visit and, particularly, for those who live there.

Similarly, the city of Frankfurt symbolizes an unnatural, inhospitable, and overall unhealthy environment, contributing to illness of all things – the body, mind and spirit. Frankfurt, in its dense urbanity, represents congestion, stuffiness and an unnatural environment and lifestyle for Heidi. Her longing for the natural, comparatively untouched, alpine landscape is reflected early on in her time in Frankfurt – her first morning living in Clara's house:

> She jumped out of bed and dressed quickly … She ran from window to window frantically, trying to open them, like a wild bird in a cage, seeking a way through the bars to freedom. She felt sure that if she could see what was outside, she would find grass somewhere, green grass with the last snow just melting from it …. After a while she gave up (Spyri, 1880/2000, 79).

Following this, she interrupts Clara's lessons while hurrying downstairs to open the door in order to hear 'fir trees rustling' – only to find the passing of light carriage wheels along the cobblestone road. After one of the servants opens a window for Heidi, pointing out the church tower as a place to 'see over the whole valley', Heidi executes her plan to see the natural world by seeking and finding the church tower. After being disappointed with the view from the top of the church tower – '… a sea of roofs, chimneys and towers …' she – punished for leaving the house without permission and without telling anyone.

Heidi's homesickness spurs her interest in going back to the Alps, although this is met with reproval from the unkind house servant, Miss Rottenmeier. Heidi's response to her describes the contrast between the two worlds of Frankfurt and her alpine home:

> I want to go home because while I'm here Snowflake [one of the mountain goats] will be crying, and Grannie will be missing me too. And here I can't see the sun saying goodnight to the mountains. And if the hawk came flying over Frankfurt he'd croak louder than ever because there are such a lot of people here being horrid and cross, instead of climbing high up where everything's so much nicer (Spyri, 1880/2000, 98).

The lack of hospitality and empathy felt by Heidi whilst in Frankfurt suggests again how the city symbolizes an unhealthful place. Since being scolded for trying to go home, a depressive change comes over Heidi; she now understands that she will have to stay in Frankfurt for a long time, perhaps forever. She begins to lose her appetite and grow pale. The reason for Heidi's mournful, heavy-hearted behaviour, although detected, is only expressed when she is alone, as she fears being called 'ungrateful' by the servants, yet once again:

> When she was alone in her quiet room at night, she often lay awake for hours, thinking of home and the mountains, and when she fell asleep at last, it was to dream of them so vividly that she woke in the morning expecting to run joyfully down the ladder from the loft – and found herself, after all, still in the big bed in Frankfurt, so far away. The disappointment of that awakening often made her cry miserably, burying her face in her pillow so that no one should hear (Spyri, 1880/2000, 116).

The few opportunities to get out of the house were equally unsatisfying: '… in the town Heidi scarcely knew whether it was winter or summer…. Now she only went out-of-doors when Clara was feeling well enough for a drive…. they saw plenty of people and beautiful houses, but not a blade of grass or a flower or a tree, and no mountains' (Spyri, 1880/2000, 125–6). Frankfurt is symbolized as a desert, replete of any health-enhancing characteristics. Heidi longs for the pure air and the unspoiled landscapes of the Alps, together with the company of Peter, the goat herder, his grandmother 'Grannie', and most of all, her grandfather and his hut. Heidi is clearly out of place in Frankfurt, having no sense of place, place identity or attachment to the city, interpreted and understood as impacting her failing health.

It is not until Dr Classen finds her sleepwalking that all learns of her homesickness. When asked about what she dreamt about, Heidi reveales her longing for home:

> I dream every night that I'm back with Grandfather and can hear the wind whistling through the fir trees. I know in my dream the stars must be shining brightly outside, and I get up quickly and open the door of the hut – and it's so beautiful. But when I wake up I'm always still here in Frankfurt (Spyri, 1880/2000: 137).

The doctor's prescription is to send Heidi home at once. Although the house servant who accompanies her back to the Swiss Alps thinks her home 'only half-civilized', Heidi is exuberant to be back home, reflecting her strong sense of place: 'I'd a million times rather be with Grandfather on the mountain than anywhere else in the world' (Spyri, 1880/2000, 150). The mountain peaks seems to greet her like old friends. The importance of place identity for Heidi in Spyri's novel is reflected in the return of her health when back in her alpine home, away from the vice and corruption of Frankfurt. The mountains are seen as Heidi's 'personal home' (Dossey et al., 1989: vii), providing an integrative network of physical, spiritual and psychological factors merging together to promote the creation of a healing and/or healthy place. The mountains' physical, emotional and spiritual healing powers, as experienced by Heidi, Dr Classen and Clara – each to varying degrees – will now be explored. Physical healing will be first discussed, followed by the emotional and, finally, the spiritual.

Physical Health

Mountains symbolize strength and in so doing, strong physical health. What was already considered more than a century ago, when the English on their continental tours discovered the stimulating powers of sun and snow, is acknowledged today as a medical fact: a stay in high altitudes exerts an influence on physical health (Auer, 1982). The high altitudes inflict deep sleep, like that experienced by Heidi her first night at home in her loft: '… Heidi did not stir. She slept soundly all night long, satisfied through and through. She was home again. She had seen the sun setting on the mountains. She had heard the wind whistling through the fir trees' (Spyri, 1880/2000, 157). Similarly, Clara sleeps and eats like never before in the alpine environment: 'As long as I can remember, I've only eaten because I had to. Everything always tasted so of cod-liver oil, and I used to wish I didn't have to eat at all. And here, I can hardly wait for your Grandfather to bring my milk' (Spyri, 1880/2000, 242).

Although somewhat encouraged by the fate of her wheelchair (thanks to the envy of Peter, the goat herder), Clara takes her first steps up in the mountain meadow. In less than four weeks, Clara is well and able to walk. Encouraged by Heidi to see the beautiful flowery meadow, Clara revels in all that has happened while gazing at all the lovely colours and smelling the poignant scents of the high pasture. The combination of the best of the alpine environment – clean air, goat's milk, an aesthetically beautiful natural environment, and nutritious, albeit simple food, together with the encouragement and loving care provided by Heidi and Uncle Alp – physically heal Clara.

The healing powers of the natural environment of the Alps are again reiterated in Heidi's explanation to Peter of why her grandfather's goats – named Daisy and Dusky – are at odds in their new stall down from the mountain in the town of Dorfli: 'Grandfather said they were feeling like I did in Frankfurt... Because they'd never left the high pasture before, and you don't know what that's like, Peter' (Spyri, 1880/2000, 207) The strong sense of place felt by Heidi, and less so by Dr Classen and Clara, suggest that the beautiful Swiss Alps are representative of a truly therapeutic landscape.

Emotional Health

Spyri's *Heidi* also illustrates mountains as symbolizing emotional health. The care, concern and sensitive nursing given freely by mountain people, exemplified by Heidi's grandfather, symbolize the cyclical and recurrent nature of the goodness produced by living a rural way of life. Uncle Alp not only befriends Dr Classen, guiding him on long walks high in the mountains while sharing his knowledge of the unique flora and fauna, but prioritizes the health of first, Heidi (upon her return from Frankfurt), and then that of Clara, when visiting.

> Uncle Alp grew very attached to his little guest [Clara] and tried to find something new every day to make her better. He took to going off in the afternoons high up on the mountaintop to look for special plants and herbs … (Spyri, 1880/2000, 244).

Heidi's profound positive effect on other people, due to her natural ability to expound hope and joy, is seen to be due to this clean, simple lifestyle. All who know Heidi are positively affected by her optimism and goodwill, particularly Grannie, the grandmother of Peter, the goat herder. Grannie's feelings for Heidi are illustrated best in her conversation with Uncle Alp:

> I've something in my heart, I'd like to say to you … If I've ever done you any harm, don't punish me by letting Heidi go away again, while I'm still above ground. You don't know what she means to me' (Spyri, 1880/2000, 170).

The healing experienced by Dr Classen is not purely physical, as it has an emotional element. The crisis of losing his daughter appears in a different light while in the alpine environment, likely influenced by the kindness shown by Heidi and her grandfather. Both Dr Classen's emotional and mental spheres are positively affected as a result of visiting the Alps. He leaves the mountains with renewed hope, likely re-evaluating his personal aims and life projects: 'This is certainly a wonderful place for sick minds as well as bodies. Life seems really worth living again!' (Spyri, 1880/2000, 199).

Clara similarly finds a new emotional perspective when away from the enclosed, stuffy world of Frankfurt. This is illustrated well when up in the mountain meadow with the goats:

> To Clara this strange new experience was very exciting. To be here, all by herself, and out of doors in such a beautiful place, with this little goat eating so trustfully out of her hand – was all so delightful. She had never expected to know such happiness, and it gave her a new idea of what it must mean to be like other girls, well and free, to run about and to help people, instead of always having to be the one who sat still and was waited on. The thought seemed to add an extra radiance to the scene, and a deeper glow to her own happiness (Spyri, 1880/2000, 251–2).

The authenticity of the Swiss Alps, as enhanced by Uncle Alp and Heidi, contrasts with the unauthentic world of Frankfurt. The hazards of living in a busy, grey city far away from nature are made worse when Heidi thinks about the unconcerned, nonchalant servants working in the Sesemann household. When Dr Classen asks her to come back with him to Frankfurt, Heidi thinks of the '… town with its many tall houses and cobbled streets, and thought of Miss Rottenmeier and Tinette [servants]. "It would be nicer if you came back to us," she said hesitantly' (Spyri, 1880/2000, 198). Luckily for Heidi, Dr Classen agrees.

Spiritual Health

Historically, mountains have long represented spiritual well-being, given their great height and perceived proximity to a heavenly existence. Although the novel has been criticized for its attempt to preach piety and morality, the underlying theme of the book is that God (not we) knows what is best for us. The mountains in Heidi's world represent God in his purest form. For Heidi, the mountains bring peace and one of the highest expressions of a spiritual life – being one with nature. This is best expressed when she returns home from Frankfurt:

> Everything seemed even more beautiful than she had expected … It was so lovely, Heidi stood with tears pouring down her cheeks, and thanked God for letting her come home to it again. She could find no words to express her feelings, but lingered until the light began to fade and then ran on (Spyri, 1880/2000, 154).

Nature as God is again reiterated by the pastor when Uncle Alp returns to town to attend church, spurred on by Heidi after a long time away: 'Neighbor, your mountains have been a good church to you, and brought you to mine in the right frame of mind. You've made me very happy' (Spyri, 1880/2000, 168). The symbolism of mountains as God's house (a church) again suggests the Alps as a sacred landscape. Similarly, Clara's description of the Alps upon her arrival suggests the presence of a higher power: 'It's heavenly, … simply heavenly… I wish I could stay here forever' (Spyri, 1880/2000, 229). These examples illustrate the perception of mountains as symbolizing a connection to the spiritual.

The clear mountain skies allow the bright starts to twinkle at night, symbolizing for Heidi a direct connection with God:

> … the stars came twinkling out one by one as she climbed up to the hut, sending her a greeting out of the sky. She stopped sometime to gaze up at them, feeling a deep peacefulness in her heart, and said a little prayer of thanks (Spyri, 1880/2000, 277).

This is discussed again when Heidi describes to Clara why the stars twinkle:

> [the stars] are up in heaven and know that God looks after us all on earth so that we oughtn't really ever to be afraid, because everything is bound to come right in the end. That's why they nod to us and twinkle like that (Spyri, 1880/2000, 336).

When sharing the hayloft with Heidi, Clara cries: 'it feels as if we were riding in a sort of carriage right into heaven' (Spyri, 1880/2000, 236).

Heidi's most enlightened experiences take place in the 'high meadow', illustrating a serious place for meeting the spiritual dimension of the Alps. Heidi's enthusiasm for the high meadow brings her to encourage Dr Classen and Clara to experience it as well. For Heidi, the high meadow provides her with a psychosomatic experience, where her feelings of well-being, elevation and uplifted state of mind teeter on euphoria:

> Heidi led [Dr Classen] the way to her favorite spot from which she could look down on the distant valley, so green below them, and up to the great mountains where the eternal snows sparkled in the sunlight. The grey rock of the two towering peaks rose majestic against the strong blue of the sky … The hawk was soaring overhead in ever-widening circles, making no sound. Heidi's eyes were happy as she gazed out at the beautiful things she loved so well … (Spyri, 1880/2000, 192).

Auer (1982, 18) summarizes the spiritual rewards of an alpine environment: 'He who knows how to open his mind and his heart in the mountains and to the mountains will be richly rewarded.' The healing experiences of Heidi, Dr Classen, and Clara represent the rewards of fully experiencing life high in the mountains.

Conclusion

Johanna Spyri, who grew up in Switzerland, first published *Heidi* more than a century ago. At that time, doctors knew little and their patients less; medical knowledge was inexact. The possibility of Clara's cure was more believable at that time than it would be today; the most invigorating of mountain air and the best of goat's milk could scarcely cure the sort of illness that would keep Clara in a wheelchair. High altitude medicine, on the other hand, has developed rapidly in the last few decades, with some suggesting that 'a person suffering from a latent disease is very likely to discover the actual source of his illness' (Auer, 1982, 9).

While extending the idea of therapeutic landscapes as lived, everyday places, the interpretation of the children's classic *Heidi* builds on the previous work which interprets literary works using the therapeutic landscape as an analytical approach. As with the rural–urban contrasts depicted in Balzac's *The Country Doctor* (Tonnellier and Curtis, 2005), the hazardous world of city life in Frankfurt (where physical, emotional and spiritual illness appears to be rampant) is contrasted with a magical alpine world – where the physical, emotional and spiritual aspects of health and wellness are enhanced. The novel thereby presents a 'utopia' of sorts and symbolizes a traditional therapeutic landscape. The fantasizing of the Alps presents an ideal which certainly may not be felt by all who visit them; such images need to be kept in check as they may raise expectations that may not be met.

As noted earlier, Spyri's *Heidi* reflects its times, as the novel was influenced by the Romantic movement in Europe, which embraced nature. This is in contrast to a century earlier, where fear was the response to 'nature in its wilder, more threatening expressions ... high, barren mountains, dangerous oceans, savage storms on land or sea, black and perilous night-time' (Charlton, 1984, 41). The perception of wild nature in eighteenth-century Europe is described by Charlton (1984, 42) as having a 'sense of repulsion, displeasing irregularity or, at best, disinterested boredom'. The idealized view of alpine environments depicted in *Heidi* thereby reflects changes in cultural attitudes toward nature in Europe, illustrating that these attitudes are contingent or context-dependent.

Acknowledgments

Thanks to the late Paul Simpson-Housley, a good friend and mentor who first inspired this work and to whom this work is dedicated. Gratitude is also extended to Wil Gesler, who kindly provided a review of the chapter, making excellent suggestions for its improvement.

References

Auer, F. (1982), *When Men and Mountains Meet: Guidelines for a Stay in High Altitudes* (Condraw SA: Stampa Romontscha).

Baer, L.D. and Gesler, W.M. (2004), 'Reconsidering the concept of therapeutic landscapes in J D Salinger's *The Catcher in the Rye*', *Area* 36 (4): 404–13.

Charlton, D.G. (1984), *New Images of the Natural in France: A Study in European Cultural History* (Cambridge, UK: Cambridge University Press).

Dossey, B., Keegan, E., Kolkmeier, L. and Guzzetta, C. (1989), *Holistic Health Promotion: A Guide for Practice* (Maryland: Aspen).

Gesler, W.M. (2000), 'Hans Castorp's journey-to-knowledge of disease and health in Thomas Mann's *The Magic Mountain*', *Health and Place* 6: 125–34.

Gesler, W.M. (2003), *Healing Places* (Lanham: Rowman & Littlefield).

Gilbert, E. and Simpson-Housley, P. (1997), 'Places and spaces of dislocation: Lady Oracle's Toronto', *The Canadian Geographer* 41: 235–47.

Preston, P. and Simpson-Housley, P. (eds) (1994), *Writing the City: Eden, Babylon and the New Jerusalem* (New York: Routledge).

Simpson-Housley, P. and Norcliffe, G. (eds) (1992), *A Few Acres of Snow: Literary and Artistic Images of Canada* (Toronto: Dundurn Press).

Spyri, J. (2000), *Heidi* (New York: Alladin) (original work published in 1880).

Tonnellier, F. and Curtis, S. (2005), 'Medicine, landscapes, symbols: "The Country Doctor" by Honore de Balzac', *Health and Place* 11: 313–21.

PART 2
Therapeutic Geographies for Special Populations

Chapter 6

Making Clean and Sober Places: The Intersections of Therapeutic Landscapes and Substance Abuse Treatment

Geoffrey DeVerteuil, Robert D. Wilton and Shaun Klassen

Introduction

In this chapter we are concerned with the nature of environments established for the treatment and rehabilitation of people with drug and alcohol addictions. Specifically, we examine the strategies treatment programs employ, and the challenges they face, in their efforts to help individuals gain control of their drinking and drug using. We do this by drawing on qualitative data collected from three treatment programs in Winnipeg, Canada. The analysis of this data is concerned with the locations of addiction treatment programs and the ways in which the surrounding environment is thought to influence the process of addiction treatment in both positive and negative ways. The analysis also illustrates the ways in which internal program environments and external contexts interrelate in important ways. The external environment – for example, the character of the local neighborhood – may shape a program's regulations with regard to the monitoring of clients' movements to and from the site. At the same time, the organizational ethos of the program – the daily routines and sense of place cultivated at the site – is often aimed at altering the ways in which individual clients relate to the external environment upon their graduation from treatment.

We employ the concept of therapeutic landscapes to frame our argument that the environment matters in terms of substance abuse treatment. In so doing, we aim to make three contributions to the evolving therapeutic landscape literature. First, we hope to extend health geographers' understanding of marginalized groups' access to, and experiences within, therapeutic landscapes. Second, we consider the extent to which treatment programs as therapeutic spaces have the potential for both positive and negative impacts on client populations. For example, our analysis suggests that facilities may sometimes be forced to operate in urban environments that are decidedly un-therapeutic in nature. In response, agency staff may develop tactics to moderate or avoid the unhealthy aspects of local contexts. Finally, our analysis sheds further light on the geographies of addiction treatment, a relatively neglected topic within the health geography literature.

We begin by briefly reviewing work on therapeutic landscapes, before turning to consider changing conceptions of addiction/dependency/abuse and the evolution of addiction treatment systems in North America. Using a Winnipeg (Canada) case study, we then outline the nature of our study and the data collection process. The remainder of the chapter is given over to an analysis of the environments of addiction treatment programs, the locational dynamics of which are structured by three intra-urban zones in Winnipeg: inner core, outer core and suburb. Particular attention is given to the extent to which each of these intra-urban zones presents different challenges and opportunities for substance abuse treatment.

Therapeutic Landscapes and Marginality

Gesler's (1992) concept of therapeutic landscapes represented an innovative geographical metaphor with which to understand the relationship between place and health. In the past decade, the concept has been extended by a number of health geographers (see Williams, 1999; Kearns and Collins, 2000; Andrews and Kearns, 2005). One recurring theme has been the attraction of such therapeutic places as commodities, and the motivations of consumers to seek out places such as summer camps, holiday destinations, remote wilderness and fitness facilities (Tyson, 1998; Palka, 1999; Andrews and Kearns, 2005). Reflecting the more traditional concerns of health geographers, therapeutic landscapes have also been understood as places of health care and health promotion (for example, homes for the elderly, private health care facilities, children's camps, and so on), again often structured by commercial interests (Kearns and Barnett, 1999; Kearns and Collins, 2000) and willingly experienced by clients (Geores and Gesler, 1999; Williams, 2002; Andrews, 2003). Less attention has been paid to how the concept intersects with our understanding of more marginal, less explicitly commercial spaces of care, including homeless services, mental health drop-ins and addiction treatment programs (but see Conradson, 2003; May, Cloke and Johnson, 2005; Wilton and DeVerteuil, 2006).[1] A common thread in these studies is the extent to which such organizations often have relatively little choice over the kinds of built and social environments in which they operate, making the tasks of creating and sustaining a therapeutic setting all the more difficult.

In this chapter, we are specifically concerned with the challenges facing addiction treatment programs as they work to create therapeutic settings in different urban environments. We suggest that the therapeutic landscape concept can be usefully applied here to consider the characteristics of neighborhood settings and their effects on the programs' operation. As we suggested earlier, the analysis affords the opportunity to further explore the complexity of therapeutic environments, particularly the extent to which they may impact health in both positive and negative ways (see Baer and Gesler, 2004; Wakefield and McMullan, 2005). As Wilson

1 Not all of these cited references make use of the therapeutic landscape concept. May et al. are interested in the extent to which service programs are drawn into broader governmental strategies aimed at controlling street homelessness. Conradson explores the capacity for organizations to constitute spaces of care, suggesting that the 'landscape' concept may not fit well with the small-scale, interpersonal environments that exist within organizations.

(2003) noted, earlier work on therapeutic landscapes tended to focus exclusively on the healthful character of specific environments (also Williams, 2002). While understandable, the absence of attention to less positive outcomes works against an understanding of places as characterized by varying degrees of therapeutic potential. Related to this – and of particular importance when considering the experience of socially marginalized populations – is the recognition of the political complexity of such environments. The identification and common-sense understanding of environments as 'therapeutic' may be shaped to a significant degree by questions of power and position (see Kearns and Collins, 2000; Wilton and DeVerteuil, 2006). While mental health professionals may have conceptualized the asylum as an instrumentally therapeutic environment, the people confined there likely had a much more nuanced understanding of its capacity for care, control and coercion.

Substance Abuse Treatment and the Role of the Built/Social Environment

Individuals dealing with addiction, as well as those in treatment, can be considered a socially marginalized population. They are often constructed as threats to decent society, labeled as non-productive, and frequently blamed for their addiction (Takahashi, 1996). While links between drug use and crime may have some basis in fact, the widespread public antipathy towards people with addictions and the organizations designed to assist them has major implications in terms of social geography and facility location.

What little research there is on the purported links between addiction treatment and the built/social environment suggests that treatment ideally requires milieus that are architecturally and medically appropriate and, critically, removed from drinking and drug cultures (Timko, 1996; Grosenick and Hatmaker, 2000). Interestingly, Gesler (1992, 738) made specific reference to addiction in his initial paper on therapeutic landscapes, suggesting that:

> Most alcoholics have low self-esteem, in part because they feel no identity with particular places. Often places represent failure, threats, or feelings of not being wanted. Therapy for alcoholics might usefully include establishment of refuges, places with positive images, where identity could be established.

In a comparison of treatment facilities in isolated rural settings and urban environments, Weinberg and Koegel (1995, 230) found that there was an advantage to being 'cocooned in a more isolated, bounded, and intense treatment community.'

Among addiction treatment providers, there is more ambivalence about the role of the environment. On the one hand, service providers frequently point to the failure of clients who tried to use 'the geographic cure' to deal with addiction. This term refers to the belief that a move away from familiar haunts and addict friends to a new place entirely can provide a basis for recovery. Service providers contend that such a move, while offering temporary relief, does little to contend with the emotional and psychological issues that lead people to drink and use drugs in the first place.[2] On the other hand, some service providers – for example, those adopting a 'social model' approach to treatment (see Wilton and Deverteuil, 2006) – have developed an activist orientation toward the social environment, arguing that the creation of healthy neighborhoods may be an intrinsic part of the recovery process.

In practice, access to the more isolated treatment environments described by Weinberg and Koegel (1995) is often limited. Where addiction treatment is a marginal component of the public health system, only those individuals able to pay for private services are able to enter settings that approximate an idealized therapeutic location. Private rehabilitation clinics are often sited in geographically attractive physical environments, and deliberately removed from the temptations of the city. Individuals who cannot pay for services must often seek treatment in very different settings. Geographical research has shown that undesirable land uses, such as non-profit and publicly funded addiction treatment facilities, are often ghettoized in central city neighborhoods with limited social and environmental amenities (Dear and Wolch, 1987). For people attempting to recover from addiction, these neighborhoods may constitute distinctly unhealthy or risky landscapes, not least because the boundary between the therapeutic environment – the treatment facility – and the community-at-large is highly permeable (Weinberg and Koegel, 1995). The immediate environment may thus act as an impediment to recovery, with clients finding it difficult to insulate themselves from people and places associated with the sale and use of drugs and alcohol. The situation may be most problematic for precariously housed people dealing with addiction. These people often cluster in downtown Skid Row areas, where there is access to cheap housing, food and rehabilitation centers, but also to liquor stores and drugs – a troublesome co-location dynamic. However – and this is a key point – there is evidence that programs in these neighborhoods do succeed in helping some people on the road to recovery. That is, despite the challenging or 'unhealthy' characteristics of surrounding neighborhoods, addiction treatment programs work hard to create and sustain therapeutic environments of sufficient integrity and duration to facilitate recovery for *some* clients.

Of course, the question of success in the context of addiction treatment is a complex one. The past half-century has seen significant change in the conceptual tools used to talk about people's problematic relationships with alcohol and other drugs. The work of Jellinek and others at Yale University in the postwar period led to the growing popularity of 'addiction' defined in terms of loss of control and the potential for harm to oneself or others (Valverde, 1998). Over time, however, the conceptual limitations of Jellinek's model meant addiction was eclipsed by a language of substance 'abuse' and 'dependency' in official medical definitions. At the same time, the development of harm reduction approaches has led to a questioning of complete abstinence as essential to the 'success' of service provision. In the context of this research, we conceptualize success in terms of helping individuals to: (1) stop drinking and/or using others drugs, or at least reduce consumption; (2) begin to address the causes of their using/drinking; and (3) develop a set of psychological,

2 This information is drawn from comments made by a number of treatment providers interviewed in the broader study from which this chapter is drawn.

social and material resources to assist them when they leave the program. Existing research and first-hand accounts by service providers and clients emphasize that recovery from addiction is a difficult task, fraught with frustration and the potential for relapse. In this sense, the question of time as a criterion for success – how long a person remains abstinent – is also difficult to answer.

In an earlier paper (Wilton and DeVerteuil, 2006), we argued that service providers work to ensure the potential for successful treatment by maintaining a highly structured internal program environment for clients, while at the same time developing strategies to cope with problematic elements of the external environment. The earlier research was based on a single case study of a city neighborhood in a US city. In the current project, we extend this analysis to include multiple programs operating in distinct settings. This offers an opportunity to explore variations in the environmental opportunities and challenges facing service providers, as well as the strategies providers develop to cope with the different settings. For the remainder of this chapter, we use data collected from three facilities in Winnipeg to examine these very issues.

Methods

Studies of therapeutic landscapes have primarily been approached using intensive, qualitative or mixed methodologies, including small-scale surveys of agencies (Kearns and Collins, 2000), detailed visits of single sites (Conradson, 2003), in-depth, unstructured interviews with small purposive samples (Wilson, 2003), and content analysis of historical and documentary sources (Geores and Gesler, 1999). Our primary focus is also qualitative, as we recognize that exploration of therapeutic landscapes requires sustained interaction with the sites themselves as well as the occupants of these sites (here both program staff and clients).

Site Selection

As part of a larger project in Canada, we gathered data using both intensive participant observation and in-depth interviews with staff and clients at treatment programs in Winnipeg, Manitoba (n=7), Hamilton, Ontario (n=5) and Toronto, Ontario (n=13). Our principal focus was on programs with a residential component, although some programs with day treatment regimens were also included, given the recent emphasis on the (cost) effectiveness of non-residential treatment options. For the purposes of this chapter, we focus only on data from the Winnipeg research.

Information on the system was accessed through the Contact Community Information and Addictions Foundation of Manitoba Directory (2006). From this inventory of facilities, we used a purposive sampling strategy (Curtis et al., 2000) to identify a sample of participant facilities in Winnipeg, with a locational cross-section of facilities that ranges from those in downtown to more suburban locations, as well as aiming for a cross-section by program focus (gender, treatment philosophy, church-based and secular, aboriginal and non-aboriginal). Location was particularly important, as we expected that facilities in stigmatized, central areas would face different locational challenges when compared with suburban and semi-rural facilities. In the research, we were not concerned with exclusive private retreats, a choice that was based to a significant extent on our interest in the challenges facing public programs with limited resources and limited capacity to control location and site characteristics.

Data Collection

Within Winnipeg, the directors of the seven addiction treatment programs were interviewed about the physical and socio-cultural character of the facility, its relationship with the surrounding neighborhood, the challenges faced in assisting clients toward recovery, as well as changes to the program and neighborhood over time. We also attempted to interview frontline staff and one or more program clients to obtain multiple perspectives on the character of facility spaces and surrounding environments. Given the organizational setting of the data collection, we were dependent on the willingness of the director as gatekeeper in each to allow access to staff and clients. Staff members from all seven programs were interviewed (n=24), along with clients at three programs (n=12). All interviews were recorded, transcribed in full and analyzed. Observation at three sites provided a second form of data. Again, the opportunity for observation varied in large part because access had to be negotiated with program directors. In some instances, observation occurred during a day-long site visit. At the other end, researchers spent one or more weeks at some facilities, interacting with clients and staff, and sitting in on treatment sessions. Observation of facility space and the immediate neighborhood offered insight into the ways that staff and clients used and adapted the local environment. Extensive field notes were taken, and these notes were also fully transcribed for analysis. Our analytical approach was largely thematic in nature, seeking common and contrasting cases across the seven sites.

Analysis

The environments of the sampled Winnipeg facilities are themselves structured by broader intra-urban patterns. Our aim here is to delineate those areas within the city that reflect different social and built environments. Figure 6.1 locates *all* treatment facilities in the city of Winnipeg by bed, overlaid by three zones – the inner core, outer core and suburbs.[3] These zones were determined using population density, housing age and household income from the 2001 census (see DeVerteuil et al., 2007 for more details). Core census tracts had population density and pre-1946 housing age values above Winnipeg city medians (2,719 person per km^2 and 8.3 per cent respectively). However, census tracts in the inner core were poorer, with household income values below the Winnipeg median ($44,321), while those in the outer core

3 Treatment facility mapping was based on a variety of sources – the Yellow Pages, studies by the Addictions Foundation of Manitoba, and the Manitoba Centre for Health Policy.

were substantially more affluent. The suburb area included tracts whose proportion of pre-1946 housing and population density were below median city values.

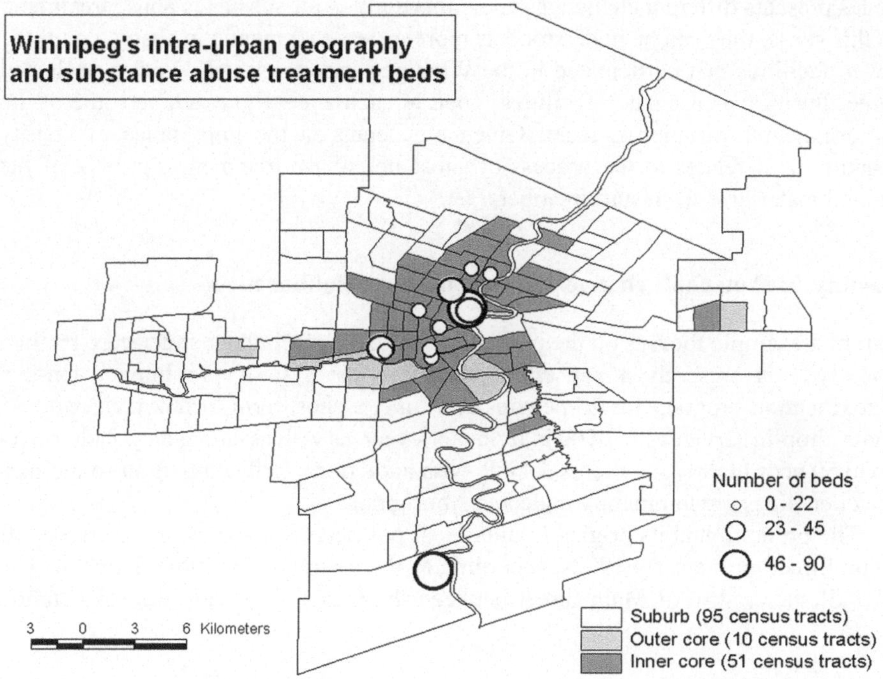

Figure 6.1 Winnipeg's intra-urban geography and substance abuse treatment beds

The resulting monocentric pattern shows a largely impoverished inner city and relatively prosperous suburbs. As the figure also shows, there is disproportionate clustering of facilities in the inner/outer core (74 per cent of beds with only 35 per cent of Winnipeg's population). Other associated facilities that people in recovery might also rely upon, including psychiatric beds, licensed mental health facility beds, and rooming home units, also clustered disproportionately in the core. This should come as no surprise, as these sorts of facilities are typically channeled to unpromising and stigmatized social/built environments, a fact that may work directly against the creation of therapeutic landscapes for people seeking treatment for addiction. Moreover, conditional zoning is currently required for the siting of neighborhood 'rehab homes,' making it difficult to relocate (City of Winnipeg, 2004).[4] Using a per capita (per 1,000 persons) measure, the inner core contained four times more

4 These homes are defined as buildings used for boarding or other residential accommodation plus penal or other mandatory supervision or treatment of not more than 12 persons, in which supervision or treatment is not provided to any persons not residents in the rehab home, and that complies with the provincial's government's licensing requirements.

psychiatric beds, three times more licensed mental health beds, and 28 times more rooming home units than the suburbs.

Our interest, in this chapter, is the extent to which each of these intra-urban zones presents different challenges and opportunities for substance abuse treatment. In this sense, they can be understood as more or less therapeutic in character. Of the seven facilities that participated in the Winnipeg-based research, we have selected three illustrative case study facilities – one in each zone – to examine some of the environmental variation evident. Given our focus on the importance of facility location as it relates to the process of substance abuse treatment, the bulk of the quoted material is from staff members.

Facility A: 'You can't change a person in that neighborhood'

Our first example focuses on the particular challenges facing one inner-core facility. Facility A is presently a key entry point into the system, providing a ten-day detoxification program for 25 persons at a time, a short-term 'drunk tank' with 15 spots, drop-in services for 2,500 individuals a year, as well as emergency shelter with up to 50 beds nightly for clients dealing with addiction.[5] Reflecting its surroundings, the clientele is predominantly male and Aboriginal.

The program had its origins in municipal politicians' desire to 'keep drunks off Main Street' in the early 1970s, according to the founding Executive Director. The two-kilometer strip of Main Street between the Canadian Pacific Railroad station

Figure 6.2 Image of Facility A, July 2005

5 These figures are taken from the organization's 2005 annual report.

to the north and the Canadian National Railroad station to the south is historically referred to as Winnipeg's 'hotel row.'

At its peak, the Main Street hotel strip was the location of upwards of sixty hotels 'and because of the high concentration of bars, pedestrians were never beyond the aroma of booze that wafted through the windows and doors of the hotels' (Artibise 1977, 62). Over time the area lost much of its original commercial luster and came to be known as Winnipeg's Skid Row. Faced with declining economic fortunes, the remaining local businesses became fed up with the rising presence of alcoholics on their doorsteps and banded together to provide core funding and establish a drunk patrol for the area in 1972, helping to found Facility A.

With no clear mandate other than clearing the local streets of drunks, a founding staff member relates some of the difficulties in the early days:

> At night we were stacking them [drunks] in there like cordwood … a friend of mine who worked at the [hospital] had given us some beds … so we just put the beds in there and that was the only thing in there … the toilet was plugged … there wasn't any running water … so this guy died … I pulled up that morning and the coroner said: 'close this joint down'. And that hit the newspapers. And when that hit the newspapers the Province came in. They gave us money to rent the upstairs of the building we were in. This friend of mine in the hospital gave us old hospital beds and that's how we started crashing the really sick ones and that was really the beginning of the detox.

Unfortunately, many other helping facilities were pressured to locate in this environment, usually because of Not In My Backyard opposition as well as the proximity to client populations who were themselves confined to marginal parts of the city. Originally a detox for alcohol, Facility A added a drop-in centre and a more permanent emergency shelter in the 1980s.

Critically, Facility A does not provide abstinence-based treatment programs; rather, its focus is on harm reduction. However, the close proximity of the very different activities occurring at Facility A and the permeability between the specific spaces designated for these activities – particularly between the detox program and the drop-in – makes even the goal of harm reduction a difficult one to achieve. As one of the crisis workers mentioned:

> I would like to see the detox off site. The simple reason is that the drop-in and the detox are so close together that when people are in the detox, the drop-in clients will be hanging around the windows watching them and sometimes that gives them the little nudge to leave and … reuse.

These conflicting objectives facing frontline staff have been further strained by a tendency to 'dump' unwanted clients there as a last resort – or, increasingly, as a first resort to save money. The Downtown Business Improvement Zone (BIZ) is currently seeking to bring inebriated and/or disruptive individuals directly to Facility A, especially since a new hockey arena opened downtown.

Facility A must deal with the fact that the boundary between program space and neighborhood is also highly permeable. The staff noted that the program is far too immersed in the local milieu of drugs and drinking, despite the agency's move to a less prominent location a block from Main Street in 1983. Its surroundings remain stigmatized and devalued, saturated with other helping facilities, bars and taverns, and frequent drug dealing. Many of the clients hail from the local area and in this sense the program is not able to provide a real 'haven' or 'respite' (see Weinberg and Koegel, 1995). Again, the crisis worker stated:

> The neighborhood itself is not conducive to their well-being. It's the people that are down here that aren't conducive to their well-being, because all their friends are here, so if they're trying to quit and they go out on the streets and their friends are there ...

Faced with these challenges, the agency's strategy has been to encourage clients into the system through the detox program or the emergency shelter, and then provide opportunities for them to move out of the core area to more suitable treatment environments. As the founding Executive Director stated emphatically: 'you can't change a person in that neighborhood,' while a post-treatment councilor argued that: 'the best way for success is to stay away from Main Street, because the only reason to be down here is to look for trouble.' For this reason, the current Director has expressed a desire to move the detox to a separate location away from the Main Street locale. However, staff members realize that community opposition would make such a move difficult. This example speaks to the ways in which constraints such as limited resources and NIMBY sentiments limit the capacity of programs to select locations with arguably more therapeutic potential.

At the same time, staff recognize that there is some benefit to not severing all ties with the inner-core neighborhood, because clients are likely to return there after treatment. This speaks to a geographical tension between the therapeutic potential of an environment removed from proximate temptation and stress, and the reality that clients may have limited income and residential choice following treatment. Particularly for people who are homeless, completion of treatment may mean a return to an environment replete with risks and triggers, such that the coping strategies and techniques learned in the controlled environment of a treatment program may be that much more difficult to implement.

Facility B: 'Close to Downtown, but not in it'

Facility B is an abstinence-based, 12-bed women-only organization that has been located in an upgrading neighborhood near downtown Winnipeg since 1966. On average, about 130 women pass through the program every year, most from the Winnipeg area. Many of the women rely on social assistance for support, and most are dealing with addiction to multiple drugs. The neighborhood surrounding the program is heterogeneous, with recent development creating a mix of high-rise buildings, new condos, old mansions, dilapidated apartments and several 'funky' commercial streets with an active nightlife.

Facility B offers treatment geared toward behavior change through an intensive four-week program. The facility follows the social model and works to cultivate an all-encompassing therapeutic community that attempts to structure and isolate its clients from their former peer groups/urban environments in order to change their behavior (see Wilton and DeVerteuil, 2006 for more information on social model programs). Facility staff explicitly recognize the challenges facing women in recovery, and directs attention to the broader context of women's lives in the context of the treatment program. As the Director commented:

> We believe in women helping women. Our program content is geared towards looking holistically at women. Certain they've all come for the same reason, and that is to look at their addiction, but our process looks at everything from self-esteem to guilt and shame, to assertiveness training, relationship issues, and parenting.

During the program, the staff attempts to instill a 'home-like' atmosphere, something that may work to replicate middle-class norms (see Veness, 1994). The nature of the program as a women-only space where clients are encouraged to focus on themselves may offer an important break from the multiple responsibilities confronting women as partners and mothers.

Figure 6.3 Image of Facility B, September 2006

With regard to the nature of the local environment, staff suggested that the program's location offered the best of both worlds – or, as one staff member stated: 'close to Downtown, but not in it.' The program's location beyond the inner core provided some protection from challenges confronting Facility A. At the same time, the pleasant yet relatively anonymous and somewhat transient nature of the local environment made it possible for the program and its clients to maintain a low profile and blend in with the neighborhood. The house manager described how these characteristics serve the program:

> A residential neighborhood … I don't think many people in the area even know who we are and that's a good thing, we do have a lot of anonymity in the neighborhood. There has never been … and I've been here twenty years … we've never had any problems with neighbors, with people not wanting us to be here.

The manager also suggested that the neighborhood is very convenient to a range of amenities, including a number of nearby AA (Alcoholics Anonymous) meetings, as well as shopping and entertainment.

The neighborhood is not free from risk. Staff noted that the area's transient and heterogeneous character attracts some drug dealing. Program regulations require the women to remain on site for the first 24 hours of their stay at the facility, in part to prevent local access to drugs. At the same time, staff members recognize a similar geographical tension to that identified at Program A. Many of the women who come to the program will eventually have to return to their old neighborhoods, where they may confront the same triggers that they faced before treatment. Recognizing this, the staff strives to balance a commitment to providing a safe haven for women – an environment that serves as a respite from previously negative environments – with the recognition that too much sheltering may ultimately be doing clients a disservice. Interestingly, staff members use the heterogeneous character of the neighborhood around the program as a training ground for their clients. As one staff member noted:

> It's practice for people to walk past the liquor commission and not go in … they learn that they can do that … you can't wrap people up in cotton wool for 4 weeks and then have too protected an environment and then you send them back home and they have to walk past the bar, the liquor commission … they may as well practice doing it here …

This approach is a strategy aimed at making positive use of a neighborhood containing what might be seen as 'unhealthy' influences. More generally, this example raises interesting questions about what constitutes a therapeutic landscape. With supports in place and a 'safe' home base to return to, the women clients' experiences in this neighborhood may actually help them to stay clean and sober. Finally, once women have left the program, there is encouragement to stay in contact, as the manager noted: 'What we encourage is that women who are able to access our continuing care program come back for 2 hours a week on Wednesdays. We ask them to commit to at least 3 months, and depending on their needs, we would be looking at asking them to participate from anywhere from a year, maybe 2 years.'

Facility C: '[The clients] like the isolation'

Facility C is positioned in a more suburban location. The program is based on a therapeutic community model. As one staff member explained:

> I think if you're talking about therapeutic community, you're talking about everybody pitching in. Everybody is part of the change. Everybody is part of the environment. The environment becomes what people put into it. It's basically a community within a community ... In a therapeutic community everyone is accountable to everybody else ... everybody has to adhere to the peer pressure to being accountable, to not lying, to attending group sessions.

Peer-surveillance is a critical component of changing one's behavior, and most of the staff members are former addicts themselves. The facility accepts men (two-thirds of all clientele on average), women, and families with children; at any one time, there are 100 adults and 12–15 children on-site. Clients on average spend a month at the facility, although there is no firm time limit. Seventy-five percent of the clients at Facility C are Aboriginal, which is not uncommon for treatment facilities in Winnipeg.

The building was originally an orphanage surrounded by substantial grounds. The treatment program took control of the property in 1971, before much of the surrounding residential development had been built. A senior staff person argued that it would be impossible to open a treatment program at this location now, given the composition of the surrounding community and the more restrictive zoning that limits the locations of rehab housing. In the context of the surrounding residential area, Facility C works to carefully manage its image, striving to be a good neighbor. The facility no longer accepts individuals with a violent offender designation, and there is now 24-hour on-site supervision that monitors the program grounds and a neighboring elementary school.

Figure 6.4 Image of Facility C, April 2005

As a strategy to facilitate integration, Facility C limits the movement of new clients within the surrounding neighborhood. During their first month in program, clients are restricted to a tightly bounded area near the facility, and the new arrivals must be accompanied by a more senior client at all times. As one of the counselors noted: 'During their first 30 days, we want people to really integrate into our community and not have a lot of outside factors ... we don't allow the immediate personal phone calls ... we don't allow people to handle their own money and things like that.' After the first month, clients can go beyond the bounded area, but must be escorted by a more senior client. After 90 days, clients may come and go as they please, subject to certain restrictions. These regulations are really designed to cocoon clients within the therapeutic community constituted at the facility. At the same time, however, staff acknowledge that the regulations create feelings of isolation and homesickness, leading some clients to abandon the program before the end of the first week.

Thinking about the neighborhood context, the surrounding population is relatively stable and middle-class. Staff and clientele talked about the area as comprised of predominantly white, employed suburban homeowners. In one sense, this means that the area can be viewed positively as an environment that presents few temptations to individual clients at the facility in terms of access to drugs.[6] However, the purportedly therapeutic nature of this safe, middle-class community also raises some feelings of alienation and institutionalization among clients, particularly those of Aboriginal heritage, which may shorten their residential stay. To combat these feelings, Facility C provides extensive Aboriginal cultural services, including sweat lodges. Nonetheless, this speaks to the complexity of the 'client' identity in terms of race and class, and the extent to which clients may feel, and/or be marked as, 'out of place' in this suburban setting.

Lastly, staff members at Facility C acknowledge the tension between the positive aspects of an isolated therapeutic site, and the recognition that clients cannot remain isolated indefinitely. Many of the clients are poor Aboriginal men and women from the city's inner core, and they often have few choices to return to this area after treatment. The Executive Director commented: 'They [the clients] like the isolation. Most of them really prefer not to be downtown, but we also know that when they split from here, that's the first place that they head to is downtown, because that is what they do know.' Recognizing the difficulty of effecting change in the material conditions of clients' lives, including income, housing and employability, Facility C must also navigate between sheltering clients and helping them to maintain ties to people and places. In addition, some clients are able to avoid a return to a problematic family or social environment by accessing Facility C's transitional housing, which is located in the same neighborhood, while attending treatment. Former clients may also continue to use the Aboriginal cultural services.

6 We are not suggesting that there is no drug and alcohol abuse in the area. The point is that drug and alcohol use is more likely to take place in private settings, while the stability and policing – both informal and formal – of the area restricts public access to/dealing of drugs.

Discussion and Conclusions

As we suggested at the outset, several factors – including a lack of resources and public opposition – act to constrain publicly funded treatment programs in their locational decision-making. One result is that programs may be forced into less than ideal locations where ready access to drugs and alcohol work against treatment objectives. We have examined the extent to which different neighborhood settings impact on the therapeutic potential of addiction treatment programs, as well as the strategies used by program staff to deal with the challenges and opportunities of these neighborhood settings. Since the facilities did not divulge their success rates, this irksome issue is only indirectly addressed in the analysis. Nonetheless, each facility clearly detailed the impacts of the immediate built and social environment on the process of treatment, which ultimately overlaps with successful treatment outcomes.

Several themes emerge from the case studies. First, they point to the ways in which facility location and the character of the local environment influence the therapeutic potential of treatment programs. Not surprisingly, they suggest that challenges and opportunities vary significantly between neighborhoods, reflecting broader intra-urban patterns. Social and built environments matter, but in distinct ways in different locales. The challenges facing the inner-city facility are extreme: a stigmatized and devalued locale, the local saturation of helping facilities, ready access to drugs and alcohol, as well as an inability to separate clients from the social networks and spatial locales associated with using/drinking. In this context, there is no way to sustain a real haven or respite. Staff attempt to use the facility as a point of contact and stabilization, with an opportunity to access services in more appropriate locations. By contrast, the outer-core location is in some ways well-suited to the women's treatment – not too isolated from clients' former neighborhoods, but distant enough to provide women with space for themselves, away from family responsibilities and familiar locales. Finally, the challenges facing the suburban facility were quite different from those confronted in the inner core. Facility C has the capacity to cocoon clients, effectively cutting them off from the outside world in order to focus on recovery. At the same time, the suburban location and strict regulations of the program, coupled with the extent to which the Aboriginal clients feel out of place in a largely white, middle-class neighborhood, produce un-therapeutic outcomes in the form of homesickness and isolation.

Second, the case studies point to the different strategies used by staff in their efforts to create and sustain what they consider to be therapeutic sites for addiction treatment. Staff members work to carve out spaces of recovery that exist somewhat independently of their surroundings. Clearly, this is easier in some cases than others. Regulations governing movement and contact with others, as well as peer monitoring, are designed to integrate clients into a programmatic environment conducive to recovery. At the same time, staff at all three sites recognized that the boundary between the programmatic environment and the world beyond is permeable. Downtown, the boundary is frequently at risk with the proximity of people and activities associated with risk. Yet even in more controlled environments staff recognized that the social and material realities confronting clients mean that many will have to return to environments that are less conducive to recovery. One response is to try to provide clients with resources or tactics to confront risks – in effect to use the local environment as a training ground. This suggests that a therapeutic environment does not have to be one in which clients are entirely removed from risk, providing they have ready access to decent housing and appropriate supports. At the same time organizations like Facility A continue to be constrained in their ability to find what they consider to be appropriate locations for treatment programs.

One broader implication of this work is the recognition that discussions about the therapeutic potential of different sites need to be adequately contextualized in time and space. What we mean here is that the ability of organizations like those in Winnipeg to have a positive impact on the health of clients is inextricably connected to the environments clients occupied before and after they entered and completed the program of treatment. This is not to suggest that there are not more or less effective programs; rather it is a recognition of the interconnectedness of place and an attempt to draw attention to the fact that effective 'health care' – in this instance, addiction treatment – is bound up with and dependent upon, for example, the provision of safe, affordable housing and economic opportunities for people in recovery from addiction.

References

Addictions Foundation of Manitoba (AFM) website, Addictions Foundation of Manitoba, <http://www.afm.mb.ca>, accessed 3 August 2006.

Andrews, G. (2003), 'Placing the Consumption of Private Complementary Medicine: Everyday Geographies of Older People's Use', *Health and Place* 9: 337–49.

Andrews, G. and Kearns, R. (2005), 'Everyday Health Histories and the Making of Place: The Case of an English Coastal Town', *Social Science and Medicine* 60: 2697–2713.

Artibise, A. (1977), *The History of Canadian Cities – Winnipeg, An Illustrated History* (Toronto: James Lorimer and Company Publishers).

Baer, L. and Gesler, W. (2004), 'Reconsidering the Concept of Therapeutic Landscapes in J D Salinger's *The Catcher in the Rye*', *Area* 36: 404–13.

City of Winnipeg (2004), *Plan Winnipeg 2020: Zoning Bylaw 6400/94 and 100/2004* (Winnipeg: City of Winnipeg).

Conradson, D. (2003), 'Spaces of Care in the City: The Place of a Community Drop-in Centre', *Journal of Social and Cultural Geography* 4: 507–25.

Curtis, S., Gesler, W., Smith, G. and Washburn, S. (2000), 'Approaches to Sampling and Case Selection in Qualitative Research: Examples in the Geography of Health', *Social Science and Medicine* 50: 1001–14.

Dear, M. and Wolch, J. (1987), *Landscapes of Despair* (Princeton NJ: Princeton University Press).

DeVerteuil, G., Hinds, A., Lix, L., Walker, J., Robinson, R. and Roos, L. (2007), 'Mental Health and the City: Intra-urban Mobility among Individuals with Schizophrenia', *Health and Place* 13: 310–23.

Geores, M. and Gesler, W. (1999), 'Compromised Space: Contests Over the Provision of a Therapeutic Environment for People with Mental Illness', in A. Williams (ed) *Therapeutic Landscapes: The Dynamic Between Place and Wellness* (Lanham, MD: University Press of America) pp. 99–121.

Gesler, W. (1992), 'Therapeutic Landscapes: Medical Issues in Light of the New Cultural Geography', *Social Science and Medicine* 34 (7): 735–46.

Grosenick, J. and Hatmaker, C. (2000), 'Perceptions of the Importance of Physical Setting in Substance Abuse Treatment', *Journal of Substance Abuse Treatment* 18: 29–39.

Kearns, R. and Barnett, J. (1999), 'Auckland's Starship Enterprise: Placing Metaphor in a Childrens' Hospital', in A. Williams (ed.), *Therapeutic Landscapes: The Dynamic Between Place and Wellness* (Lanham, MD: University Press of America) pp. 169–99.

Kearns, R. and Collins, D. (2000), 'New Zealand Children's Health Camps: Therapeutic Landscapes Meet the Contract State', *Social Science and Medicine* 51: 1047–59.

May, J., Cloke, P. and Johnson, S. (2005), 'Re-phasing Neoliberalism: New Labour and Britain's Crisis of Street Homelessness', *Antipode* 37: 703–30.

Palka, E. (1999), 'Accessible Wilderness as a Therapeutic Landscape: Experiencing the Nature of Denali National Park, Alaska', in A. Williams (ed.), *Therapeutic Landscapes: The Dynamic Between Place and Wellness.* (Lanham, MD: University Press of America) pp. 29–51.

Takahashi, L. (1996), 'A Decade of Understanding Homelessness: From Characterization to Representation', *Progress in Human Geography* 29: 291–310.

Timko, C. (1996), 'Physical Characteristics of Residential Psychiatric and Substance Abuse Programs: Organizational Determinants and Patient Outcomes', *American Journal of Community Psychology* 24: 173–92.

Tyson, M. (1998), *The Healing Landscape: Therapeutic Outdoor Environments* (New York: McGraw-Hill).

Valverde, M. (1998), *Diseases of the Will: Alcohol and the Dilemma of Freedom* (Cambridge, UK: Cambridge University Press).

Veness, A. (1994), 'Designer Shelters as Models and Makers of Home: New Responses to Homelessness in Urban America', *Urban Geography* 15: 150–167.

Wakefield, S. and McMullan, C. (2005), 'Healing in Places of Decline: (Re)imagining Everyday Landscapes in Hamilton, Ontario', *Health and Place* 11: 299–312.

Weinberg, D. and Koegel, P. (1995), 'Impediments to Recovery in Treatment Programs for Dually Diagnosed Homeless Adults: An Ethnographic Analysis', *Contemporary Drug Problems* 12: 193–236.

Williams, A. (1999), 'Introduction', in A. Williams (ed.) *Therapeutic Landscapes: The Dynamic Between Place and Wellness* (Lanham, MD: University Press of America) pp. 1–11.

Williams A. (2002), 'Changing Geographies of Care: Employing the Concept of Therapeutic Landscapes as a Framework in Examining Home Space', *Social Science and Medicine* 55: 141–54.

Wilson, K. (2003), 'Therapeutic Landscapes and First Nations People: An Exploration of Culture, Health and Place', *Health and Place* 9: 83–93.

Wilton, R. and DeVerteuil, G. (2006), 'Spaces of Sobriety/Sites of Power', *Social Science and Medicine* 63: 649–61.

Chapter 7

Anxious Subjectivities and Spaces of Care: Therapeutic Geographies of the UK National Phobics Society

Joyce Davidson and Hester Parr

> We think of landscapes of healing as an ideal, a potential, something to strive for. Of course, what is ideal for one person is not for another, so we must allow for difference. (Gesler and Kearns, 2002, 137).

Introduction: Extending Therapeutic Geographies

How are therapeutic spaces constituted as emotionally supportive geographies? How do they incorporate different kinds of emotional encounter? How do these encounters contribute to a distinctive ethos of care and caring in relation to particular groups? How do we accommodate difference into landscapes of healing? These questions and others frame this chapter, as we highlight the dynamic properties of 'non-physical', therapeutic encounters that constitute the caring networks of the UK National Phobics Society (NPS). Here we consider the special case of people who experience phobias and anxieties and who may be reluctant to seek out help or access therapeutic resources that are located in both physical spaces beyond the home and social spaces characterized by face-to-face engagement. The NPS provides alternative modalities of care and support in the form of telephone, print and virtual communication and we describe, compare and interpret these below.

Such concerns immediately need to be located within an expansive understanding of what might constitute therapeutic geographies. In this chapter, the term 'therapeutic' refers to the *potential* healing power available through social and spatial engagements between people who have experienced similar kinds of psychological difficulties. Therapeutic healing here does not refer to *certain* relief or cure from specific psychological conditions or difficulties for those that share their experiences, but rather refers to the emotional benefits of co-identification and support. This implies an explicitly relational understanding of what constitutes therapeutic dimensions of co-caring without co-presence. In this instance, this support occurs remotely and through disembodied encounters: ones that embrace, and offer access to, particular emotional geographies (Davidson et al., 2005). In taking this line, our arguments risk running counter to (at least) two lines of enquiry in recent writing on 'therapeutic landscapes' as it is currently constituted in human geography.

The therapeutic landscapes literature owes much to the pioneering work of Gesler (1992; 1993; 1996; 1998; 2003) in which he outlines the social, physical and symbolic factors that contribute to how 'the healing process works itself out in places' (Gesler, 1992, 743). Although Gesler's conceptualization also allows for healing in diverse types of 'situations, settings, localities and milieus' (ibid), it is the analysis of distinctive (nameable) places that has most characterized this work, particularly with respect to how places (for example, Lourdes and Bath) hold an 'enduring reputation for achieving physical, mental and spiritual healing' (Gesler, 1993, 71). Although in tune with a humanistic version of post-medical or health geography (Kearns, 1994), others have critiqued this work for being overly focused on taken-for-granted healing properties of nature and places (Wilson, 2003; Andrews, 2004; Lea, 2006), without due attention paid to the range of different spatialities (Doel and Segrott, 2003) and relations (Conradson, 2005) that must be more-or-less enrolled – and which may sometimes fail to be enrolled – in the practical accomplishment of 'healing'. In this latter regard, Conradson argues that a focus on therapeutic landscape *experiences* is instructive, in that these emerge 'through a complex set of transactions between a person and their broader socio-environmental setting' leading to 'an individual [...] becoming imbricated within a different set of relations' (2005, 346 and cited in Lea, 2006, 6). As Lea (2006) also argues, such relations may involve different kinds of anxieties, performances, practices and embodiment that configure healing and landscape connections as forever contingent. Notable in this literature is the *physicality* of the environments and experiences involved. Here the landscapes and geographies at stake are tangible and touchable; the contingent therapeutic experiences which might emanate from them are still largely embodied to some extent (although see Doel and Segrott, 2003).

In creatively engaging with notions of therapeutic landscapes, recent commentators have been keen to expand the boundaries of what constitute both the therapeutic and therapeutic geographies as they emphasize the complicated processes of subjectification bound into healing outcomes (Lea, 2006; Andrews, 2004). In this work the *practices* that might be constitutive of therapeutic experience come more directly to the fore, as well as human and non-human relations and encounters (and see McCormack, 2003; Milligan et al, 2004; Parr, 2007). Lea's (2006) work, for example, emphasizes both human *and* non-human participation in creating therapeutic places and outcomes. She is explicit about how this approach:

> begins with a consideration of the ways in which simply being in place might act upon the body, but then moves to […] consider […] the ways in which 'nature' is not a passive entity, but rather an active participant in modes of embodied engagement with the retreat place (Lea, 2006, 3).

While this approach does much to extend understanding of the non-passive nature, subjectivity, encounter and the relational constitution of the therapeutic, it still relies on very tangible embodied engagements. As Andrews (2004, 315) has recently pointed out, the therapeutic landscapes literature has relied on 'physical locations within which a person needed to be located in order to experience a therapeutic effect', and indeed he questions 'the necessity of physical places and co-presence' in terms of therapeutic experience. In calling for further attention to the 'non-physical and imagined', Andrews' argument begins to connect with other geographical writing that has also emphasized the power of safe spaces for the healing of self and body which is constructed via different modalities of communication (Bondi with Fewell, 2003).

Taking this critique a little further, and responding to the call for further investigation into 'non-physical' therapeutic experience, this chapter will discuss the implications and experiences of remote networks of support for and between people with phobias and anxieties. In discussing the therapeutic implications of such networks, we are explicitly understanding the therapeutic as a *process* involving mutual support and recognition, which can be facilitated through and with a variety of 'matter' and materials (paper, virtual technology, telecommunications). In evaluating these networks, we are being quite specific with our aims and ambitions: we seek to understand them as rather different modalities of therapeutic and emotional identification and engagement, revealing of the diverse properties that form the experience of support in non-clinical groupings (Segal et al., 1993; Radin, 2006). In thinking through the specificities of these modalities of support as they are organized by the NPS, we are also contributing to a discussion about a specific 'ethos of care'. In this case this ethos of care is emergent in response to the particular needs of NPS members, and one that seeks to be responsive and flexible with regards to the dynamic nature of anxieties. In discussing these concerns we do not reject the importance of non-human properties, place or landscape in therapeutic experiences, but rather deliberately choose to focus here on the supportive and potentially therapeutic environments that are inculcated via remote networks of communication. We deliberately use the term 'therapeutic geographies' rather than 'landscapes' in order to specify our distinctive focus here, as well as argue that such a term better encompasses the *range* of spaces, places and localities which might be enrolled in relational healing experience. We also, however, employ the term 'therapeutic environment', which in this chapter is used to refer to the specific healing potential inculcated in a particular context, in this case one enabled by print, telephone or internet.

Anxiety Disorder and the Construction of Therapeutic Geographies

The National Phobics Society (NPS) serves a client group who experience extreme social and spatial exclusion as a result of debilitating anxiety disorders including phobias, and generalized and specific anxieties. On joining the NPS, members have access to physically situated self-help groups and a wide range of subsidized therapy services in various locations around the UK. These services are important and widely used by many of the 6,000-strong membership. However, negotiating access to and participating in these therapeutic localities requires the bodily presence of users in spaces that are at once highly personal and emotionally charged. The *immediate* experience of such therapeutic encounters and the perceived lack of protective 'distance' can lead to precisely the kind of exposure that many phobics desperately try to avoid. The public (and so often problematic) situation of such services also renders them inaccessible to many NPS members. Each individual does, however, have access to alternative communicatively mediated supportive spaces that provide a network of social – and potentially supportive – relations, within which the individual can situate themselves.

Drawing on extensive ethnographic involvement with the NPS,[1] we argue that the salient features constituting these communicatively mediated therapeutic landscapes are best revealed through understanding the nature of these kinds of social *contact* that require no shared *physical* presence, contact that is (technologically) negotiated. Interview and survey responses reveal that relations established through telephone, print newsletters and computer are re-envisaged spatially by participants in such a way that they come to provide some*where* for members to go, a 'place' for confidential communication with sympathetic and trusted others, who are (mostly) kept a respectful (and safe) distance from the vulnerable self. While not without risks, these mediums contribute to the provision of more *inclusive* versions of therapeutic geographies. In bringing different kinds of emotionally supportive space into being, the NPS can be seen to provide for and be inclusive of difference (cf. Gesler and Kearns, 2002, 137).

The NPS and Spaces of Care: Anxious Subjectivities

The NPS operates from a base in the Zion Community Resource Centre, in the Hulme district of the city of Manchester in the North of England. It is a voluntary support organization, formed in 1970 by an agoraphobia sufferer, with the intention

---

1 There are two separate studies which contributed empirical materials to this case study. The first was funded by the UK National Health Service (RDO/35'12), and set out to collect and examine service users' experiential accounts of their phobic disorders and their perception of statutary and non-statutary service provision in the UK. Service users were recruited via their membership of the NPS, and 40 individuals (5 male, 35 female) took part in in-depth individual interviews, which were audio-tape recorded, fully transcribed and anonymized with respondents' permission. Interviews took 'place' in one of three ways as chosen by respondents: by telephone (34); face-to-face meeting at respondent's home (5); face-to-face meeting at NPS (1). Interviews lasted between 35 minutes and 2½ hours, with the majority approximately 1 hour in length. Ethnographic methods were also intermittently employed in the NPS office over a period of three months, and ten service providers, recruited via NPS and contacts with health professionals, were interviewed to provide complementary data regarding service provision for phobia sufferers.

In addition, the NPS formed part of a wider study funded by the UK Economic and Social Research Council (RES-000-27-0043) on the internet and mental health problems. In this study the empirical materials were gathered from an extensive on-line internet user survey designed with a combination of both closed and open-ended questions and completed by 78 respondents. The survey was advertised on four UK mental health discussion forums, including the National Phobics Society (www.phobics-society.org.uk). Quotations from survey responses are used in anonymized form. The survey in the study was followed up by telephone interviews with five respondents which each lasted between 30 and 60 minutes, were tape recorded and transcribed. Interviews were also conducted with managers or moderators of each web-based forum, and all materials were coded and analysed. These methods were supplemented with on-line overt ethnography in each of the sites, including participation in various discussion threads and live chat-rooms. All quotations are from NPS members.

of helping those affected by anxiety disorders of all kinds. The primary aims and objectives of the NPS are broadly similar to those of other health support groups (Allsop et al., 2004; Powell, 1994). The organization's 'user-led' philosophy is also consistent with that of health-related self-help groups, which are '[f]ounded on the principle that people who share a disability have something to offer each other that professionals cannot provide (Segal et al., 1993, 705). The NPS is explicit in placing exceptional value on sufferers' own experience and expertise: 'Our belief is that *those who have experienced anxiety disorders are best placed to provide support for other sufferers* because they are able to truly understand the impact these conditions have on people's lives' (www.phobics-society.org.uk, 2006).

The anxious qualities of the life-worlds of NPS members differ according to individual context and circumstance, not least, the form of anxiety from which they suffer (Smith and Davidson, 2006). In their struggle to make sense of disordered socio-spatial experience, some sufferers describe feeling marginalized and misunderstood, dislocated and 'cut off' from the rest of the world. Others describe an agony of proximity rather than distance, of feeling too intimately and inescapably located. In all cases, when confronted with an imagined or actual phobic object, anxiety's embodied and mindful affects work to disrupt protective boundaries of sufferers' subjectivities, and something feels desperately wrong with the world and their place within it (Davidson, 2000). Anxiety – particularly in its intensified manifestation as panic – is never a predictably internal or external affect. Rather, it takes place at the intersections between selves and spaces, feeding off of yet transforming both. That is to say, environments or objects *make* the sufferer anxious at the same time as their anxiety *makes* that environment or object more frightening. The mechanisms of such viciously anxious cycles are at least partially understood and articulable by the sufferers, yet never manageable.

Given the complex phenomenology and curious placement of panic and anxiety (Davidson, 2003a), what, then, might constitute a therapeutic environment for those enduring its affects? Sufferers' accounts suggest that what is required is a psycho-social intervention that disrupts the cycle of continually re-established anxious interactions. To be therapeutic, an intervention must somehow short-circuit the anxious-self-fulfilling process, and interject something qualitatively different, a sense of support, co-identification or *care* that enables less anxious affective relations to be imagined, experienced and even tentatively established. Therapeutic space should ideally s(m)ooth and strengthen fragile edges of the self, restore a sense of calm, coherence, and comfortable connection, and so enable less emotionally fractious relations. In the following section, we analyse three different modalities within the NPS ethos of care in terms of their therapeutic potential as well as their potential dangers as geographies of support.

Telephone Support Services

The NPS receives an average of 50 calls per day to helplines staffed between 9am and 9pm by volunteer support workers.[2] These trained volunteers are themselves

2 A number of other services are provided by telephone, including a psychiatric pharmacy helpline, a psychology information line, Cognitive Behavioural Therapy and counselling, all provided by professionals. Our focus here is particularly on *peer* support

NPS members whose own prior or current experience of anxiety renders them well placed to engage empathetically with callers. Staff offer information and advice relating to coping strategies, available therapy services, or often, simply reassurance and a sympathetic ear (Boehm et al., 1991). They are capable of hearing and indeed sharing experiences in a non-hierarchical, open communicative style, and this co-identification configures a potential source of and space for support where members can 'speak to somebody who sort of understands [...] Because some people, and my husband's one of them, he, he just says, you know, just *go* to the dentist' (Judy).

Telephone helpline services might be considered qualitatively similar in some ways to physically situated self-help groups in their intention to provide social and emotional support through contact with someone experiencing similar difficulties (Galinsky et al., 1997; Lazar and Erera, 1998). While the person with whom you are communicating is not physically by your side, neither is the sense of physical connection entirely absent. The person you speak with is *there*, at the other end of a telephone connection which draws you together, if not 'here' at least 'now' (Teare et al., 1995). This immediacy of response has a form of sensorial, spatial affect. It brings a feeling of *presence* to interactions with another who, though not with you 'in the flesh', is partially embodied and brought close, arguably grounded near by the weight of their voice. 'People *need* people to support them' (Rhona), and this modality provides a 'life-line' for many.

While the *immediacy* of the telephone is potentially therapeutic, this same quality is not, however, without danger for vulnerable subjectivities. Penny, for example, draws attention to some potential risks and anxieties when she expresses her own reluctance to present the painful details of her experience in a forum that feels insufficiently distanced: 'I couldn't phone them because that was too stressful for me, I couldn't phone and talk about it because that was too close to home.' Melanie, too, describes feeling unable to articulate her own experience (of vomit phobia) by phone, though she also suggests that any modality of encounter that required her to verbalize her fears would be dangerous and painful and so 'best' avoided: 'I mean I didn't used to speak on the phone [...because] when it got really bad and I would just be speaking and I'd just think oh my God I'm going to be sick and so I'd have to put the phone down. So I'd just, it's just easier not to speak.'

For many others, however, helplines provide some of the positive aspects of face-to-face contact in a way that is more convenient as well as less (or at least *differently*) risky. This forum lacks some of the temporal pressure associated with fixed and limited sessions typical of physically situated support groups: 'they can call when it suits' (NPS manager). Many members use the service more than once, and helplines can serve as a conduit through which anxiety is repeatedly, even familiarly, off-set. Helpline staff come to know what their members need, sometimes even on an individual basis, as the NPS manager illustrates: 'If you said to me, "this is so and so", I would *know*. Like we had X on the phone from [an English town] this morning. I know that person, what she wants, cause I've been in the office three or four times when she's called before, and that's really important.' Nicky explicitly

---

offered by telephone, as a modality of encounter that enables co-identification between participating individuals.

refers to this informed affective stance in terms of an 'ethos', and for regular callers, the familiarity, as well as absence of seemingly prevalent patronizing language, unhelpful assumptions and stereotypes, is a crucial part of the caring context that enables fractured and fractious boundaries to be reworked in relation to others.

There are obviously certain potentially therapeutic qualities associated with face-to-face support groups that this modality of encounter lacks (Davidson et al., 2000). However, while it certainly *matters* that the telephone masks the language of the body, the intricacies of expression in face, hand and eye, the consequences are not necessarily negative. For some, the absence of full physical presence provides freedom from visual and other immediate sensory distractions, *clarifying* communications by creating space to focus on the meanings of words (Galinsky et al., 1997). It may even be therapeutic to *imagine* the effects of one's own voice on space that embodies one's projections positively, to be free from anticipation or recognition of less than caring expressions and gestures. Saying what one really feels in a relatively controlled communicative context can be emotionally liberating and even empowering: 'Knowing that one's contributions are anonymous can lead to increased self-disclosure, honesty and intimacy [...] and participants may feel freer to discuss taboo subjects' (Lazar and Erera, 1998, 185).

Callers value the opportunity to 'blow off steam' and often bemoan the lack of support and understanding from friends, family and physicians, without fear of repercussion for their standard of care or of damaging relationships that may already be strained by anxiety (Davidson, 2007). However, this freedom to complain, to express frustration or desperation, clearly has implications for the (also anxious) individuals on the receiving end of callers' vocal presentations of emotional pain. Staff work to understand and sympathize with such frustrations, but bearing the brunt is not without emotional costs. On the telephone, some communicate their anxieties in ways far more demanding than might be the case if they took *time* to consider their comments (as in writing a letter) or with *regard* to the emotionally responsive face of another, visibly upset by their encounter.

In such an inherently relational modality, risks are not borne by one side alone: and the helpline forum is not always or entirely emotionally protective. Nadia, who considers her phobia of frogs to be uncommon and relatively unknown, feels she took a considerable risk by disclosing and exposing her fears, even at a 'distance', by phone:

> I did phone one phobic agency, I can't remember which one it was because there was so many I did contact and the woman did kind of laugh. [...] It wasn't helpful at all because it wasn't, it had been a big step for me to eventually think right, I'm gonna do something about this, and then [...], it didn't help (Nadia).

Fiona describes a similar emotionally damaging encounter through a (non-NPS) helpline when she was feeling 'really, really bad' about her (far more common) phobia:[3]

3 While interviewees offered no negative evaluations of NPS helplines, their recruitment through this organization might be seen to mitigate against direct criticism.

> And I spoke to the man who's in charge [... and] he said 'what are you, an arachnophobic?' And I said 'no, no I actually have a phobia of wasps so it's, I know, it's a bit embarrassing'. And, and he just laughed [...] he was so rude, he was so horrible and thoughtless (Fiona).

At their best, telephone support services involve confidentiality, compassion, trust and respect, and interviewees describe having somewhere to *go* because of them, having a place and person to turn to when they need to 'talk it over' (Elaine). The accessibility and reliability of this connection can serve to strengthen the emotional fabric of members' lives. However, its immediacy and anonymity poses risks as well as benefits. Over the phone, in order to tell how another is feeling, you literally have to be *told*. The lack of available interpersonal cues and non-verbal communication creates a space of encounter that, while emotionally liberating and therapeutic for some, can be felt *care-free* in a far less positive sense by others.

Newsletters

Although not everyone has experience of telephone support services, every NPS member receives a quarterly newsletter that comprises a very different modality of encounter from that considered so far. Like telephone support, newsletters are typically experienced in the sufferers' own homes, and can be engaged with at a time of their own choosing. We want to suggest that these too represent a circulation of the caring ethos characteristic of the NPS, but can such disembodied and distanced communication really configure a *context* of care? Can solitary consumption of print media ever constitute a relational therapeutic environment for identification and engagement? Clearly, visual interaction with the written word can involve less active or immediate response to the other/s involved. The imagined author can be 'engaged' fairly passively, their contribution dipped into from a distance or dismissed in a seemingly care*free* manner.

Somewhat ironically, it is perhaps this sense of control over the medium's capacity to affect that makes it so potentially therapeutically valuable for its anxious users. We have already seen how crucial a sense of agency can be in otherwise disordered life-worlds, and newsletter contributors know this first-hand. The similarly positioned anxious authors offer advice without making demands. They present information and invitations, but reader response and degree of involvement with the content encountered is entirely voluntary, unseen and pressure-free. This sense of being 'in charge' looms large in sufferers' accounts, but it does so in a way often tempered by an air of connection and care for that charge, of co-identification and also belonging. Members suggest that the regular appearance of newsletters in their homes serves as a tangible reminder of and conduit to a substantial network of support:

> You read articles in there that you can identify with these people with all their different problems, in the newsletters. They're very helpful, you don't feel like you're on your own (Amanda).

The newsletter is entitled *Anxious Times* and recent issues have included articles on subjects as diverse as exercise and mental health, auricular acupuncture, and computer-aided therapy programmes. In addition to 'expert'-authored information and advice, newsletters present personal accounts of, for example, living with agoraphobia and of experiencing hypnotherapy. Authors communicate a sense of trust in their own and other sufferers' judgement, and this is commented upon frequently and valued highly by members who frequently describe feeling dis-empowered by the lack of support available elsewhere. Sufferers learn and derive strength from each others' experience, as with Jase, who describes benefiting from accounts of contributors 'who say "oh I took Kava Kava and it really, it really has helped me", or something like that' (Jase).

Members also author reviews of relevant products such as hypnotherapy tapes and self-help manuals, some of which become available for subsidized purchase through the NPS. Harry is among those who have found this aspect of newsletter support particularly valuable: 'I mean my saviour has been two £10.00 books from the National Phobics Society, when I've spent thousands on various therapies.' The therapeutic benefits of all kinds of writing from personal experience are well known (Bolton et al., 2004). Committing one's thoughts to paper for distribution through a therapeutic network where they are likely to be received sympathetically can be especially empowering. Articulating an emotionally difficult experience can help off-load and lessen its weight, and the sense that the process of disclosure might benefit others can be therapeutic too (Pennebaker and Seagal, 1999). Mary describes communicating her successful experience of hypnotherapy to staff at NPS, and how 'they were pleased for me and asked me to write that [newsletter] article [...] And I, you know I was really pleased, I just wanted someone, I thought well if I can help just one person.'

Although the written (at least printed) word might be thought to lack some of the spontaneity of the spoken, perhaps it also lacks some of the risks associated with making 'off-the-cuff' remarks that might lead to later regret. The published account tends to be considered and thoughtful, a relatively measured and sensible performance of the anxious self in a format that is fixed and formalized. In print, members (re)present their anxieties in an apparently manageable form, through coherent narratives of, for example, achievement or warning. Whatever their aim, authors aspire to communication in *relation*. By being open to the personal touch and expression of a voice embodied by print, two-dimensional narratives can be brought to colourful life, configuring spaces that might interact and overlap with others in ways potentially (if not intrinsically) therapeutic (Conradson, 2005). Indeed, we might see newsletter reading and writing as 'observable, knowable and distributed community practices' (Laurier and Whyte, 2001, 1):

> The newsletters are helpful. And I don't know, it's just nice to know this is something that a lot of people have got, because before I didn't think that people really had phobias, do you know what I mean? (Melanie).

> I think they're informative, they tell you where to find help. The other side of it is the emotional thing where you suddenly think, 'oh I'm not alone, I'm not the only idiot in,

> you know, England who is, is suffering'. [...] And that was a lovely feeling actually. (Rhona).

That 'emotional thing', the 'lovely feeling' that emerges from print to enter one's immediate experience can clearly be important and therapeutic. However, the reader's active and imaginative engagement is key to the configuration of emergent shared space, and less positive constructions can also transpire from this modality.

An active (and especially anxious) imagination has obvious dangers, and these can be multiplied through exposure to representations of other anxious subjectivities. Even carefully constructed accounts of another's suffering can trigger similarly damaging emotions in oneself. Both reading and writing about personal experience have the potential to be painful as well as positive (Parr, in press; Wright and Chung, 2001) and so, for some, are activities best avoided: 'Perhaps it sounds selfish but I didn't want to hear about other people's phobias' (Kate). The dangers of exposure to emotional triggers incorporated in the written word are perhaps more pronounced in relation to the internet (see below; also Dias, 2003; Parr, in press).

Newsletter 'contacts pages' provide opportunities to explore differently mediated engagements with other NPS members, inviting readers to take a chance on different kinds of connection. As Melanie explains, 'there's a contact section, I've made some friends through that [so] now I've got people who I can talk to that are having the same thing. Whereas even though people around me try to understand, they don't really.' For others, that support network exists only in an imaginative sense; that is to say, they might not *actually* contact anyone – most members do not. While some find it reassuring to know that they *could* (Brown et al., 2004; Rappaport, 1993), engagement inevitably involves risks and others, such as Judy, communicate their sense that interacting through this modality requires courage: 'I like the contact page, although *I've never actually been brave enough to, to contact anybody*.'

The Internet

> There is a strong link with all of us because of our phobias, and I was very nervous at first about not being able to fit in, and most new members feel the same way, but once I posted my hello, I got such nice welcoming messages back, I felt welcome from the start. People are so nice and non-judgmental, they are knowledgeable and willing to help and they made me feel a part of a group after being on the outside for so long (Kim).

For an increasing number of NPS members, online services serve a similar function to newsletters, facilitating contact, advice and support (see Fox, 2001; Parr, 2002; Powell et al., 2003, for further discussion of internet support groups). However, concerted efforts have been made in recent years to expand and improve members' online services, as the NPS recognizes the potential value of such resources for a client group reluctant to 'come out' in places perceived to be public (Davidson, 2003b, 2005). Online member services are password protected, creating the sense that only community members can participate. There is an e-contact (or e-pals) list, a bulletin board and a chat-room, available 24 hours for general use, and with weekly chat sessions on specific anxiety conditions, where members can have greater chance of more focused and guided exchanges. A great deal of research and literature has recently become available about the potential benefits and pitfalls of online support services for those with health problems (see Parr, forthcoming), but our purposes for evaluation are specifically to question whether and the extent to which they are capable of enabling access to new therapeutic geographies for people isolated by anxious subjectivities.

Interviewees show that online forums provide some*where* – metaphorically at least – to comfortably incorporate an aspect of their identity that they are normally under pressure to hide. In these supportive online spaces, members learn how to look at their behaviour in a way that is often constructive and supported, although there are also risks in this regard (see below). In the following account, Aileen reflects on her own use of NPS chat-rooms:

> It's a place where people can meet up without particular, for areas that aren't talked about a lot in society and aren't appreciated. Um, and you can, I would be more likely to visit an Internet site than I would a face to face meeting [because of …] just, I suppose the embarrassment factor.

For Aileen, it is the *meaning* of the support space, the emotional and relational rather than physical qualities that matter (Korpela and Hartig, 1996). In terms of the capacity for online services to constitute therapeutic geographies of support, members articulate their value as intense textual spaces in which the anxious self is laid bare, reconstituted and understood through shared correspondence. Here, then, online forums and chat-rooms, for example, do not provide access to passive therapeutic environments in which relief from suffering occurs automatically, but rather engenders a process of self-understanding through shared experience that *may* inculcate positive effects:

> I am much more aware of how and why I feel like I do. I am also aware that it is not just me feeling the way I do, so the sense of isolation and wondering why I was the only one feeling this way has been replaced by knowledge and understanding. By knowing more about how you feel, and perhaps why, you can then start to tackle it (Steve).

Here, Steve argues that knowledge and understanding may provide access to relief from anxiety, but through non-passive actions, perhaps ones also previously tried and tested by other online NPS members. Emotional identification, shared senses of suffering, swapping experiences of recovery, all contribute to a dynamic online therapeutic environment in which anxious subjectivities are constantly articulated, sometimes on a daily basis in ways that create regular support. For people usually isolated by their mental health problems, this comprises a potentially therapeutic environment considered worth striving for (Gesler and Kearns, 2002). The notion that therapeutic geographies involve online mutual effort and activity in order to create the potential for healing underlines their *contingent* properties. Here, as in the case of telephone and print media, there is no stable (natural) entity or treatment or process that is mutually assumed as therapeutic between members. In writings on therapeutic geographies, undue emphasis may sometimes be placed on positive and risk-free experiences of a potentially healing milieu. In online environments, characterized by dynamic self and co-writing of anxious subjectivities and of the daily struggles of living with phobic disorders, such risks should be acknowledged, as Debbie reminds us below:

> [the content of other member's emails] can influence and worsen your fears, for example someone could post that they are scared of eating mayonnaise – then you could take on that fear yourself (Debbie).

The assumed distance, anonymity and transience characterizing other kinds of online relationships do not necessarily feature in chronic illness forums in quite the same way as elsewhere on the web (Radin, 2006). The dense writing of disruptive thoughts and emotions on NPS forums exposes other members to difficult intimacies, especially if posted from regular users of forums, where senses of responsibility and norms of reciprocity mean that the content of posting must be noted and addressed in ways that challenge notions of technological 'drive-by relationships' (Puttnam, 2000, 177). Occupying online forums as potentially healing spaces is, then, an ambivalent endeavour, as some 'do feel part of an online community, but whether I feel comfortable there is still to be decided' (Mark).

These caveats do not detract from the argument that online services – particularly these forums and chat-rooms – may comprise effective therapeutic geographies. A supportive space, clearly, need not be physical or risk-free. Potential therapeutic geographies can thus be seen to be constructed from non-proximate and iterative co-writing: as Warin et al. (2000, 1871) explain, '[t]he spatial boundaries of social relationships, of open communication and non-hierarchical encounters are equally as important as the physical spaces to people's experience of care'.

Conclusion: Different Therapeutic Geographies?

In this chapter we have laid out a basic argument for the need to consider therapeutic geographies as ones inclusive of non-physical environments in which the potential for healing is evidenced through contingent and distanced social relations. Using the case study of the NPS, we have attempted to show how telephone, print and online encounters constitute contingent therapeutic environments in which social support, psychological well-being and risk are all encountered. Our findings suggest two further speculative arguments which we will elaborate in conclusion. Firstly, that anxious subjectivities benefit from *particular* configurations of care in ways that still 'allow for difference' (Gesler and Kearns, 2002, 137) in the co-construction of healing potential. Secondly, that in understanding the particularities of the therapeutic environments provided by NPS, we are led to re-evaluate the role of risk in therapeutic geographies and disrupt predominantly celebratory assessments of healing spaces and places.

In taking the first of the points above, it is the particularities of severe anxiety that has led the NPS to provide diverse access points to therapy and social support. In this chapter it has been the latter upon which our comments have focused. As a support group, the NPS operates through different modalities which in turn involve different social relations and therefore request different requirements of its members. In telephone, print and internet communication, members are required to engage with, and represent, their own anxiety disorders in different ways, also drawing upon the experience of others' anxious subjectivities *differently*. We have argued, for example, that each modality provides access to a potentially therapeutic environment in which anxious subjectivities are required to 'work' at being relational at differing levels. None of the modalities briefly described in this chapter clearly provide definitive solutions to anxious dissolution, and yet they do enable access to others who can be seen as *striving* for the potential of healing. NPS can thus be argued to facilitate an exploration of experience and behaviour in a way that is constructive and supportive rather than simply self-critical and painful. This provision helps build the possibility for stronger reflective and recovering identities (Francis, 1997; Pennebaker, 1997).

Secondly, in creating diverse access points to potential geographies of healing and recovery through their different modalities of co-supportive encounters, the NPS highlights how and why we should think differently about what might constitute the therapeutic. Our materials above suggest that some therapeutic environments can be considerably risky in their relational co-constitution (a notion also addressed by Conradson, 2005 and Lea, 2006). In evaluating therapeutic geographies, should we be asking further questions about these risks, or perhaps be involved in more targeted analysis of ambivalent healing relations? Criticisms of the therapeutic landscape literature might include mention of its overt celebratory nature, the emphasis on taken-for-granted and implicit therapeutic properties of places and landscapes coupled with a lack of attention to 'relational notions of selfhood' (Conradson, 2005, 337). In this chapter, we have tried to address some aspects of these deficiencies by highlighting the *contingent* nature of diverse non-physical therapeutic geographies, while emphasizing that any potential in this regard is intimately tied to *how and where* anxious selves relate. While we cannot offer definitive conclusions for how non-physical therapeutic geographies might be constructed and encountered, we can offer our evaluation of the NPS as one which highlights how therapeutic geographies are a thoroughly relational achievement. For a space to be experienced as therapeutic it has to communicate an affective stance and invitation, open up a space within problematic environments and relations, engage risk, and still exist as one where more positive experiences can emerge.

Acknowledgments

We would like to thank the UK National Phobics Society and its members for their participation in our research, and Victoria Henderson and Allison Williams for helpful comments and advice.

References

Allsop, J., Jones, K. and Baggott, R. (2004), 'Health consumer groups in the UK: A new social movement?', *Sociology of Health and Illness* 26: 737–56.

Andrews, G. (2004), '(Re)thinking the dynamics between healthcare and place: therapeutic geographies in treatment and care practices', *Area* 36: 307–18.

Bolton, G., Howlett, S., Lago, C., Wright, J. (2004), *Writing Cures: An Introductory Handbook of Writing in Counselling and Psychotherapy* (London: Brunner Routledge).

Boehm, K., Chessare, J.B., Valko, T.R., Sager, M.S. (1991), 'Teen line: a descriptive analysis of a peer telephone listener service', *Adolescence* 26: 643–8.

Bondi, L. with Fewell, J. (2003), '"Unlocking the cage door": the spatiality of counselling', *Social and Cultural Geography* 4: 527–47.

Brown, P., Zavestoski, S., McCormick, S., Mayer, B., Morello-Frosch, R. and Altman, R.G. (2004), 'Embodied health movements: new approaches to social movements in health', *Sociology of Health and Illness* 26: 50–80.

Conradson, D. (2005), 'Landscape, care and the relational self: Therapeutic encounters in rural England', *Health and Place* 11: 337–348.

Davidson, J. (2000), '"… the world was getting smaller": women, agoraphobia and bodily boundaries', *Area* 32: 31–40.

Davidson, J. (2003a), *Phobic Geographies: The Phenomenology and Spatiality of Identity* (Burlington, VT and Aldershot: Ashgate Press).

Davidson, J. (2003b), '"Putting on a face": Sartre, Goffman and agoraphobic anxiety in social space', *Environment and Planning D* 21: 107–22.

Davidson, J. (2005), 'Contesting Stigma and Contested Emotions: Personal Experience and Public Perception of Specific Phobias', *Social Science and Medicine* 61: 2155–64.

Davidson, J. (2007), 'Caring and daring to complain: An examination of UK National Phobics Society members' perception of primary care', *Social Science and Medicine*, Vol. 65: 560–71.

Davidson, J., Bondi, L. and Smith, M. (eds) (2005), *Emotional Geographies* (Burlington, VT and Aldershot: Ashgate Press).

Dias, K. (2003), 'Virtual Sanctuary: Geographies of Pro-Anorexia Websites', unpublished PhD thesis, University of British Columbia.

Doel, M.A. and Segrott, J. (2003), 'Beyond belief? Consumer culture, complementary medicine, and the dis-ease of everyday life', *Environment and Planning D* 21: 739–759.

Fox, N. (2001), 'Use of the internet by medical voluntary groups in the UK', *Social Science and Medicine* 52: 155–6.

Francis, L. (1997), 'Ideology and interpersonal emotion management: redefining identity in two support groups', *Social Psychology Quarterly* 60: 153–71.

Galinsky, M. J., Schopler, J. H., Abell, M. D. (1997), 'Connecting group members through telephone and computer groups', *Health and Social Work* 22: 181–8.

Gesler, W.M. (1992), 'Therapeutic landscapes: medical geographic research in light of the new cultural geography', *Social Science and Medicine* 34: 735–46.

Gesler, W.M. (1993), 'Therapeutic landscapes: theory and a case study of Epidauros, Greece', *Environment and Planning D* 11: 171–89.

Gesler, W.M. (1996), 'Lourdes: healing in a place of pilgrimage', *Health and Place* 2: 95–105.

Gesler, W.M. (1998), 'Bath's reputation as a healing place', in R. Kearns and W.M. Gesler (eds) *Putting Health into Place: Landscape, Identity and Wellbeing* (Syracuse: Syracuse University Press) pp. 17–35.

Gesler, W.M. (2003), *Healing Places* (Lanham, MD: Rowman & Littlefield).

Gesler, W.M. and Kearns, R. (2002), *Culture/Place/Health* (London: Routledge).

Kearns, R. (1994), 'Putting health and health care into place: An invitation accepted and declined', *Professional Geographer* 46: 111–15.

Korpela, K. and Hartig, T. (1996), 'Restorative qualities of favourite places', *Journal of Environmental Psychology* 16: 221–33.

Laurier, E. and Whyte, A. (2001), '"I Saw You": Searching for Lost Love via Practices of Reading, Writing and Responding', *Sociological Research Online* 6:1, <http://www.socresonline.org.uk/6/1/laurier.html>, accessed 19 February 2007.

Lazar, A. and Erera, P. I. (1998), 'The telephone helpline as social support', *International Social Work* 41: 89–101.

Lea, J. (2006), 'Retreating to nature: rethinking "therapeutic landscapes"' unpublished paper available from the author: Department of Geography and Earth Sciences, University of Glasgow, Glasgow G12 8QQ.

McCormack, D. (2003), 'An event of geographical ethics in spaces of affect', *Transactions of the Institute of British Geographers* 28: 488–507.

Milligan, C., Gatrell, A. and Bingley, A. (2004), '"Cultivating health": therapeutic landscapes and older people in northern England', *Social Science and Medicine* 58: 1781–93.

Parr, H. (2002) 'New body-geographies: the embodied spaces of health and illness information on the Internet', *Environment and Planning D* 20: 73–95.

Parr, H. (2007), 'Mental health, nature work and social inclusion', *Environment and Planning D* (in press).

Parr, H. (forthcoming), *Mental Health and Social Space: Towards Inclusionary Geographies?* (Oxford: Blackwell).

Pennebaker, J.W. (1997), *Opening Up: The healing power of expressing emotion* (New York: Guilford Press).

Pennebaker, J.W. and Seagal, J.D. (1999), 'Forming a story: the health benefits of narrative', *Journal of Clinical Psychology* 55: 1243–54.

Powell, J., McCarthy, N. and Eysenbach, G. (2003), 'Cross-sectional survey of users of internet depression groups', *BMC Psychiatry* 3: 19.

Powell, T.J. (1994), *Understanding the Self-help Organization: Frameworks and Findings* (London: Sage).

Radin, P. (2006), '"To me, it's my life": Medical communication, trust, and activism in cyberspace', *Social Science and Medicine* 62: 591–601.

Rappaport, J. (1993), 'Narrative studies, personal stories, and identity transformation in the mutual help context', *Journal of Applied Behavioral Science* 29: 239–56.

Segal, S.P., Silverman, C. and Temkin, T. (1993), 'Empowerment and self-help agency practice for people with mental disabilities', *Social Work* 38: 705–12.

Smith, M. and Davidson, J. (2006), '"It makes my skin crawl…" The embodiment of disgust in phobias of "nature"', *Body and Society* 12: 43–67.

Teare, J.F., Garnett, C.R., Coughlin, D.G., Shannon, D.L. and Daly, D.L. (1995), 'America's children in crisis: adolescents' requests for support from a national telephone hotline', *Journal of Applied Developmental Psychology* 16: 21–33.

Warin, M., Baum, F., Kalucy, E., Murray, C. and Veale, B. (2000), 'The power of place: space and time in women's and community health centres in South Australia', *Social Science and Medicine* 50: 1863–75.

Wilson, K. (2003), 'Therapeutic landscapes and First Nations Peoples: an exploration of culture, health and place', *Health and Place* 9: 83–93.

Wright, J. and Chung, M.C. (2001), 'Mastery or mystery? Therapeutic writing: a review of the literature', *British Journal of Guidance and Counselling* 29: 277–91.

Chapter 8

Preventive Applications of the Therapeutic Landscapes Concept in Urban Residential Settings: A Quantitative Application

Matthias Braubach

Introduction

This chapter explores how the therapeutic landscape (TL) concept can be applied to support our understanding of the health influencing characteristics of urban areas and residential neighbourhoods. It gives a preventive dimension to the concept of TL by applying it within a daily life setting instead of a health (care) setting, and considers the same characteristics that are deemed 'therapeutic' as having a potential preventive value as well. Based on a recent WHO survey on housing and neighbourhood conditions and health, this chapter attempts to provide empirical data to support the preventive application of the TL concept.

Urban Residential Settings, Health and Well-being

Depending on the location in space, the health status of the population varies. Countries as well as cities show considerable differences in health measures which can not only be analysed by spatial units and geographical areas, but are often directly caused, influenced or aggravated by these areas. Empirical research has shown that especially within smaller areas (such as neighbourhoods), health differences can be extreme (for example Cohen et al., 2003; Macintyre et al., 2003; Stafford and Marmot, 2003; Mackenbach and Howden-Chapman, 2002). It has been argued that

> … individual or household level socio-economic factors can explain most of these area differences in health. On the other hand, other studies hypothesise that ecological or environmental effects on health exist, independent of individual or household level factors (Chandola, 2001, 105).

One of the main concerns in current research on the relationship between neighbourhood quality and health or well-being is therefore targeted on identifying the association between: neighbourhood quality, perception of physical and social characteristics of the residential area, and the health status of residents (Wilson et al., 2004).

Although urban planning processes are – and always have been – strongly informed by public health arguments,[1] a variety of stressors and health risks still exists in residential environments (WHO, 1997). In addition, it has been found that architects and building professions often show less interest in evaluating the performance of their buildings than in the development of new projects, and that there are still considerable challenges in implementing scientific knowledge in the area of urban and residential design and health impacts (Brown, 2003). Issues of residential neighbourhood effects on health and well-being have received increasing interest in recent years, with an attempt to identify the health impact of contextual and/or environmental factors. The debate between contextual or environmental factors and compositional factors represents a perfect example to explore the relevance of place-based approaches to health.

Despite the diversity of residential areas, and the uniqueness of local problems, there are five general features of urban areas which may have a direct or indirect influence on health, and which may be of relevance for any neighbourhood (Macintyre and Ellaway, 2000). These over-arching features are: (1) physical characteristics of the environment shared by all residents; (2) healthy environments in general, referring to the conditions of, and the functionality of, home, work, school or recreational settings; (3) services provided to support the daily life of residents; (4) socio-cultural features of neighbourhoods; and (5) reputation of the neighbourhood. These five features indicate that urban residential quality covers a variety of characteristics, including physical and social characteristics as well as perceptive dimensions. The health effects of urban residential settings therefore strongly resemble the concept of TL as defined by Gesler (1992) and Williams (1998).

Therapeutic Landscapes – A Useful Approach for Urban Health

The TL concept is aimed at explaining the mechanisms by which '... certain places or situations are perceived to be therapeutic' (Gesler, 1992, 735). Based on developments in cultural geography, Gesler (1992) and Williams (1998) have worked towards an improved understanding of the relation between spatial settings and human health.

The concept of TL is based upon three elements: natural and built environments, social interactions, and the symbolic meaning of those elements for an individual (Gesler, 1992). Further, TL have been defined as

> ... those changing places, settings, situations, locales, and milieus that encompass both the physical and psychological environments associated with treatment and healing, and the maintenance of health and well-being (Williams, 1998, 1193).

1 Relevant paradigms were, for example, the garden city model, the functional model (Charta of Athens), the car-adapted city or the city of short distances, and the compact city.

This definition combines several dimensions. On the input side, the TL concept deals with both physical and psychological spaces, with psychological space being an individually filtered reflection of the physical space (based on Ley, 1981). Gesler defined this reflection as the 'negotiated reality' of place (1992, 742). On the output or effect side, it features both mental and physical health status and the more general construct of well-being. The definition thereby postulates that individual well-being and comfort are shaped by: (a) characteristics of physical as well as psycho-social environments, and (b) subjective perceptions and evaluations of these surroundings by the individual (Figure 8.1), which encapsulates all five health-relevant features of urban residential environments as listed above. The dialectic between environment and environmental perception, jointly creating the output of health and well-being, may be expressed through reduced quality of life and decreased residential satisfaction as well as stress, mental imbalance or direct health effects.

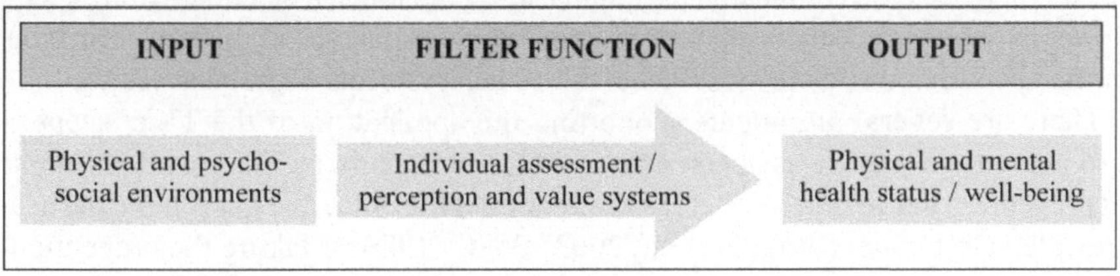

Figure 8.1 A model of place effects on health

The TL concept is partially reminiscent of the environment-behaviour-relationship work initiated in the late 1970s and early 1980s (for example, see Rapoport, 1982) or the research carried out on place and place attachment (see Relph, 1976 and Tuan, 1980). Both research areas are closely related to neighbourhood and urban and residential research which aims at identifying the impact of the built environment on humans and as an example carries out post-occupancy evaluations of buildings or built areas (Dieckmann et al., 1998) in order to assess how the users/residents use the building or the area and whether this use corresponds to the original planning objectives. Creating a bridge to health research, the TL concept provides various opportunities for applications within the field of environmental quality and health, and given its strong emphasis on perception, it could especially represent a suitable way forward for focusing on mental health conditions.

From Therapeutic to Preventive Landscapes

The desire to 'put health into place', as brought forward by Kearns and Gesler (1998), aimed at identifying or even quantifying place-related impacts on individual health and well-being. It attempted to describe those mechanisms by which places can influence human beings, and to identify those places that provide positive and health-supportive outcomes. While many scholars have chosen to focus on a therapeutic use of these places, as the TL concept suggests, recent work carried out by researchers such as Ellaway and Macintyre (2000), Cattell (2001), Stafford and Marmot (2003), Evans (2003) and Latkin and Curry (2003), has shown that characteristics of places and settings may also have high relevance for ill health and mental disorders. Many of these publications have raised the question of how – in addition to compositional factors – contextual factors such as neighbourhood/environmental characteristics (Diez Roux, 2001), or collective factors such as social values/attitudes (Gatrell, 1997), may impact on individual health and well-being in both positive and negative ways.

In this context, the expression 'putting health into place' can be understood in a much more active way, calling for creating environments that support health and well-being (Srinivasan et al., 2003; Sime, 1986). The objective of such an understanding would have to be the *intentional* integration of salutogenic characteristics into places, providing healthy spatial settings to individuals as a means of primary prevention.[2] Therefore, the concept of TL is relevant for all population groups, beyond those defined as ill; it is also relevant for places used or inhabited by humans. Within urban residential settings, the TL concept may be especially important for those persons that spend most of their time in their neighbourhood and are most vulnerable to health threats within that setting.

There are several arguments supporting the application of the TL concept in a more preventive way. It has been acknowledged by many public health researchers that prevention has to receive increased attention if the objective to promote the health of people is to be met (Stokols et al., 2003; WHO, 1986). Making the prevention of ill health a priority, ideas and lessons learned – irrespective of their background or origin – must be transferred and applied prospectively. This is also valid for the TL concept, as health effects can also be gained with preventive objectives. Can it help individuals not only to recover from ill health, but also support and maintain the health status of healthy people in order to prevent ill health in the first place?

In conclusion, this chapter claims that the TL concept is too valuable and too relevant to be only seen in therapeutic terms. In order to assess its preventive potential, this chapter uses the example of urban residential settings as one suggested avenue to consider. It is argued that for people's everyday environments, we need to literally 'put health into place' by intentionally shaping and planning health-supportive characteristics in spatial settings. The development of such preventive landscapes is, most of all, characterized by an intentional integration of health-sustaining and therapeutic elements into places. In addition, providing such preventive landscapes in the context of residential settings would be an ideal way to implement the objectives of public health, which have been defined as: (a) the 'reduction of social and health inequalities', and (b) the 'striving for health-sustaining environments' (McMichael and Beaglehole, 2000, 495).

---

2 The concept of salutogenesis (Antonovsky, 1987) aims at identifying those factors that make or keep people healthy. As various health risks and exposure do not always and inevitably lead to manifestation of disease, the theory of salutogenesis postulates that next to the exposure to pathogenic factors, the presence or absence of salutogenic factors is a major cause for the expression of health effects. In other words, conditions making people sick can be counteracted and compensated by health-supportive conditions. Antonovsky therefore advocated that health research should not only be restricted to factors leading to ill-health, but also identify factors promoting and supporting health.

Preventive Landscapes and Linkages to Health Concepts

The preventive application of the TL concept in urban residential settings is based on the idea that residential landscapes and housing environments could be provided with elements and characteristics that keep people healthy. In this context, the concept of preventive landscapes is close to the idea of 'supportive environments', as put forward by Bistrup and the Danish National Board of Health in 1991. The term 'supportive environments' was created to define the dimension of positive and health-sustaining effects of housing and community environments, one of the topics of the 3rd International Conference on Health Promotion held in Sweden in June 1991. The briefing book on housing and community environments (Bistrup, 1991) featured a variety of characteristics of urban residential environments, such as: noise, air quality, safety issues, crowding, waste and pollution, traffic, and green open spaces. It also identified various health impacts of these conditions. Although many researchers have dealt with similar issues in a more scientific manner (for example, Cohen et al., 2003; Evans, 2003; Macintyre et al., 2003; Stafford and Marmot, 2003; Mackenbach and Howden-Chapman, 2002), this briefing book remains a most useful example for the health-sustaining effects of well-designed community environments. In this context, the objective of 'preventive landscapes' is to identify those characteristics of residential settings that keep people healthy.

In parallel with supportive environments, the concept of preventive landscapes also borrows from theories on environmental preference and the restorative effects of natural landscapes versus urban residential settings (Van den Berg et al., 2003). Evidence from various studies indicates that the view of natural scenery positively affects mental capacity and physiological indicators such as mood states and stress hormones (Kaplan, 1995; Ulrich et al., 1991). The findings on the restorative effects of natural spaces are helpful for preventive landscapes in that they can provide evidence of the healthiness of natural spaces as islands of peace and tranquillity within the built environment.

Preventive Landscapes in Urban Residential Settings: Quantitative Proof for a Qualitative Concept

The spatial and environmental characteristics of urban residential settings provide the physical and social context for a variety of impacts on health and well-being. Looking at the relevant literature, urban and residential health research mostly deals with the mental and social effects of the physical and built environments (Evans, 2003; Jackson, 2003; Twiss, 2003). In urban planning and landscape architecture, numerous publications also emphasize the meaning and importance of public and social places (Thompson, 2002), green and open spaces, and vegetation in urban residential settings (Attwell, 2000; Botkin and Beveridge, 1997). Urban ecology is a major field within urban planning, and concepts such as the urban green structure aim at improving the liveability of cities (Bergen Jensen, 2000), including ecological as well as urban and recreational functions. Urban design discusses issues of accessibility and participation in trying to make cities integrative spaces for all groups of society (Imrie, 2000), while healthy city networks focus on the health effects of urban systems (Tsouros, 2000).

Still, a variety of stressors and risk factors can affect health and well-being in urban residential settings, and it is the objective of urban planners to provide adequate and balanced living conditions to each resident. Urban residential characteristics can, depending on their quality, decrease or increase well-being (Van Kamp et al., 2003); examples include the exposure to background stressors such as pollution, graffiti and litter (Van Poll, 1997). Urban and neighbourhood characteristics modify exposure to residential and traffic noise (Stansfeld et al., 2000) and affect feelings of individual security and perception of safety (Austin et al., 2002). Furthermore, a number of characteristics have an impact on quality and use of urban residential spaces, including: the provision of local amenities such as parks, playgrounds, shops, public facilities and recreation sites (Thomson et al., 2003), the accessibility to public transportation and opportunities for cycling and walking (Emery et al., 2003), and adequate maintenance of private and public properties. Stress, decreased residential satisfaction and health constraints most often arise from a lack of one or several of these necessary features (Halpern, 1995). Severe limitations in well-being result from the collision of too many overlapping expectations towards residential environments, possibly leading to social tension and conflicts (Thompson, 2002).

Data and Methods

The empirical results presented in this chapter indicate the degree of residential satisfaction and health variation that can be associated with the quality of urban residential characteristics. The results are based on a study of housing and health in eight European cities coordinated by the World Health Organization Regional Office for Europe (WHO LARES – Large Analysis and Review of European Housing and Health Status (WHO, 2006)). The project was carried out in 2002–2003 and involved the collection of data on the residential conditions and individual health status of 8,519 residents in the cities of Forlì (Italy), Vilnius (Lithuania), Ferreira (Portugal), Bonn (Germany), Geneva (Switzerland), Angers (France), Bratislava (Slovakia) and Budapest (Hungary). The cities merely represent a convenience sample aiming at covering various European housing stocks. Therefore, results can neither be representative on a national, nor on international scale. As well, the survey data represent the residential conditions in urban settlements only and are not useful for indicating the conditions in rural settings.

The data were collected by interview teams through the use of three survey tools: a face-to-face household questionnaire on the perception of residential conditions by the households, an inspection check list for quality of residential conditions, and an individual self-administered health questionnaire. The survey tools covered a number of different urban residential characteristics, such as housing conditions (that is, heating and ventilation systems, air quality, layout, noise, hygiene equipment) and the quality of the housing environment. The health questionnaire collected data on subjective health perceptions, mental health, functional limitations and the prevalence of self-reported and diagnosed diseases. The database therefore enables a variety of investigations linking residential and housing conditions with a number of health outcomes. In addition, it provides opportunities for identifying the comparative effect of selected urban residential stressors on health. Full details of the methodology is described by Bonnefoy et al. (2003) and found on the WHO Regional Office for Europe website (www.euro.who.int/Housing/activities/20020711_1).

Table 8.1 Overview of dependent variables used for analysis

Outcome variables	Dependent variable	Source	Original variable format	Dichotomized format for analysis
Satisfaction	Residential environment satisfaction	Residents	5-point scale (very good – very bad living area assessment)	Very good/good assessment versus average, bad or very bad assessment
Health	Self-rated health	Residents	5-point scale (very good – very bad subjective health assessment)	Very good/good health versus average, bad or very bad health

As the dependent variables (Table 8.1), satisfaction with the residential environment and self-rated health were selected, providing composite indicators on residential conditions (Amerigo and Aragones, 1997; Anderson and Weidemann, 1997) and health status (Dunn, 2002; Gold et al., 1996).

The results presented are based on multiple logistic regression models using various residential environment characteristics as independent variables. The independent variables were selected on the basis of a literature review on residential environments and health, and the results of a correlation test using the Spearman rank correlation coefficient. The chosen parameters reflect the most relevant residential and urban stressors and covered:

- greenery, public spaces and play areas (vegetation along streets; recreational areas for children; place to sit and relax);
- noise exposure (traffic noise; surrounding area noise);
- public safety (assessment of residential area safety at night; perception of safety in the home);
- upkeep and maintenance (graffiti; amount of litter);
- transportation issues (connection to city centre).

For the survey, the data for these residential and urban stressors was – with few exceptions – taken from the residents and are, therefore, their perceptions of housing and neighbourhood conditions. Table 8.2 provides information on the format of the data used for the analyses and indicates whether the data is provided by the residents or – in two cases – by the interviewer as a part of the inspection.

Table 8.2 Overview of independent variables used for analysis

Urban residential characteristic	Independent variable	Source	Original variable format	Dichotomised format for analysis
Greenery, public spaces and play areas	Vegetation along streets	Interviewer	Binary variable (yes – no)	Original format was applied
	Recreational areas for children	Residents	3-point scale (yes – to some extent – no)	Yes versus to some extent and no
	Place to sit and relax outside of dwelling	Residents	Binary variable (yes – no)	Original format was applied
Noise exposure	Traffic noise	Residents	5-point scale (not at all – extreme)	Not at all/slightly versus moderately, strongly and extremely
	Surrounding area noise	Residents	5-point scale (not at all – extreme)	Not at all/slightly versus moderately, strongly and extremely
Public safety	Safety perception in residential area at night	Residents	3-point scale (yes – to some extent – no)	Yes versus to some extent and no
	Feeling of safety in the dwelling	Residents	5-point scale (strongly agree – strongly disagree)	Strongly agree/agree versus neither/nor, disagree and strongly disagree
Upkeep and maintenance	Graffiti on residential buildings	Interviewer	4-point scale (no – one or two – three to five – six and more)	No versus all other
	Amount of litter and trash in residential environment	Residents	5-point scale (very dirty/littered – not at all dirty/littered)	Not at all/slightly dirty littered versus all other
Transport issues	Connection to city centre	Residents	3 options (no – yes for some residents – yes for all residents)	Yes for all residents versus no and yes for some residents

Before analyses, the original variables were adapted and transferred into dichotomous format as shown in the tables above. For each outcome variable, a regression model was computed. All regression models included the variable 'city' as a fixed effect in order to control for the variation of the eight cities.

To adjust for the confounding influences of known health determinants related to individual characteristics, four variables were selected as confounding variables to be controlled for. They include: 'age group' (child – adult – senior), 'gender' (male – female), 'socio-economic status' (high SES – low SES) and 'functional limitations' (yes – no).

Each regression model was run twice. In the first, the impact of the selected urban residential characteristics on: (a) residential environment satisfaction, and (b) self-rated health was tested. In the second run, the four confounder variables were controlled for, testing the stability of the associations found by the first models. In addition, the direct impact of residential environmental satisfaction on self-rated health was analysed by a third model.

Results

Table 8.3 presents the results of descriptive cross-tabulations with chi-square tests, showing the essential features of the sample and the ratio of residents reporting low satisfaction with the residential environment and low self-rated health for each independent variable.

The cross-tabulations show that the individual characteristics such as age, gender, SES and functional limitations have only little meaning for the expression of residential environment satisfaction – only SES provided a significant variation of satisfaction (25–29.4%). However, compared to the degree of health variation related to SES (28.7–49.6%), its impact on residential environment satisfaction seems rather modest. The largest degree of health variation, however, is found for 'functional limitations' (33.6–76.7%) and 'age' (11.1–75%).

All residential characteristics are significantly associated with both low residential environment satisfaction and low self-rated health status, except for 'vegetation along streets' which is only significant for health status. The strongest association with low residential environment satisfaction is found for residential noise exposure, as annoyance by surrounding area noise (44.6% with low satisfaction) and traffic noise annoyance (39.1% with low satisfaction) show the highest impact on residential quality. Looking at the impact on low self-rated health, the strongest impacts are indicated by individual-level characteristics and score highest for persons with functional limitations (76.7% reporting low health status), and elderly residents (75% with low health). For residential characteristics, the highest percentages for low self-rated health were found for the perception of not being safe in the dwelling (48.9%) and surrounding area noise (47.3%).

Turning to the multivariate analyses, Table 8.4 shows that for all selected characteristics of the urban residential environment, negative assessments are consistently related to an increased likelihood of low residential environment satisfaction (model 1). This association is very stable as all variables remain

Table 8.3 Bivariate analysis for low residential environment satisfaction (RES) and low self-rated health

Characteristic	n	Percentage with low RES	Chi² p value	Characteristic	n	Percentage with low self-rated health	Chi² p value
Gender			0.853	Gender*			0.000
Male	3927	27.1		Male	3927	34.9	
Female	4459	26.9		Female	4473	41	
Functional limitation			0.910	Functional limitation*			0.000
Yes	846	27		Yes	846	76.7	
No	7408	27.1		No	7466	33.6	
SES*			0.000	SES*			0.000
Low SES score	3691	29.4		Low SES score	3687	49.6	
High SES score	4386	25		High SES score	4403	28.7	
Age group			0.101	Age group*			0.000
Child (<18)	1571	28.8		Child (<18)	1569	11.1	
Adult (18–65)	5538	26.9		Adult (18–65)	5555	37.1	
Senior (>65)	1303	25.3		Senior (>65)	1296	75	
Play areas for children*			0.000	Play areas for children*			0.000
Not enough areas	3836	35.1		Not enough areas	3811	43	
Enough play areas	4435	20.1		Enough play areas	4465	34	
Place to sit and relax outside of dwelling*			0.000	Place to sit and relax outside of dwelling*			0.000
No such place exists	2667	37.1		No such place exists	2656	46.7	
Place exists	5527	22		Place exists	5544	34.1	
Vegetation along streets			0.475	Vegetation along streets*			0.000
Not present	3067	27.5		Not present	3062	43.5	
Present	5345	26.7		Present	5358	35	

Traffic noise perceived as annoying*			0.000	Traffic noise perceived as annoying*			0.000
Annoying traffic noise	1724	39.1		Annoying traffic noise	1731	43.8	
No annoying noise	6160	24		No annoying noise	6202	36.2	
Surrounding area noise perceived as annoying*			0.000	Surrounding area noise perceived as annoying*			0.000
Yes – noise is annoying	543	44.6		Yes – noise is annoying	545	47.3	
No annoying noise	7302	26		No annoying noise	7348	37	
Graffiti on residential buildings*			0.000	Graffiti on residential buildings*			0.009
Graffiti exist	2638	35.3		Graffiti exist	2636	39.8	
No graffiti	5713	23.2		No graffiti	5723	37.2	
Litter and trash in residential environment*			0.000	Litter and trash in residential environment*			0.000
Rather dirty and littered	3057	35.5		Rather dirty and littered	3058	43.5	
No litter and trash	5273	22.2		No litter and trash	5273	34.9	
Connection to the city centre*			0.000	Connection to the city centre*			0.000
Not well connected	933	34.5		Not well connected	939	46.3	
Well connected	7397	26.3		Well connected	7394	37.1	
Feeling of safety in residential area at night*			0.000	Feeling of safety in residential area at night*			0.000
Not feeling safe at night	3411	36.8		Not feeling safe at night	3411	44.3	
Feeling safe at night	4888	20.2		Feeling safe at night	4891	33.3	
Feeling of safety in the dwelling*			0.000	Feeling of safety in the dwelling*			0.000
Not feeling safe in dwelling	1763	34.6		Not feeling safe in dwelling	1772	48.9	
Feeling safe in dwelling	6093	25		Feeling safe in dwelling	6134	35	

* = significance at 95% level.

significant after controlling for individual characteristics (model 2). Even more, controlling for individual characteristics strengthens the associations found for eight of the ten residential environment variables. The results of this full model indicate that low residential environment satisfaction largely depends on environmental conditions while individual factors play a very limited role.

The relevance of the ten residential environment characteristics for the overall low level of satisfaction shows a balanced pattern. Odds ratios range from 1.30 to 1.89, and four of the variables (play areas for children; surrounding area noise; litter and trash; and connection to the city centre) have an odds ratio of at least 1.80. It is therefore not possible to identify a dominant parameter.

Table 8.4 Logistic regression results – examining the impact on low residential environment satisfaction

	Model 1: Residential environment characteristics (REC) only		Model 2: REC and individual factors	
Characteristic	Odds ratio for low RES	CI	Odds ratio for low RES	CI
Play areas for children				
Not enough play areas	1.82*	1.60–2.06	1.81*	1.59–2.06
Place to sit and relax outside of dwelling				
No such place exists	1.68*	1.47–1.92	1.70*	1.48–1.96
Vegetation along streets				
Not present	1.44*	1.26–1.65	1.50*	1.30–1.72
Traffic noise perceived as annoying				
Yes	1.67*	1.46–1.91	1.74*	1.51–1.99
Surrounding area noise perceived as annoying				
Yes	1.66*	1.34–2.05	1.80*	1.45–2.23
Graffiti on residential buildings				
Graffiti exist	1.30*	1.13–1.49	1.30*	1.13–1.50
Litter and trash in residential environment				
Dirty and littered area	1.78*	1.57–2.01	1.82*	1.61–2.07
Connection to the city centre				
Not well connected	1.84*	1.54–2.20	1.89*	1.57–2.27
Feeling of safety in residential area at night				
Not feeling safe at night	1.65*	1.46–1.86	1.66*	1.46–1.89

Characteristic	Model 1: Residential environment characteristics (REC) only		Model 2: REC and individual factors	
	Odds ratio for low RES	**CI**	**Odds ratio for low RES**	**CI**
Feeling of safety in the dwelling				
Not feeling safe in dwelling	1.31*	1.14–1.52	1.32*	1.14–1.54
Socio-economic status				
Low SES			0.96	0.84–1.09
Age group (compared to elderly)				
Adults			1.06	0.88–1.27
Children			1.37*	1.10–1.72
Gender				
Male			1.04	0.92–1.17
Functional limitation				
Yes			0.87	0.71–1.07

REC: residential environment characteristics; RES: residential environment satisfaction
* = significance at 95% level.

Table 8.5 presents the results of the regression model using low self-rated health as the criterion variable. Model 1, covering the residential environment characteristics only, indicates that four of these variables fail to provide a significant influence on the variation of low health status (play areas for children; place to sit and relax outside of dwelling; graffiti on residential buildings; and connection to the city centre). For the six variables with a significant increase of likelihood of low self-rated health, the safety parameters show the strongest impact, followed by the noise exposure parameters.

Controlling for individual characteristics (model 2) strongly modifies the impact of the residential environment variables. Two variables (no vegetation along streets; annoyance by surrounding area noise) are no longer significant. 'Not feeling safe in the dwelling' and 'Not feeling safe in the residential area at night' still provide the largest explanation for low self-rated health (OR 1.60 and 1.58), although their odds ratios have decreased in comparison to model 1. Exposure to traffic noise remains the only significant noise variable (OR 1.34), showing a stable association with low self-rated health. Finally, residents living in dirty and littered areas are more likely to report low health status, although the odds ratio is lower than in model 1 (OR 1.26).

The results indicate that low self-rated health depends much more on individual characteristics than was the case for low residential environment satisfaction. Age and functional limitations show the strongest odds ratios (OR 0.04 for children compared to elderly, and OR 6.36 for persons with functional limitations) and are therefore identified as the most important health determinants. However, the results for low SES (OR 1.43) and men (OR 0.73) are of the same magnitude as the results for the four significant residential environment characteristics (OR range from 1.26–1.60), showing that residential environment quality independently contributes to self-rated health.

Table 8.5 Logistic regression results – examining the impact on low self-rated health

	Model 1: REC only		Model 2: REC and individual factors	
Characteristic	**Odds ratio for low health**	**CI**	**Odds ratio for low health**	**CI**
Play areas for children				
Not enough play areas	0.98	0.85–1.12	1.07	0.91–1.25
Place to sit and relax outside of dwelling				
No such place exists	1.08	0.95–1.22	1.05	0.91–1.21
Vegetation along streets				
Not present	1.22*	1.08–1.38	1.07	0.93–1.23
Traffic noise perceived as annoying				
Yes	1.40*	1.19–1.68	1.34*	1.10–1.65
Surrounding area noise perceived as annoying				
Yes	1.40*	1.08–1.80	1.32	0.98–1.76
Graffiti on residential buildings				
Graffiti exist	0.83	0.66–1.04	0.85	0.65–1.10
Litter and trash in residential environment				
Dirty and littered area	1.32*	1.10–1.58	1.26*	1.02–1.56
Connection to the city center				
Not well connected	1.01	0.82–1.25	0.93	0.73–1.18
Feeling of safety in residential area at night				
Not feeling safe at night	1.64*	1.41–1.90	1.58*	1.33–1.87
Feeling of safety in the dwelling				
Not feeling safe in dwelling	1.75*	1.43–2.13	1.60*	1.28–2.00
Socio-economic status				
Low SES			1.43*	1.25–1.63
Age group (compared to elderly)				
Adults			0.19*	0.16–0.23
Children			0.04*	0.03–0.05
Gender				
Male			0.73*	0.65–0.83
Functional limitation				
Yes			6.36*	5.06–7.98

REC: residential environment characteristics
* = significance at 95% level.

Using low residential environment satisfaction as a composite assessment of the quality of the residential area, Table 8.6 indicates that it has a small impact on the variation of low self-rated health (OR 1.09, CI 0.98–1.22; model 1). However, model 2 shows that the individual characteristics of the residents have masked a significant relationship between residential environment satisfaction and low self-rated health, as the odds ratio increases to 1.15 (CI 1.02–1.31) when adjusted for age, gender, SES and functional limitations. Therefore, the perceived quality of the residential environment shows a small but significant association with the subjective health perception of residents, independent of individual characteristics.

Table 8.6 Logistic regression results – examining the impact of low RES on low self-rated health

	Model 1: RES only		**Model 2: RES and individual factors**	
Characteristic	**Odds ratio for low health**	**CI**	**Odds ratio for low health**	**CI**
Residential environment satisfaction				
Low RES	1.09	0.98–1.22	1.15*	1.02–1.31
Socio-economic status				
Low SES			1.46*	1.30–1.65
Age group (compared to elderly)				
Adults			0.20*	0.17–0.24
Children			0.04*	0.03–0.06
Gender				
Male			0.73*	0.65–0.81
Functional limitation				
Yes			6.42*	5.25–7.86

RES: residential environment satisfaction
* = significance at 95% level.

Discussion

The results support the argument that social and environmental problems in the urban residential neighbourhood are significantly associated with low residential environment satisfaction and low self-rated health. This finding, based on data from eight European cities, corresponds to the results of other research on the health impact of neighbourhood conditions (Howden-Chapman, 2004; Drukker et al., 2003; Dunn, 2002). This study thereby suggests that, irrespective of the city, the relationships between contextual or environmental factors, residential environment satisfaction and self-rated health seems to be universal. To some extent, residential environment satisfaction may even provide a useful proxy indicator for potential health effects of living places; this is an important finding as there are a variety of residential environment satisfaction studies that do not investigate the interaction with health.

The fact that all selected parameters of the residential environment are found to be significantly linked to variation in residential environment satisfaction, before as well as after controlling for individual level factors, indicates that the perception of satisfaction is directly shaped by the quality of the given residential environment. These results correspond to other studies (Wilson et al., 2004; Türkoglu, 1997) and provide evidence that the individual level factors have comparatively little to do with the assessment of residential environment quality.

The largest increase of likelihood of low residential environment satisfaction is found for neighbourhoods that are not well connected to the city centre; there are two possible explanations for this. First, it could be hypothesized that a low degree of integration, together with difficulties reaching the city centre, may have a direct impact on the perception of residential environment quality. On the other hand, poor integration of residential areas could be taken as an indicator for isolated and neglected areas, in which the quality of the residential environment is decreased due to other residential features. Irrespective of the mechanism, the results show that a good connection to and integration of residential areas are essential for avoiding neighbourhood problems affecting residential quality.

Noise from the surrounding area, including noise sources such as cafes and bars, backyards or playgrounds, was perceived as even more annoying than traffic noise, which is commonly seen as the largest problem (Sharp, 2002). This is possibly due to an increased degree of adaptation to traffic noise, while area noise is more irregular by nature, making it harder for residents to adapt.

Visual impression (litter and trash, graffiti, vegetation along streets) and functional capacities (recreational areas for children, place to sit and relax) of the close residential environment are identified as meaningful features of neighbourhoods and show that residential areas have to fulfil a variety of expectations. Strongest results were found for 'litter and trash' and 'play areas for children', indicating that a provision of local amenities and clean, maintained areas are among the most relevant expectations of residents. Similarly, increased likelihood for low satisfaction ratings is found for a lack of safety in the residential area at night as well as in the home. The perception of fear at night in the neighbourhood especially leads to strongly decreased satisfaction, as it is known to have a powerful impact on the daily life of residents and can strongly affect social behaviours (Keane, 1998). In total, the results confirm that residential environment satisfaction represents a suitable assessment of the quality of residential environments, as it reflects the impact of a variety of neighbourhood characteristics.

The increased health relevance of individual level factors (Table 8.5, model 2) marks the most visible difference on health amongst the regression models tested. However, there are still four residential environment characteristics with an independent impact on self-rated health (traffic noise, litter and trash, safety in residential area at night and safety in home). The impact of traffic noise indicates that noise exposures are capable of affecting the health perception of residents, which is supported by both empirical evidence and by biological mechanisms linking noise exposure with health outcomes (such as sleep disturbance, mental health symptoms, and hormonal changes) (Stansfeld and Matheson, 2003).

Dirty environments were associated with a decreased health perception in this study. Although direct links cannot be ruled out and may be especially relevant for mental health effects, it seems more likely to assume that the association between litter and self-reported health is the result of residential selection. In this case, littered areas with a low level of maintenance may then be inhabited by less privileged and therefore less healthy population groups. This would define housing conditions as a health inequality parameter as suggested by Evans and Kantrowitz (2002), and may lead to an aggravation of pre-existing health problems.

Issues of safety are the most important residential environment features for subjective health perception, showing that fear and perceived danger – representing a very personal and emotional threat to health – are much more linked with perceived health than any other residential feature. Not feeling safe in the dwelling even leads to a stronger increase of poor health than of low residential environment satisfaction. Although there is a variety of mechanisms linking the perception of safety problems and fear of crime with health (Green et al., 2002; Chandola, 2001), it is possible that some of the persons reporting not feeling safe in this survey have health perceptions that makes them more vulnerable, and thereby more likely to perceive fear.

In total, the results show that the quality of the residential environment does have some impact on health, although this is only valid for selected variables and is less expressed than the impact of individual level factors. This suggests that various health outcomes, although mostly a product of individual determinants, can still be influenced by environmental conditions (Macintyre et al., 2002). In this context, already published results of the LARES survey have indicated that green and socially stable residential environments are associated with increased physical activity and decreased obesity rates (Ellaway et al., 2005), while fear of crime has a limiting impact on the amount of physical activity – especially so for women (Shenassa et al., 2006). However, the results also show that the impact of a given residential environment parameter on residential environment satisfaction does not predict its relevance for self-rated health, as only one of the residential environment characteristics that provided the highest odds ratios for low residential environment satisfaction is also significant for poor health (dirty and littered area).

Finally, the extreme odds ratio provided by persons with functional limitations (OR 6.36) cannot be left without comment. The strong association of functional limitation with low health shows that, depending on the hypothesis tested, it could be extremely important to control for this specific risk group feature when looking at spatial differences of health (Wainwright and Surtees, 2003). With the global population showing a 10 per cent prevalence of some form of handicap, often leading to a severe limitation of health, functional limitations could represent a valid confounder for the spatial patterns of selected health outcomes.

With four out of ten residential environment characteristics significantly associated with poor health perception, residential environment satisfaction is becoming increasingly attractive as a composite indicator of residential environment quality, and it is justified to ask about its direct health relevance. Table 8.5 shows that, controlling for individual level factors, residential environment satisfaction provides a significant impact on self-reported health status. This impact is less strong than for individual characteristics such as age, gender, SES or functional limitations, but it exists – significant, independent, and based on more than 8,000 individuals. Based on the assessment of individual residential environment features, and on the composite assessment of residential environment satisfaction, it can therefore be confirmed that:

> neighbourhoods have a small but consistent impact on health status (Howden-Chapman, 2004, 165).

Conclusion

Although urban planning and neighbourhood development projects are often based on a variety of health-relevant objectives, they have yet to develop an improved understanding of how their practices can impact on human well-being (Dannenmaier, 1995). In this context,

> … one of the major challenges is to bridge the divide between the environmental quality/wellbeing/quality-of-life specialists and the players who make urban policy and who shape our physical and social environments – in other words the engineers, planners, architects, service delivery specialists, etc. (Brown, 2003, 85).

In this chapter, it has been suggested that urban and residential neighbourhood planning has both the capacity and the mandate for developing health-supportive settings, promoting the general life satisfaction and health status of individuals. The TL concept is suggested as a suitable conceptual framework for the health-effectiveness of such settings, in the context of urban residential neighbourhoods. Applying the TL concept in a preventive manner provides the possibility to shape and develop urban residential areas, offering a rich diversity of characteristics and functions that are capable of providing restorative and healthy benefits such as tranquillity, functional diversity, comfort and safety. Based on the data analysis and evidence from the literature, key aspects for such characteristics and spatial functions could, for example, include: availability of green and open spaces acting as quiet and comforting islands within the stress of urban life; general protection of urban residential neighbourhoods from noise and other urban exposures; improvement of the perception of public safety through adequate lighting or thoughtful urban planning; increased maintenance of urban and residential spaces; and adequate access to (public) transportation opportunities to avoid 'disconnection' of urban districts.

It is suggested that by offering adequate access to such health-supportive features of urban residential environments, health-relevant exposures to physical, social and mental challenges are reduced, while negative feelings such as stress, aggression and lack of control are counteracted and their effects mitigated. In other words, the residential environment perception of the inhabitants needs to be prioritized to shape a residential experience that provides psychosocial support and satisfaction instead of stress and frustration.

Preventive urban residential landscapes, adapted to the needs and preferences of humans and aimed at providing suitable spaces for realizing personal priorities and needs, could therefore complement the TL approach through a prospective implementation of healthful elements. The results presented show that in urban residential settings, such health-sustaining elements do exist and have a measurable impact on residential environment satisfaction and health. If adequately provided in urban residential areas, many of these features could improve the quality of life of residents, representing a strong and preventive contribution to population health.

References

Amerigo, M. and Aragones, J.I. (1997), 'A theoretical and methodological approach to the study of residential satisfaction', *Journal of Environmental Psychology* 17: 47–57.

Anderson, J.R. and Weidemann, S. (1997), 'Developing and utilizing models of resident satisfaction', in G.T. Moore and R.W. Marans (eds) *Advances in Environment, Behavior, and Design*, vol. 4 (New York: Plenum Press) pp. 287–314.

Antonovsky, A. (1987), *Unraveling the Mystery of Health. How People Manage Stress and Stay Well* (San Francisco: Jossey Bass).

Attwell, K. (2000), 'Urban land resources and urban planting – case studies from Denmark', *Landscape and Urban Planning* 52: 145–63.

Austin, D.M., Furr, L.A. and Spine, M. (2002), 'The effects of neighborhood conditions on perceptions of safety', *Journal of Criminal Justice* 30: 417–27.

Bergen Jensen, M. (2000), 'Green structure and sustainability: developing a tool for local planning', *Landscape and Urban Planning* 52: 117–33.

Bistrup, M.L. (1991) *Housing and community environments – how they support health. Briefing book for the Sundsvall Conference on Supportive Environments 1991.* (Copenhagen: National Board of Health).

Bonnefoy, X.R., Braubach, M., Moissonnier, B., Monollbaev, K. and Robbel, N. (2003), 'Housing and health in Europe: preliminary results of a pan-European study', *American Journal of Public Health* 93 (9): 1559–63.

Botkin, D.B. and Beveridge, C.E. (1997), 'Cities as environments', *Urban Ecosystems* 1: 3–9.

Brown, A.L. (2003), 'Increasing the utility of urban environmental quality information', *Journal of Landscape and Urban Planning* 65: 85–93.

Cattell, V. (2001), 'Poor people, poor places, and poor health: the mediating role of social networks and social capital', *Social Science and Medicine* 52: 1501–16.

Chandola, T. (2001), 'The fear of crime and area differences in health', *Health and Place* 7: 105–16.

Cohen, D.A., Mason, K., Bedimo, A., Scribner, R., Basolo, V. and Farley, T.A. (2003), 'Neighborhood physical conditions and health', *American Journal of Public Health* 93 (3): 467–71.

Dannenmaier, M. (1995), 'Healing gardens', *Landscape Architecture* 85/01: 56–8.

Dieckmann, F., Flade, A., Schuemer, R., Strohlein, G. and Walden, R. (eds) (1998), *Psychologie und gebaute Umwelt. Konzepte, Methoden, Anwendungsbeispiele* (Darmstadt: Institut Wohnen und Umwelt).

Diez Roux, A.V. (2001), 'Investigating neighbourhood and area effects on health', *American Journal of Public Health* 91 (11): 1783–89.

Drukker, M., Kaplan, C., Feron, F. and van Os, J. (2003), 'Children's health-related quality of life, neighbourhood socio-economic deprivation and social capital. A contextual analysis', *Social Science and Medicine* 57: 825–41.

Dunn, J.R. (2002), 'Housing and inequalities in health: a study of socio-economic dimensions of housing and self reported health from a survey of Vancouver residents', *Journal of Epidemiology and Community Health* 56: 671–81.

Ellaway, A. and Macintyre, S. (2000), 'Social capital and self-rated health: support for a contextual mechanism', *American Journal of Public Health* 90: 988.

Ellaway, A., Macintyre, S. and Bonnefoy, X. (2005), 'Graffiti, greenery, and obesity in adults: secondary analysis of European cross sectional survey', *British Medical Journal* 331: 611–12.

Emery, J., Crump, C. and Bors, P. (2003), 'Reliability and validity of two instruments designed to assess the walking and bicycling suitability of sidewalks and roads', *American Journal of Health Promotion* 18 (1): 38–46.

Evans, G.W. (2003), 'The built environment and mental health', *Journal of Urban Health* 80 (4): 536–55.

Evans, G.W. and Kantrowitz, E. (2002), 'Socioeconomic status and health: the potential role of environmental risk exposure', *Annual Reviews of Public Health* 23: 303–31.

Gatrell, A.C. (1997), 'Structures of geographical and social space and their consequences for human health', *Geografiska Annaler* 79B (3): 141–54.

Gesler, W.M. (1992), 'Therapeutic landscapes: medical issues in light of the new cultural geography', *Social Science & Medicine* 34 (7): 735–46.

Gold, M., Franks, P. and Erickson, P. (1996), 'Assessing the health of a nation: the predictive validity of a preference-based measure and self-rated health', *Medical Care* 34: 163–77.

Green, G., Gilbertson, J.M. and Grimsley, M.F.J. (2002), 'Fear of crime and health in residential tower blocks', *European Journal of Public Health* 12: 10–15.

Halpern, D. (1995), *Mental Health and the Built Environment: More than Bricks and Mortar?* (London: Taylor & Francis).

Howden-Chapman, P. (2004), 'Housing standards: a glossary of housing and health', *Journal of Epidemiology and Community Health* 58: 162–8.

Imrie, R. (2000), 'Responding to the design needs of disabled people', *Journal of Urban Design* 5 (2): 199–220.

Jackson, R.J. (2003), 'The impact of the built environment on health: an emerging field', *American Journal of Public Health* 93 (9): 82–83.

Jones, M. (1991), 'The elusive reality of landscape. Concepts and approaches in landscape research', *Norsk Geografisk tidsskrift* 45: 229–44.

Kaplan, S. (1995), 'The restorative benefits of nature: towards an integrative framework', *Journal of Environmental Psychology* 15: 169–82.

Keane, C. (1998), 'Evaluating the influence of fear of crime as an environmental mobility restrictor on women's routine activities', *Environment and Behavior* 30 (1): 60–74.

Kearns, R.A. (1993), 'Place and health: towards a reformed medical geography', *Professional Geographer* 45 (2): 139–47.

Kearns, R.A. and Gesler, W.M. (1998), *Putting Health into Place* (Syracuse: Syracuse University Press).

Latkin, C.A. and Curry, A.D. (2003), 'Stressful neighbourhoods and depression: a prospective study of the impact of neighbourhood disorder', *Journal of Health and Social Behaviour* 44: 34–44.

Ley, D. (1981), 'Behavioral geography and the philosophies of meaning', in K.R. Cox and R.C. Golledge (eds) *Behavioral Problems in Geography Revisited* (New York: Methuen) pp. 209–30.

Macintyre, S. and Ellaway, A. (2000), 'Ecological approaches: rediscovering the role of the physical and social environment', in L.F. Berkman and I. Kawachi (eds) *Social Epidemiology* (Oxford: Oxford University Press).

Macintyre, S., Ellaway, A. and Cummins, S. (2002), 'Place effects on health: how can we conceptualise, operationalise and measure them?' *Social Science and Medicine* 55: 125–39.

Macintyre, S., Ellaway, A., Hiscock, R., Kearns, A., Der, G. and McKay, L. (2003), 'What features of the home and the area might help to explain observed relationships between housing tenure and health? Evidence from the West of Scotland', *Health and Place* 9: 207–18.

Mackenbach, J.P. and Howden-Chapman, P. (2002), 'Houses, neighbourhoods and health', *European Journal of Public Health* 12: 161–2.

McMichael, A.J. and Beaglehole, R. (2000), 'The changing global context of Public Health', *Lancet* 356: 495–9.

Rapoport, A. (1982), *The Meaning of the Built Environment. A Nonverbal Communication Approach* (London: Sage Publications).

Relph, E. (1976), *Place and placelessness* (London: Pion).

Sharp, D. (2002), 'Silencing cities', *Journal of Urban Health* 79 (2): 162.

Shenassa, E.D., Liebhaber, A. and Ezeamama, A. (2006), 'Perceived Safety of Area of Residence and Exercise: A Pan-European Study', *American Journal of Epidemiology* 163 (11): 1012–17.

Sime, J.D. (1986), 'Creating places or designing spaces?' *Journal of Environmental Psychology* 6: 49–63.

Stafford, M. and Marmot, M. (2003), 'Neighbourhood deprivation and health: does it affect us all equally?' *International Journal of Epidemiology* 32: 357–66.

Stansfeld, S., Brown, B. and Haines, M. (2000), 'Noise and health in the urban environment', *Reviews on Environmental Health* 15: 43–82.

Stansfeld, S.A. and Matheson, M.P. (2003), 'Noise pollution: non-auditory effects on health', *British Medical Bulletin* 68: 243–57.

Stokols, D., Grzywacz, J.G., McMahan, S. and Phillips, K. (2003), 'Increasing the health promotive capacity of human environments', *American Journal of Health Promotion* 18 (1): 4–13.

Thompson, C.W. (2002), 'Urban open space in the 21st century' *Landscape and Urban Planning* 60: 59–72.

Thomson, H., Kearns, A. and Petticrew, M. (2003), 'Assessing the health impact of local amenities: a qualitative study of contrasting experiences of local swimming pool and leisure provision in two areas of Glasgow', *Journal of Epidemiology and Community Health* 57: 666–7.

Tsouros, A.D. (2000), 'Why urban health cannot be ignored: the way forward', *Reviews on Environmental Health* 15: 267–71.

Tuan, Y.F. (1980), 'Rootedness versus sense of place', *Landscape* 24: 3–8.

Türkoglu, H.D. (1997), 'Residents' satisfaction of housing environments: the case of Istanbul, Turkey', *Landscape and Urban Planning* 39: 55–67.

Twiss, J. (2003), 'Community gardens: lessons learned from California Healthy Cities and Communities', *American Journal of Public Health* 93 (9): 1435–38.

Ulrich, R.S., Simons, R.F., Losito, B.D., Fiorito, E., Miles, M.A. and Zelson, M. (1991), 'Stress recovery during exposure to natural and urban environments', *Journal of Environmental Psychology* 11: 201–30.

Van den Berg, A.E., Koole, S.L. and van der Wulp, N.Y. (2003), 'Environmental preference and restoration: (How) are they related?' *Journal of Environmental Psychology* 23: 135–46.

Van Kamp, I., Leidelmeijer, K., Marsman G. and de Hollander, A. (2003), 'Urban environmental quality and human well-being. Towards a conceptual framework and demarcation of concepts; a literature study', *Landscape and Urban Planning* 65: 5–18.

Van Poll, R. (1997), *The perceived quality of the urban residential environment. A multi-attribute evaluation* (Roermond: Westrom Drukkerij).

Wainwright, N.W.J. and Surtees, P.G. (2003), 'Places, people, and their physical and mental functional health', *Journal of Epidemiology and Community Health* 58: 333–9.

WHO Regional Office for Europe (1986), *Ottawa Charter for Health Promotion* (Copenhagen: World Health Organization Regional Office for Europe) <http://www.euro.who.int/AboutWHO/Policy//20010827.2.>

WHO Regional Office for Europe (1997), *Green cities, blue cities. Local authorities, environment and health. Briefing pamphlet series 18* (Copenhagen: World Health Organization Regional Office for Europe).

WHO European Centre for Environment and Health, Bonn (2006), *Large Analysis and Review of European housing and health Status. Preliminary overview* (Copenhagen: World Health Organization Regional Office for Europe) <http://www.euro.who.int/Document/HOH/LARES_results.pdf>

Williams, A. (1998), 'Therapeutic landscapes in holistic medicine', *Social Science and Medicine* 46 (9): 1193–1203.

Wilson, K. (2003), 'Therapeutic Landscapes and First Nations peoples: an exploration of culture, health and place', *Health and Place* 9: 83–93.

Wilson, K., Elliott, S., Law, M., Eyles, J., Jerrett, M. and Keller-Olaman, S. (2004), 'Linking perceptions of neighbourhood to health in Hamilton, Canada', *Journal of Epidemiology and Community Health* 58: 192–8.

Chapter 9

A 'Family-Friendly' Place: Family Leisure, Identity and Wellbeing – The Zoo as Therapeutic Landscape

Bonnie C. Hallman

This chapter explores how the zoo, as a specific site for family leisure, can be considered a therapeutic landscape, particularly for those at the family formation/transition to parenthood stage in the life course. The zoo landscape is understood as a cultural landscape imbued with meanings reflective and constitutive of social norms regarding family life in contemporary Western society. These norms figure prominently in the social construction of a sense of family through the building and maintenance of family connection and cohesion via shared quality time spent in family leisure activities.

The chapter begins with an examination of the relationship between family leisure, parental/family identities and wellbeing. Drawing on interdisciplinary research on gender, family status and leisure, the association of 'purposive leisure' (Shaw and Dawson, 2001) and 'leisure-based parenting' (Such, 2006) with contemporary notions of successful parenthood and family life is presented, connecting 'success' to notions of family wellbeing. The discussion continues with an examination of zoos as places of family leisure characterized by a widespread and largely unexamined identification as family-friendly recreation destinations. Informed by literature on the relationship between place identity and self/group identity formation, attention centres on the use of zoos as stages or backdrops for the public performance of family life and of parenthood as identity, practised through interaction and communication behaviours made possible or encouraged in zoo environments. It is argued that where a degree of fit is perceived between personal and group identities (good parent/good families) and the apparent place identity of the leisure environment (the family-friendly zoo), a sense of wellbeing can be engendered for those who seek/accept this identity, thereby constituting the zoo as a therapeutic landscape.

The chapter ends with a presentation of an illustrative analysis of family zoo photographs. Family photographs taken at the zoo are understood to be both a product of family leisure behaviour, and 'souvenirs' capturing the interaction, connection and cohesion of the family group sought, and engaged in, through the family recreation experience of the zoo visit. Such family photography is widely recognized as both a process of inscribing meanings about family and familial space, and as tangible representations of idea(l)s of family functioning and identity (Bourdieu, 1990; Chambers, 2003).

Family Leisure and Wellbeing

> People practice family – artfully, creatively, with intention – in local settings, ... And their intentions and craft are formed within material and discursive contexts that shape and channel those efforts ... (DeVault, 2000, 501).

In contemporary conceptualizations of healthy family life, family leisure is considered integral to the emotional labour of constructing a positive family life. It is understood to assist in the building of connection and cohesion amongst family members, thus creating, maintaining and valuing group identity as a 'well-functioning' family. Typically, family leisure is considered as occasions that are happy, mutually beneficial and enjoyable for all family members, and thereby contribute to wellbeing (Shaw, 1992; Daly, 1996; Haworth, 2003). This connection between leisure and family wellbeing is well captured in the modified adage 'the family that plays together, stays together'.

The importance and value placed on families spending quality leisure time together, and significantly, private family time in public leisure settings for young families with small children, is stressed in North American culture. Numerous contemporary parenting and lifestyle self-help books (for example Partow and Partow, 1996; Ellis, 1999; Baicker-McKee, 2003; Ehman et al., 2005), television programmes (for example SuperNanny, Nanny 911, Shalom in the Home), and articles in parenting and women's magazines extol the virtues of family outings and vacations. Emphasis is placed on: how quality leisure time improves children's mental, physical and moral/spiritual development; strategies for making family outings fun and educational for children of all ages; and the importance of family togetherness as a cultural, sometimes religious, value (Daly, 1996; Shaw and Dawson, 2001). This culturally embedded view of family leisure time tends to emphasize only positive aspects and tends to reify nuclear, traditional family ideology (Shaw, 1992) while at the same time normatively constructing family recreation as an integral practice in successful parenting and therefore a key practice in the development of a positive family life.

Constructions of Parenthood through Leisure

The view of family life outlined above fails to recognize that family time with children is far more likely to be experienced as a combination of 'care-work' (associated with producing and sustaining a fulfilling and positive family life) *and* leisure by women (with the boundary between leisure and work often blurred), and more often as leisure alone by men, though increasingly as a central part of their identities as fathers (Zabriskie and McCormick, 2003; Maher, 2005). According to Kelly's (1983) model of leisure behaviour 'relational and role determined leisure', such as family outings and vacations, is differentiated from intrinsically motivated individual leisure that characterizes the recreation of adults prior to parenthood. This categorization acknowledges the tangible and emotional work involved in planning and producing recreational family activities.

Research consistently points to the strong sense of purpose that parents attach to family participation in public leisure activities and their determination and commitment to this aspect of family life, such that leisure-based parenting plays a central role in the recreational lives of parents with young children (Samuel, 1996; Maher, 2005; Such, 2006). For some time this has been recognized amongst mothers, where pragmatic, activity-focused interactions with their children can become closely tied to a positive maternal identity, as they are the ones most often identified as responsible for suggesting, planning and enacting purposive family leisure activities. Studies have indicated that mothers report consciously organizing family leisure such as zoo trips in order to build family cohesion, and recognize that they do so at least in part to guard against the possibility of family dysfunction (DeVault, 2000; Howard and Madrigal, 1990; Turley, 2001).

Family leisure has also become a key component of men's roles as fathers, a trend reflective of contemporary cultural changes in the dominant image of fatherhood (Kay, 2006), which '... now highly values active participation that facilitates child development and family functioning ... [this] supercedes traditional leisure goals of personal enjoyment and development and relaxation' (Such, 2006, 119). Recent research asserts that contemporary mothers *and* fathers increasingly define recreational family time as a (willingly fulfilled) duty to promote family functioning, togetherness and wellbeing; however there may be key differences in how this is thought of and accomplished. The orientation of fathers is described as a desire to 'be with' their children, a perspective somewhat different from the discourse of care, emotional work and responsibility that Such (2006) identifies as 'being there' for their children amongst mothers interviewed in her research.

Therefore, family leisure activities such as trips to zoos provide a context for family recreation behaviour, intimately linking leisure and parenthood for both mothers and fathers. Those recreational spaces identified as positive, safe and educational reflect back to parents the very values of family emotional connection and togetherness they are both inspired by, and seeking to transmit to their children through quality family leisure time. It is to a discussion of the identification of zoos as an important and interesting example of such 'family-friendly' places of leisure, and the association of this place identity with family/parenthood identities, that this chapter now turns.

Identifying Zoos as Family-Friendly Therapeutic Landscapes

> Place identity is '... an interpretation of self that uses environmental meaning to symbolize or situate identity' (Cuba and Hummon, 1993, 112).

Zoos are a specific type of built environment imbued with place-specific personal, social and cultural meanings (Hallman and Benbow, 2006; Mullan and Marvin, 1999), which collectively form a place identity through which personal and group identities are constructed, maintained and sometimes transformed, as individuals develop their sense of self in the world and their connectedness to place and environment. People use places to communicate qualities of the self to themselves and to others. Moral and value judgments are transferred to specific places, which thereby acquire a personality or quality with which they are readily identified and labeled (Williams, 1999a). Individuals and groups synthesize – even internalize

– such subjective meanings ascribed to the objective reality of a place (Hagerstrand, 1982; Pred, 1984). Relph (1976, 55) described this as existential insidedness: '… the most fundamental form of insidedness … in which a place is experienced without deliberate and self-conscious reflection, yet is full of significance.'

Thus, we can understand zoos as socially defined public places inscribed with meanings about private family life, where families can enact or perform their particular 'family fantasy' (Aitken, 1998). Parents, their children, and sometimes other family members (for example, grandparents) inhabit the zoo landscape together, immersed in a discourse of family life that metaphorically swirls around them (DeVault, 2000), shaping their expectations of, and behaviours within, the zoo landscape. Where there is a sense of fit or continuity between the zoo place identity and the self/group identities of parents and family groups, then the zoo can be considered a therapeutic landscape for a given family. Such continuity between self and environment is a significant contributor to emotional health and subjective wellbeing as one feels to be 'in place' (Williams, 1999a).

The Zoo: 'A Good Family Place'

Every year more than 140 million Americans visit zoos, with over 80 per cent of those visitors coming in family groups with children aged 12 and under (American Zoo and Aquarium Association, 2006; Tomas et al., 2003). According to Morey & Associates (2004), responses to their national survey of American zoo and aquarium patrons indicated that over 65 per cent of households visiting a zoo did so 'for a family activity' (the most common response). In research conducted at Assiniboine Park Zoo in Winnipeg, Canada, the vast majority of visitors (over 70 per cent) reported in open response questions that they considered zoos a 'classic family outing', a 'good family place' and/or a 'good place to bring the kids' (City of Winnipeg Community Services Department, 2004). These findings reflect the persistent popularity of zoos as destinations for family leisure outings. This is despite other prevalent public discourses critical of zoos specifically and of captive wild animals in general, such as the various groups that constitute the animal rights movement.

Such descriptions of zoos as 'good for families' are based on a place identity understanding so culturally embedded as to be largely unquestioned and nearly automatic in its recognition. Such identification appears to strongly reflect Relph's (1976) notion of existential insidedness; the zoo is simply 'read' as family-friendly without deliberate consideration. This extent of 'insidedness' is apparent in the use of zoos as referents in popular culture. They simply would not resonate with the public if the zoo environment was not unquestioningly understood as a 'good place' for children and families in the dominant culture. Figure 9.1 is an example of popular media reflecting this culturally embedded, largely unproblematized association of zoos with a positive depiction of 'mom, dad and the kids' family life.

The comic strip depicted here is from the enormously popular Baby Blues series, launched in 1990 and published in some 900 newspapers and with an estimated 40 million fans worldwide; this series follows the trials and tribulations of the MacPhersons as they raise a contemporary young family (Kirkman and Scott, 2004).

Figure 9.1 A popular culture reflection of the significance of zoos and photographs in family life. ©Baby Blues Partnership. King Features Syndicate

It portrays the whirlwind of needs and demands of young children, and the worries, joys and compromises of parenthood. Here we see the MacPhersons, parents and children, at the zoo getting their family 'jungle pix' and thus marking the event of being together at the zoo, engaged in family leisure. In turn, the millions of parents who read and identify with this strip will share the reinforcement of the idea of the zoo trip as valued family leisure, together with the memorializing of the experience in a (overpriced!) photograph. In this example, this reading of the zoo as family-friendly is even given the moral weight of a proverb!

Figure 9.2 Place identity at work. Photography by Mark Alexander. Created by Animal Instinct Advertising for Newport Aquarium

In Figure 9.2, we see another expression of the power of identification of zoos as 'family-friendly'; this is just one example of the pervasiveness of the marketing of zoos and similar locales such as aquariums, to their 'target market' – families with young children. This example particularly reinforces the role of the mother in this experience – there is no 'father-otter' depicted, however, the message is clear through the text and the stereotypical garb of the 'mom and kids' that this is a near-perfect and traditional family leisure destination. Thus, marketers and the zoo community that employs them clearly recognize the place identity of zoos as family-oriented leisure destinations, and the primary role of the mother in family recreation decisions.

In addition to media advertising, large corporations are increasingly seeking to reach young families through advertising and sponsorship within the zoo environment itself, through exhibit and facility sponsorship (for example, the Dodge Wild Earth exhibit and safari ride at the Philadelphia Zoo, complete with Durango SUVs) (Hampp, 2006). Marketers are increasingly '… realizing this is a great way to reach Middle America … typically zoos have higher attendance numbers than most pro sports teams in markets because they are open every day … drawing good crowds and obviously drawing families is a desirable target for marketers' (Hampp, 2006).

While these are only some of the possible examples that could be examined, they do illustrate the pervasiveness of the assessment by the dominant users of the zoo landscape (families with young children) and those who provide and develop these sites (the zoo community) of zoos as 'family-friendly', as well as something of the particularity of the dominant conceptualization of family. Together they (re)create the place identity of zoos as positive environments for engaging in quality family time and for building subjective wellbeing in family life.

Such a reading of the zoo as a 'landscape of wellbeing' conceptually fits within the socio-ecological model of health, in which health and wellness are seen as the desired outcomes of a complex interaction of physical, mental, emotional, spiritual, environmental and social factors (Evans and Stoddart, 1990; Gesler and Kearns, 2002). This perspective affords an opportunity to examine and reveal the myriad associations between environment and wellbeing, including examination of those spaces and places wherein behaviours deemed emotionally and socially positive by the larger community are engaged (Williams, 1999b; Gesler and Kearns, 2002). Zoos and similar family-oriented outdoor leisure destinations are understood to be, and market themselves as, sites conducive to positive family interaction. This 'place of interaction' is understood to literally create 'happy families' through behaviour not only facilitated, but also strongly identified as central to their very purpose. Thus, identity as a well-functioning family, as defined by the dominant culture, has ties to place identity and the performance of social identity through participation in activities within that place. We can think of zoos as a therapeutic landscape in the sense that they are thought of and used, especially by families with young children, to contribute to wellbeing via increased family togetherness, communication and positive memory-making. This association can be put quite simply: *for those who identify with zoos as family-friendly places to be, zoos operate as therapeutic landscapes*.

An understanding of the zoo as a cultural landscape reflective of hegemonic messages privileging the two-parent nuclear family and reinforcing traditional values and family forms means that this landscape may be anything but therapeutic for those who *do not* desire, or easily fit within, the image of family and family life ascribed to zoos. Some will find ways to challenge and change the dominant place identity of the zoo, in order to make a place for their experience and definition of family. A good example of this is those zoos such as London Zoo (Gay Sunday at the Zoo) and San Diego Zoo (Gay Day at the Zoo), where the 'family-friendly' label is challenged and redefined to be inclusive of homosexuality through the literal occupation of a place redolent with otherwise unproblematized heterosexist familism (Zoological Society of London, 2006; Pickle, 2001). Such forays into challenging perceptions about what constitutes family and who may comfortably use family-friendly places are all too limited and specific in time and place. However, for those who do accept all or part of the message of familism that is both inscribed into and reflected outwards from the social institution of the zoo, a created sense of 'being in the right place; doing the right thing' for the good of their children and the family unit as a whole may well create an important sense of wellbeing.

Capturing Family Life and Wellbeing: Evidence from Family Zoo Photographs

The zoo visit is an experience in practising family often captured in family photography. Like family leisure, family photography is widely recognized as a process of inscribing meanings about family life and familial space. In addition, family photographs, and particularly photographs of the family at leisure, are tangible representations of idea(l)s of family identity and belonging (Sontag, 1977; Bourdieu, 1990; Spence and Holland, 1991; Chambers, 2003).

Like the zoo space, photographs of families at leisure are both shaped by and reflective of the building and maintaining of a desired state of family wellbeing. Through the understood meanings of the place and behaviours captured in the photographs, and through the later viewing and sharing of the created images, a visual narrative of the continuity between a set of parent/family identities and the specific place identity is generated; this works as a depiction/performance of the wellbeing of the family to both themselves and to others (Chambers, 2003; Haldrup and Larsen, 2003; Larsen, 2005). The zoo as place contextualizes the performance of family, tying hegemonic place meanings to the photograph that captures and transmits the documented realization of and fulfillment in family life, however fragile and/or fabricated – perhaps even to the point of being a constructed fantasy (Aitken, 1998; Chambers, 2003).

In the following section, examples of family zoo photographs are introduced as illustrations of the visual narrative of family life being created in and through family leisure at the zoo. The examples are drawn from the author's research (Hallman and Benbow, in press) analysing and identifying themes in a sample of over 140 family zoo photographs, drawn from both personal family photograph collections and online photograph sites. The process of data collection involved a Google Images search under the terms 'zoo' and 'family' conducted over several days in July 2005. The initial search yielded some 728,000 hits. Of this initial sample we selected and examined the first 140 images of children and family members taken at zoos. All of these images came from publicly accessible online family photo albums. An additional ten photographs, either pre-dating this research or taken in support of other zoo-related research projects, were selected from the personal photo collection of the author. These were added to the sample based on a reading of Chambers, (2003) chapter on the analysis of family photo albums as family space, where she convincingly demonstrates that the 'insider' perspective afforded a more autobiographical approach to family photograph analysis.

Following Rose (2001), content analysis of this combined sample of zoo family photos revealed startlingly clear themes in the portrayal of families in the zoo environment. In fact, as Rose (2003), Chambers (2003) and Larsen (2005) all note in their analyses, there is a sameness to family photography in general, and to family leisure photography specifically, that verges on the boring! In this analysis of a set of family zoo photos, children and intergenerational family groups were consistently depicted with either animal exhibits as backdrops, or with significant features – such as animal statues or entrance gates – as stages for posed shots. A second group of family zoo photos depicts interaction either between family members (most often parents and young children) or between children and animals in the controlled setting of children's or petting zoo areas, capturing the (direct and indirect) performance of parenthood.

In order to avoid copyright issues, the family zoo photographic images used in this chapter are from the personal collection of the author. It is emphasized that these photographs are used to *illustrate* the themes identified in the broader analysis, and thus how family photography is reflective of the construction of self/group and place identities of contemporary families at leisure in the zoo environment.

Family Zoo Photographs: A Visual Narrative of Family Life

Using zoo exhibits and landmarks such as animal statues as photograph backdrops (as depicted in Figure 9.3) clearly places those depicted in the picture within the zoo

Figure 9.3 Placing the child, placing the family at the zoo

Figure 9.4 Depicting the happy family

landscape. Such images cannot be confused with any other setting, as the animals, animal statues or signage figure prominently, unmistakably marking the place of family leisure and reflecting a certain determination to portray parental and family status. Group images (Figure 9.4) reveal and celebrate the relationships and ties between family members, recording not only who was there but also implying – through body contact and smiling faces – that ties are cohesive and emotionally close. The prominent use of zoo entrances reflects not only a very easy way to place the group in the zoo location which, similar to zoo statues, are often unique to a specific zoo, but also speaks to their symbolic meaning. As portals between the outside world and the special 'zoo world', entrances mark the transition from their daily routines to a place and 'time out of time' (Gillis, 1996) where families can immerse themselves in the construction of positive family life associated with the zoo environment.

Photographs such as Figure 9.3 and Figure 9.4 are taken to memorialize being at the zoo, whether the trip was specifically for the benefit of children, or more broadly to be together as a family engaged in an activity that is identified as 'properly' familial in the dominant culture. The very act of creating the photograph betrays the value placed on the leisure activity and the zoo location in which its subjects are participants. The documenting of the event in the photograph becomes central to the construction of continuity between the family ideal and the family-friendly place depicted, as the photograph is a tangible record of the event, recording the (created) history of individual families.

Figure 9.5 Capturing purposive family leisure at the zoo

Family photographs depicting parents and/or adult family members interacting with children at the zoo immortalize moments of shared experience and genuine interaction and communication between child and adult. Figure 9.5 demonstrates this type of zoo family snapshot 'in progress'. It shows a father encouraging his daughter to pet a goat in a contact area within a larger children's zoo. Such opportunities for interaction between parent, child and animal are routinely identified as one of the most sought-after zoo experiences by parents (Turley, 2001; Tomas et al., 2003). However, this image is included here because it also shows the grandmother in the act of photographing this 'family moment' between father and child, and thus capturing it to relive and enjoy. Such photographs become a physical record of the meaning sought in the activities enacted in the zoo space, of the very emotional connections and family ties that make such occasions of family leisure important purposive acts in the emotional lives of parents and their families. Such images become proof of, however fragile or elusive, a moment of congruence between ideals of family life, and the reality of the family pictured. For those parents and families who identify with the dominant discourse of family life and family forms, or who seek to, this evidence of 'fit' between the ideal and actual can help to actively create a sense of subjective wellbeing, making the zoo a therapeutic landscape for them.

Constructing Wellbeing: Families and Zoos

The acts of going to the zoo for family recreation, and then capturing the 'high points' of the outing in family photos, reflect an active construction by parents of young children of a sense of family that is intimately associated with the dominant discourse on family life and structure. It is argued that by working at, and demonstrating, some type of continuity or fit between self-identity (as a parent and as a family) and place-identity (the zoo as family-friendly), parents actively construct their own sense of family wellbeing, at least for those who ascribe to, or aspire to, a traditional nuclear family ideal, and value this as constituting a happy and meaningful family life. Where this purposive work of building family generates a sense of satisfaction with family life, the public places in which family social life is enacted may function as therapeutic landscapes. It is the promise of ideal family life, repeatedly identified with and (re)created through places like zoos, which motivates family leisure involvement to such an extent that it becomes a normative part of the dominant discourse of family life in Western societies. This is nothing less than the construction of personal identity (as parent, as family) in the world through the recognition and/or adoption of place identity. In short, those who really aspire to be a 'proper family' as normatively defined, and (want to) feel good about it, are those who are in large numbers taking the children to the zoo, and making sure to take a few photographs with the zebras while they are there to mark their 'family-friendly' moment in time and place.

References

Aitken, S.C. (1998), *Family Fantasies and Community Space* (New Jersey and London: Rutgers University Press).

The American Zoo and Aquarium Association <www.aza.org/> accessed 1 July 2006.

Baicker-McKee, C. (2003), *Fussbusters on the Go: Strategies and Games for Stress-free Outings, Errands, and Vacations with Your Preschooler* (Atlanta: Peachtree Publishers).

Bourdieu, P. (1990), *Photography: A Middle Brow Art*, translated by S. Whiteside (Stanford, CA: Stanford University Press).

Chambers, D. (2003), 'Family as Place: Family Photograph Albums and the Domestication of Public and Private Space', in J. Schwartz and J. Ryan (eds) *Picturing Place: Photography and the Geographical Imagination* (London and New York: I.B. Tauris) pp. 96–114.

City of Winnipeg Community Services Department – Strategic Information Systems Services Division, and Assiniboine Park Enterprises (2004), *Assiniboine Park Zoo Survey Analysis, 2003 and 2004.*

Cuba, L. and Hummon, D. (1993), 'A place to call home: identification with dwelling, community and region', *The Sociological Quarterly* 34: 111–31.

Daly, K.J. (1996), *Families and Time: Keeping Pace in a Hurried Culture* (London: Sage).

DeVault, M.L. (2000), 'Producing family time: practices of leisure activity beyond the home', *Qualitative Sociology* 23 (4): 485–503.

Ehman, K., Hovermale, K. and Smith, T. (2005), *Homespun Memories for the Heart: More Than 200 Ideas to Make Unforgettable Memories* (Grand Rapids MI: Revell).

Ellis, G. (1999), *The Big Book of Family Fun* (Grand Rapids, MI: Revell).

Evans, R.G. and Stoddart, G.L. (1990), 'Producing health, consuming health care', *Social Science and Medicine* 31 (12): 1347–63.

Gesler, W. and Kearns, R. (2002), *Culture/Place/Health* (London and New York: Routledge).

Gillis, J. (1996), 'Making time for family: the invention of family time(s) and the reinvention of family history', *Journal of Family History* 21 (1): 4–21.

Hagerstrand, T. (1982), 'Diorama, path and project', *Tijdschrift voor Economische en Sociale Geographie* 73: 323–39.

Haldrup, M. and Larsen, J. (2003), 'The family gaze', *Tourist Studies* 3 (1): 23–46.

Hallman, B.C. and Benbow M. (2006), 'Naturally cultural: the zoo as cultural landscape', *The Canadian Geographer* 50 (2): 256–64.

Hallman, B.C. and Benbow, M. (in press), 'Family leisure, family photography and zoos: exploring the emotional geographies of families', *Social and Cultural Geography.*

Hampp, A. (2006), 'Animal attraction: advertising at the zoo', *Advertising Age,* 29 October 2006. <http://adage.com/print?article_id=112828> accessed 30 October 2006.

Haworth, J. (2003), 'Editorial: Leisure and Wellbeing', *Leisure Studies* 22: 317–22.

Howard, D.R. and Madrigal, R. (1990), 'Who makes the decision, the parent or the child? The perceived influence of parents and children on the purchase of recreational services', *Journal of Leisure Research* 22 (3): 244–58.

Kay, T. (2006), 'Where's Dad? Fatherhood in leisure studies', *Leisure Studies* 25 (2): 133–52.

Kelly, J. R. (1983), *Leisure Identities and Interactions* (London: Allen & Unwin).

Kirkman, R. and Scott, J. (2004), *Playdate: Category 5 – A Baby Blues Scrapbook* (Riverside, NJ: Andrews McMeel Publishing).

Larsen, J. (2005), 'Families seen sightseeing: performativity of tourist photography', *Space and Culture* 8 (4): 416–34.

Maher, J. (2005), 'A mother by trade: women reflecting mothering as activity not identity', *Australian Feminist Studies* 20 (46): 17–29.

Morey & Associates (2004), *National Zoo and Aquarium Attitude and Usage Survey Report*, conducted for the American Zoo and Aquarium Association (AZA).

Mullan, B. and Marvin, G. (1999), *Zoo Culture. The Book about Watching People watch animals*, 2nd ed (Urbana, IL: University of Illinois Press).

Partow, C. and Partow, D. (1996), *Families That Play Together Stay Together: Fun and Healthy TV-Free Ideas,* (Minneapolis MN: Bethany House Publishers).

Pickle, R. (2001), *Gay Day at the San Diego Zoo,* 9 February 2001 <http:www.everything2.com/index.pl?node_id=94510> accessed 9 November 2006.

Pred, A. (1984), 'Place as historically contingent process: structuration and the time-geography of becoming places', *Annals of the Association of American Geographers* 74: 279–97.

Relph, E. (1976), *Place and Placelessness* (London: Pion Limited).

Rose, G. (2001), *Visual Methodologies: An Introduction to the Interpretation of Visual Materials* (London: Sage).

Rose, G. (2003), 'Family photographs and domestic spacings: a case study', *Transactions of the Institute of British Geographers* 28 (1): 5–18.

Samuel, N. (1996), *Women, Leisure and the Family in Contemporary Society: A Multinational Perspective* (Wallingford UK: CAB International).

Shaw, S. (1992), 'Dereifying family leisure: An examination of women's and men's everyday experiences and perceptions of family time', *Leisure Sciences* 14: 271–86.

Shaw, S.M. and Dawson, D. (2001), 'Purposive leisure: examining parental discourses on family activities', *Leisure Sciences* 23: 217–31.

Sontag, S. (1977), *On Photography* (London: Penguin).

Spence, J. and Holland, P. (eds) (1991), *Family Snaps: The Meanings of Domestic Photography* (London: Virago).

Such, E. (2006), 'Leisure and fatherhood in dual-earner families', *Leisure Studies* 25 (2): 185–99.

Tomas, S., Crompton, J. and Scott, D. (2003), 'Assessing service quality and benefits sought among zoological park visitors', *Journal of Park and Recreation Administration* 21 (2): 105–24.

Turley, S.K. (2001), 'Children and the demand for recreational experiences: the case of zoos', *Leisure Studies* 20: 1–18.

Williams, A. (1999a), 'Place identity and therapeutic landscapes: the case of home care workers in a medically under serviced area', in A. Williams (ed.) *Therapeutic Landscapes: The Dynamic Between Place and Wellness* (New York: University Press of America) pp. 71–96.

Williams, A. (1999b), 'Introduction', in A. Williams (ed.) *Therapeutic Landscapes: The Dynamic Between Place and Wellness.* (Lanham, MD: University Press of America) pp. 1–11.

Zabriskie, R. and McCormick, B. (2003), 'Parent and child perspectives of family leisure involvement and satisfaction with family life', *Journal of Leisure Research* 35 (2): 163–89.

Zoological Society of London, *Gay Sunday at the Zoo*, 17 September 2006 <http://www.zsl.org/print/London-zoo/whats-on/gay-sunday-at-the-zoo,138,EV.html> accessed 9 November 2006.

PART 3
Applications in Health Care Sites

Chapter 10

Application of Concepts of Therapeutic Landscapes to the Design of Hospitals in the UK: The Example of a Mental Health Facility in London

Wil Gesler and Sarah Curtis

Introduction

Inpatient facilities for the National Health Service (NHS) in the UK have recently been updated by a wave of hospital building. Much of this new building has been funded under the Private Finance Initiative (PFI) scheme. This is a public/private partnership arrangement, under which a private contractor builds a hospital to NHS specifications and, when it is constructed, the NHS leases the facility, usually including the associated non-clinical services (catering, cleaning, and so on), from the private provider, as a site for provision of NHS-funded and -managed clinical care. This method of 'levering' private funds for infrastructure renewal has allowed the NHS to update its stock of medical facilities and has provoked considerable new debate about how best to design hospitals to meet current and future needs of the UK population. It is generally recognized that good designs contribute to better patient outcomes and improve the well-being of patients and staff. Therefore a significant effort has gone into generating a research evidence base to guide plans for new hospital design (Canter and Canter, 1979; Ulrich, 1984; CABE, 2000; Lawson and Phiri, 2003; Francis et al., 1999). The NHS in the UK has provided, and is continually improving, design evaluation tools for assessment of hospital design (NHS Estates, 2001; 2006). Especially in the psychiatric sector, this thinking is complemented by historical research that has examined the design of mental health facilities in the past and its effect on users and staff (Scull, 1979; Philo, 1989; 2004; Edginton, 1997).

The authors of this chapter are engaged in research to evaluate the hospital design process in the UK, including summarizing the various design goals of different groups of stakeholders (Gesler et al., 2003) and a case study of an acute care mental health unit (Curtis et al., 2007). This chapter discusses some ways in which our work contributes to interpretation of, and potential improvement in, hospital design, especially for mental health care. For purposes of this project, ideas about therapeutic landscapes have been derived from three theoretical strands: (1) *Cultural ecology*, (2) *structuralism and poststructuralism*, and (3) *humanism*. For a detailed explanation and discussion of these, see previous publications by Gesler (1992; 2003). The strands have been operationalized and respectively referred to in brief as (1) physical, (2) social, and (3) symbolic environments.

Findings from our Preliminary Research

In a recently published paper (Curtis et al., 2007) we report a pilot study which collected information through discussions with small groups of ex-users, nursing staff and psychiatric consultants who had experience of a new psychiatric inpatient facility in London, UK, which had been built under the PFI scheme.

Full details of the research method are provided in Curtis et al. (2007). We recruited volunteers to participate in the study. Participants took part in discussions with the researchers either individually or in groups. Each group comprised either psychiatric patients (who were currently well and were recruited through a user group, independently of the NHS), or nurses and managers at the new hospital, or consultants working there. Selected quotes from the discussions are included below and we have used notation which distinguishes between comments by users, staff (and managers) and consultants. We posed two simple questions: (1) *What specific features of [the hospital] (in terms of physical layout, activities, and so on) do you think are good for the well-being of users and staff? (2) What specific features of [the hospital] do you think are not good for the well-being of patients and staff?* Then we let the discussion flow as freely as possible. Discussions were tape-recorded and transcribed. Both principal investigators read through the transcripts several times to elicit major themes and then met to discuss their findings. We were able to interpret the discussion in terms of the three therapeutic landscapes environments mentioned above. The study contributed to theories of therapeutic landscapes by showing how these ideas might apply to an acute inpatient facility of this type and how our informants understood and interpreted these different dimensions of the setting in which care was being provided.

Our aims in interpreting the transcripts were: (a) to ascertain which specific hospital designs informants perceived to be significant in response to our two initial questions; (b) to elicit their views on how and why these features of hospital environments were important. To achieve this we employed attributional coding (Sylvester, 1998) with the aim of understanding the informants' reasoning about *how* and *why* certain hospital features were important for well-being. Attributional coding aims to interpret discourses in terms of what is said about the '*agent*' (a person, group or entity seen to *produce* an outcome) and how this impacts on the '*target*' (a person, group or entity mentioned in the *outcome*). In this case we were especially interested in what our informants had to say about physical, social or symbolic aspects of the hospital environment producing a therapeutic or non-therapeutic effect for patients, staff or consultants. Below we also relate our findings to concepts and theories found in the mental health and hospital design literature that we had previously reviewed.

Physical Environments

Therapeutic landscapes theory postulates that certain physical attributes are perceived as important for human health and well-being, and for healing. Consistent with this idea, study participants commented favourably on some aspects of the hospital design, including the amount of light the large windows let in, the bright colour scheme and curtain fabrics, the curving walls and simple way-finding in the building. More negatively, there were criticisms of the poor quality of the furniture and food. Participants also felt that maintenance was sometimes poor and commented on structural problems with the building, such as faulty window closures, which meant that some windows had to be kept permanently closed, and the lack of air-conditioning in hot weather. These findings are supported by more than three decades of research and hundreds of studies that relate specific features such as amounts of light and colour schemes to recovery rates and patient and staff well-being. The basic idea behind this work is that what people experience from their material surroundings affects their moods, emotions and behaviour (Bagley, 1974; Holahan, 1979; Reizenstein, 1982; Spencer and Blades, 1986; Hutton and Richardson, 1995).

Taken as a whole, the discussions we held emphasized that the *combination* of these physical elements was crucial for good design, rather than single elements in isolation. For example, the quotes below show that ideas about providing a well-lit, sunny environment for patients were seen as theoretically beneficial for patients, and that this influenced the hospital design and use of the space. However, as the following comments also show, this needed to be more effectively complemented with measures to control other physical aspects of the environment, such as ambient temperature.

> Group facilitator: What's the other positive thing?
>
> User: Just the design of the building. It was on a bigger scale, very light, very airy.
>
> User: … the time when I was there it was very hot and I was … asking for permission to open the window and he [member of care staff] goes 'Oh, you can only open it a [a little way].' And I said 'that's no good, I'll have to sit on the sill … and wait for the wind to blow!' And he says, 'That's it; take it or leave it.'
>
> Consultant: … actually when we moved in, I think the patients got the south side because people felt it would be good for patients because [of] the bright sunlight … But now, seeing what happens in summer, the place gets unbearably hot on the south side, the place remains remarkably cool on the north side. … The patients are really in trouble – difficulty sleeping at night. It gets too hot.
>
> Member of staff: We had great debates about air conditioning in the design process and some very dodgy arguments were made about how we couldn't incorporate it. … I don't understand how we don't have it, in the 21st century.

We also noted that *activity* within spaces was constantly emphasized, as well as just the visual appearance of the hospital space. The small garden around the hospital was appreciated, but there was a call for more garden space and a greenhouse. The quotes below suggest that our informants judged these to be important not just as visual features but as environments to be physically used, providing therapeutic physical activity and engagement.

> Consultant: … this probably was quite a barren piece of land – so they have planted a whole lot of trees, but I do find it important to have a whole lot more greenery. What you just have, if a patient came out here, is a car park to walk on. I think you can walk behind the back [but] … it's not a park, its not a lovely garden space that you can go and use, its just all quite barren and not inviting and I think that's part of a need when you in actual fact allow people to go out for walks and that they actually have a place to do that in.

> User: I think they could actually use a greenhouse on the hospital grounds, most probably at the back of it, a room where people could chill out, get rid of their anxieties. I think that would be best, like. … On the hospital premises, but not inside the main building … You would have plants in it, … people planting plants, and back of that a place to work out [get physical exercise]. On the weekend. Use their time.

Interest was also shown in the location of and access to the unit. Several felt that there were good opportunities to go off-site to use services and amenities in the local urban area and that visits from relatives and friends were facilitated. These findings about the importance of physical location and accessibility are consistent with a copious literature, especially within the field of health geography, on the importance of physical access to care, including distances travelled, and transportation availability and use (Aday and Andersen, 1974; Joseph and Phillips, 1984; Ricketts et al., 1994). Mental health literature stresses the re-integration of users into the community as an essential aspect of the treatment process (CABE, 2000; Francis and Glanville, 2001; Weaver, 2001).

However, there was also a feeling that there had been little choice over the site for the new hospital, that transportation links were poor, and that the new site, located overlooking a busy highway nearby, was in some respects unsuitable.

> User: … where else can you put it? There is nowhere else it can go I guess, but it isn't in the right place, it's too near a main road … The suicidal possibilities are obvious, you know. You get someone who loses it who just climbs over the fence and straight into the road. Slopes are never a good one either. The land slopes down like this down towards the main road. But having said that you've maybe got the added advantage of seeing a lot further. And you've got maybe an illusion of space, because you could see right over to Greenwich, so that's not a bad thing.

These comments link to discussion of the symbolic attributes of hospital locations (discussed below). Several of our informants seemed to feel that the hospital had been sited in a relatively unattractive part of the city, close to sewage and waste disposal facilities, and that this reflected the relatively low priority accorded by society to this type of hospital. These points underline the challenges, as well as advantages, of siting hospitals in densely populated urban areas, where land is costly and conflicting land uses compete for available space. These tensions were identified by the designers of nineteenth century asylum buildings, who chose rural locations for psychiatric hospitals; these tensions seem no less important today than in the past. Historically, these tensions arose partly through the social construction of mental illness and social processes seeking to marginalize people with mental illnesses. Today, we can still hear echoes of similar issues arising in debates about psychiatric hospital location.

Social Environments

Our pilot study responses also involved debates on interrelated issues concerning social environments which resonated with ideas from therapeutic landscapes theory concerning the ways that certain settings encourage healthy social interactions. Topics raised by our respondents touched on: how to achieve a balance between surveillance and security; how to reconcile the needs for privacy and for social interaction; and how to promote patient empowerment.

The first issue was the need felt by some staff for security and surveillance, set against the argument expressed by some users in favour of freedom and openness. Discussions centred on specific design features: the location of the nursing station at the hub of radial wards; vision panels in doors to allow staff to observe patients; door handles with a smooth design to prevent ligature (attempting suicide by hanging oneself) but which as a result were also difficult to grip and turn; windows that could not be opened to prevent patients jumping out; and security of access to the building. These points are consistent with previous literature that documents the conflict between the need to control and restrain the mentally ill and the opposing desire to promote human needs and individuality (Moon, 2000; Foucault, 1979; Philo, 1989; Geores and Gesler, 1999; Curtis, 2004).

While some of the medical and nursing staff tended on the whole to favour design features to facilitate surveillance and prevention of self-harm, others seemed to take the view that these were ineffective and undesirable.

> Member of nursing staff: I look at the structure and it is quite good and from the nurses station we can see all the corridors, unlike before [in the old hospital building]. We know where everything is … the way our ward is designed we could easily see what is up in any of the three corridors that we have, which is quite good.

> Consultant: [discussing whether vision panels in bedroom doors were necessary] I really feel that its irrational … it's there supposedly to reduce risk and obviously people want to make sure that the service users are safe, but I don't think there is any evidence to show that actually it does reduce risk, and it seems to me that it's a token to make people feel better, to make the staff feel that somehow or other by every hour peering through a window is somehow helping to reduce risk of self-harm or absconsion, or whatever.

> Consultant: … the door handles. I know it's a sort of small point but … they're a conical shape and extremely slippy and nobody can open them, neither staff, most of the time, except if you have … the strength to open them, and the clients can't get out of the rooms either.

> Group facilitator: It has to do with ligature, I think.

> Consultant: … Maybe it has … but it could actually be quite problematic, because if you're in a room with somebody and they are getting very irate with you and you can't get out and they can't get out either, you know – a recipe for disaster.

A second issue was the balance between the need for opportunities for privacy and withdrawal from social contact, *versus* the therapeutic value of user–staff interactions. Both users and staff felt that it was beneficial to retreat to their bedrooms, office or a quiet place. Expressions of the need for privacy in the hospital building (both for patients and for staff) can be linked to the literature on territoriality (Sack, 1986) and privacy or the desire to be in places where one can relax and not be observed (Philo and Parr, 2003). Psychiatric units may be seen by users as refuges from stressful social contact (Tomlinson et al., 1996; Parr, 1999; Pinfold, 2000). In contrast, social interactions are advocated as beneficial (Main, 1980; Manning, 1989; Moos, 1997) and this was also stressed in the discussion groups we recorded. Some participants said that the new hospital provided good 'shared' space for interactions, but more needed to be done to facilitate these through encouragement of a wider range of activities for leisure as well as formal occupational therapy classes. Several innovative projects were mentioned (for example, a group art project produced a mosaic), but some users felt there was not enough to occupy their minds, especially at weekends when there were fewer staff available to organize activities with the patients. Some of the limitations were due to the restrictions of the physical buildings, but availability and willingness of staff to superintend this type of activity seemed at least as important. This highlights the significance of the links between the functional space of the hospital and the care model and staff resources provided to support it.

> User: I don't know how much complete peace you could get. Whereas in the other place [referring to shared space in the older hospital replaced by the new building] you could get complete peace if you needed it … You need some peace and space on your own.
>
> Consultant: The other asset of the new building is all the separate bedrooms which I think is also being seen as very positive by patients because they generally, in hospitals that are older, are in shared wards and they don't actually enjoy that, they do enjoy more privacy of their own bedroom.
>
> First member of staff: I think it's very important that everybody has access to private space that they can actually say is their own.
>
> Second member of staff: Right! That goes for staff as well as users.
>
> User: Another thing that would have been useful would be to have a very small gym, put somewhere on the hospital grounds for people to use and have a shower afterwards, and go back to the wards, like. And just chill out, like. Cause there were a lot of people even like myself that was going in the wards and hitting their heads against doors and things like that and damaging themselves, and they had nothing to get their frustration out on, like. They had nothing to take it, there's no punch bag, no treadmill to run it off, no cycle to burn it off, like. Things like that would be useful for people to have in hospital. It doesn't need a massive room, just a small room would have been useful.

> User: It's OK … putting bricks and mortar together to make a building, but a building is just a building if there's no facilities in there that's active to make a person better.

The third issue had to do with rights and empowerment. Some users felt their rights to have their views respected were not always observed while they were hospitalized. There were accounts of nurses 'shouting' at patients:

> User: It was nice in there. I can't complain about it. But one thing when I was complaining about, she [one of the nurses] shout behind me, for no reason. … I might get a letter from them, the doctor complaining about it [i.e. making a complaint on the patient's behalf].

A member of staff also expressed concern that not all patients who had been hospitalized against their will would understand their rights of appeal:

> Member of staff: … they have a right to appeal against their section … I mean knowing how unwell they are, it's important that they are made aware [of this]. Particularly on the secure unit, they do come in very unwell and they think they are in prison. They think there is nothing they can do about it, and I think it's sad and … it's good that they're made aware of their rights. Cos often they come in and they think it's the end of the world.

However, some staff felt that they were disempowered relative to patients:

> Member of staff: Users seem to have more rights than staff: when there is a conflict between a user and staff member, staff are interrogated and made to feel vulnerable. Users can be aggressive.

Also, although some users and staff had been consulted in the hospital design process, they felt some of their recommendations had been ignored and some would have liked more ongoing involvement in decisions about use of the space and operation of the hospital buildings. Concerns over rights and empowerment stem from power relations operating within particular institutions like hospitals and in wider society. Foucault (1991) provides an analysis of hospitals as spaces of medical power where patients are subordinated to staff as well as society in general (see also Philo, 1989). This type of perspective is linked to the idea of 'bio-power' which refers to a form of power that regulates individual and social life, in which life itself (represented in the human body) has become the object of power (Foucault, 1991; Gastaldo, 1997).

Patient empowerment in decisions about hospital design may not always be easy to achieve through conventional consultative mechanisms. One very articulate informant, who had been a mental health service patient and had participated in the discussions about the hospital design, commented on the issue of how to consult patients who may have difficulty expressing their views.

> User: … it's good that I was eloquent enough, because my illness was about mood swings, kinda like that doesn't interfere too much with perception, it's more to do with up and down, quite simple really. But you know, for people who can't talk, they're not going to be able to get to these levels where, of understanding, or perhaps their brain just doesn't work that way.

Symbolic Environments

Three issues encapsulated respondent perceptions of landscapes which were consonant with theories about the symbolic meanings attached to aspects of landscape that are perceived to be significant for health and well-being. First was 'homeliness'. Several users and some staff commented on the homelike atmosphere of the unit, enhanced, for example, by the furnishings and the fact that people could bring some small personal belongings with them. The mental health literature includes the desirability of creating a 'homelike' atmosphere (Philo, 1989; Edginton, 1997), referring at times to specific features such as good lighting and soft furnishings (Waller and Finn, 2004). People who often feel unaccepted in public places or have insecure or uncomfortable housing may benefit from feeling 'at home' in hospital (Smith, 2000; Dyck et al., 2005). On the other hand, some staff warned that creating too homely a feeling would make the transition back into the community more difficult. In our study some participants found the hospital felt impersonal in spite of attempts to create a homely atmosphere. Some of the freedoms experienced at home were thought to be lost in the more institutional, hospital setting, for example, in terms of self-catering:

> Consultant: The staff do sometimes make breakfasts with the clients at the weekend. I suppose it would be nice if we had a larger kitchen and that the clients ... who wanted to could prepare their own meals, not all the time but when they wanted to and certainly they would have more choice and better quality of food.

A second issue was cultural identity or cultural sensitivity, expressed in terms of gender, religion, language, eating habits and health beliefs. Concern was expressed for the vulnerability of women to sexual harassment in mixed-gender wards and lack of a communal area reserved for women. Some accommodation was provided for people to practise their religion in places set aside for worship and to receive spiritual advice from visiting faith leaders, but participants discussing these points also felt that this type of provision was not adequate to meet all of a patients' needs (see Curtis et al., 2007). The large variety of languages spoken by users presented communication problems. Several users felt the food served showed a lack of awareness of cultural preferences. There is a strong tradition within health geography that recognizes the importance of cultural differences in health situations (Gesler and Kearns, 2002; Kearns, 1995). Differences in social and cultural background may lead to different perceptions of whether specific design features are therapeutic or not (Gillespie, 2002), and therefore must be accommodated by planners of health facilities (Parr and Butler, 1999).

The cultural diversity at the hospital is evident among the staff as well as the patients. One user suggested that some staff who had come to work in the UK from other countries might need more training, both in terms of language skills and in terms of their manner of speaking, in order that they be culturally acceptable to patients. Several users complained that staff had 'shouted' at them, which they found unacceptable and this user speculated that cultural differences might contribute to this problem:

> User: Some of the staff may have language problems … It may not be their fault, but … [t]he staff have been recruited from elsewhere, they come here and they don't get training in language skills … in some cultures … you can shout without meaning it … Diversity is something that has to be looked at carefully.

One of the consultants discussed the question of how to achieve an acceptable 'mix' of different cultural influences in a building like a hospital.

> Consultant: It does make it interesting, it makes the whole idea of how are you going to build something that would sort of appeal to everybody. Because I can immediately think of the Islam and their, sort of architecture – there's a lot of colour and a lot of mosaic and there's all kinds of more pretty design and whatever. So how do you sort of get all the cultures together in one building and that everybody is happy? That's pretty difficult actually.

Another symbolic dimension of the new hospital that was raised by users and some staff was the way it reflected the level of respect accorded to people with mental illness. Some staff commented that they were pleased that a new facility was being provided because it showed recognition that people with mental illness deserved this investment in improved infrastructure.

Users thought that having their own room showed respect for individual privacy and they appreciated when members of staff were friendly and considerate. On the other hand, they complained that the symbolism of the new hospital could also seem derogatory because it was located on a site that had been previously used for waste disposal. One user commented on the 'carceral' appearance of the security fencing around the hospital and felt that this should be made to appear less imposing with planting. Again we see in this comment a reference to the perceived therapeutic properties of plants as natural features, which could compensate for the degrading, prison-like aspect of the fencing. The concern being expressed was particularly for the visual impression greeting *visitors* as much as patients in the hospital, suggesting that this user was concerned about how the hospital communicated ideas about mental illness to the wider community.

> User: As you look to your right, you see this huge, green frame … Any design of future hospitals, they should not have a thing like that looking you in the face as you walk in as a visitor … That whole fence could be covered with some form of, not ivy, that stuff that spreads everywhere, but a clematis type of plant.

Others complained that they were bullied and treated as a danger to society. Concerns about respect, or lack of respect, relate to a large literature on the stigma attached to the mentally ill, who are often labelled as deviant or the 'other' and are therefore excluded or marginalized (Foucault, 1965; Smith and Giggs, 1988). Social construction and public perceptions of mental illness continue to mistake madness for badness (Moon, 2000) and keep the mentally ill at arm's length (Smith and Hanham, 1981; Dear and Wolch, 1987). These attitudes mean that the symbolic attributes of new psychiatric hospitals have particular significance and need very careful consideration.

General Issues Arising from the Pilot Study

Links between Theory, Research Evidence, and User and Staff Perceptions

The people that we spoke to in this pilot study (in all three respondent groups) certainly seemed to have their own theories about how the hospital environment related to the well-being of the patients and whether or not it helped to realise their objectives for a model of inpatient care of people with serious mental illness. We were interested to find that these ideas often accorded with well-established theoretical and empirical findings from research concerning how physical environments, social settings and symbolic attributes of places might be important for well-being. We also noted that our respondents made almost no direct reference to evidence from theoretical or empirical research that might support their views or explain their experiences. Perhaps this is because messages from research have been distilled and transmitted in 'non-academic' language before being 'absorbed' into everyday public knowledge. Alternatively, actors in particular hospital settings may form their own theories from experience about the key relationships, which are independent of scientific evidence.

The 'evidence layer' in the revised 'AEDET evolution tool' developed for NHS Estates is a helpful innovation for those needing evidence about how particular attributes of environment relate to clinical outcomes and staff and patient experiences (NHS Estates, 2006). However, it does not offer a coherent theory of how the different elements of the hospital environment work together. We would argue that the three-dimensional framework offered by therapeutic landscapes theory in geography might be an example of a more comprehensive conceptual model, fitting well with experiential knowledge, on which participants might base local work to develop therapeutic hospital environments that are beneficial to well-being.

This seems all the more important since there were elements of the design of the hospital considered in our pilot study that seemed to have been unsuccessful because planning and implementation of the hospital building project had not recognized the *combined* significance of a *range* of elements that operate together in a particular setting. For example, we described above how light levels seemed to receive a lot of attention, but a comfortable ambient temperature in hot weather was not assured in the new building. Similarly some security features, such as door handles designed to prevent self-harm, produced other, unanticipated security risks. Very importantly, to complement good functional and aesthetic physical design, it seemed crucial to create therapeutic *social relations* in the new hospital space, especially through the behaviours of staff towards patients and through the care provided, and to consider the *symbolic* features of the hospital design.

This all suggests that it is actually very challenging to design a hospital under the rules of the PFI initiative which require that contractual agreements about design are drawn up in advance of construction and provision of the hospital, which will be difficult to modify later. This demands quite sophisticated information to be able to anticipate *a priori* how all the physical, social and symbolic elements of a hospital will work together once it is built and in use. Thus there seems to be a case for more post-construction studies of hospitals to improve our comprehensive understanding of the hospital as a therapeutic setting in a holistic sense. Research of this sort would enable us to learn from the 'civic intelligence' (Elliot et al., 2004) that hospital staff and patients can contribute about successes and failures of the new hospitals constructed in the UK.

The Importance of Qualitative, Non-clinical Studies

The effects of hospital design on health and well-being have been quite comprehensively researched over the last three or four decades, often using quasi-clinical trials of the effects on clinical outcomes such aspects of physical environments as light, temperature, noise, cleanliness and views of nature, as well as the extent to which these can be controlled and varied by patients or hospital staff. Clinical outcomes are often measured in terms of factors such as lengths of hospital stays and amounts of analgesic medication taken by patients. Such research results have impressed health care planners and architects and have provided important evidence leading to specific design changes such as providing larger windows in hospital wards.

Evidence from our pilot study supports work by other authors (including geographical research by authors such as Parr (1999) and Parr and Butler (1999)) indicating that other forms of empirical research also have an important complementary role to play in informing decisions about hospital design. These include work using qualitative methods, such as participant observation, semi-structured interviews, or, as in our case, free-flowing in-depth discussions with patients and staff. Qualitative studies are often suggestive rather than (appearing to be) definitive, but because they represent the experiences, feelings and thoughts of those directly involved in health care situations, they can aid in understanding the pathways between specific design features and better health. As argued above, these strategies may also allow consideration of the complex interrelationships between different specific aspects of hospital design.

The advantages of qualitative and participative research to explore the issues raised in this chapter will be well known to many readers. However, we are unsure how far consultation processes generally used to plan hospital design make use of these types of techniques. Although there is a great deal of rhetoric in the NHS about patient and staff input into decision-making about such things as hospital design, we were told several times by participants that this seldom happened in practice. While 'experts' such as architects, academic researchers and NHS planners certainly have a vital role to play in determining hospital design, they do not experience hospital care on a day-to-day basis as our participants do. Decision-making about hospital design raises some fundamental questions about who should be able to influence the process and how 'public science' works to inform discussions about risks and opportunities of new developments for human well-being (Beck, 1999). We suggest that there are some important challenges involved in trying to allow all the stakeholders to participate fully in discussion.

Conclusion

We feel that the application of therapeutic landscapes to the study of the effects of hospital design on patient and staff health and well-being has proved very fruitful this far in our research programme. Our work has the potential to contribute to discussion about hospital design research that moves beyond discussion of specific features of physical environments and clinical goals. We have been able to link research findings from a case study to theoretical models that invoke a more balanced consideration of social and symbolic as well as physical aspects of environment, and view these in the context of wider social relations as well as within particular hospital settings. Our work also suggests causal pathways between specific design features, and patient feelings and behaviour, and demonstrates some implications for the relationships between hospital design and psychiatric models of care. Our findings underline the importance of considering hospital design holistically (in terms of the interactions between different elements of design), as well as concentrating on specific design features. These ideas need to be explored in a wider range of settings in order to determine how far our findings in this pilot project are context-dependent. This calls for similar studies in mental health units in a variety of socio-geographical contexts within the UK and elsewhere.

References

Aday, L. and Andersen, R. (1974), 'A Framework for the Study of Access to Medical Care', *Health Services Research* 9 (3): 208–20.

Bagley, C. (1974), 'The Built Environment as an Influence on Personality and Social Behavior: A Spatial Study', in D. Canter and T. Lee (eds) *Psychology and the Built Environment* (London: Architectural Press) pp. 156–62.

Beck, U. (1999), *World Risk Society* (Cambridge, UK: Polity Press).

CABE (2000), *Design Quality in PFI Projects: HM Treasury Guidance Note No. 7* (with HM Treasury). Available from: The Office of Government Commerce, Rosebery Court, St Andrews Business Park, Norwich, Norfolk, NR7 0HS.

Canter, S. and Canter, D. (1979), 'Building for Therapy', in D. Canter and S. Canter (eds) *Designing for Therapeutic Environments* (New York: Wiley).

Curtis, S.E. (2004), *Health and Inequality: Geographical Perspectives* (London: Sage).

Curtis, S.E., Gesler, W., Fabian, K., Francis, S. and Priebe, S. (2007), 'Therapeutic Landscapes in Hospital Design: A Qualitative Assessment by Staff and Service Users of the Design of a New Mental Hospital Inpatient Unit', *Environment and Planning A* (in press).

Dear, M. and Wolch, J.R. (1987), *Landscapes of Despair: From Deinstitutionalization to Homelessness* (Princeton, NJ: Princeton University Press).

Dyck, I., Kontos, P., Angus, J. and McKeever, P. (2005), 'The Home as a Site for Long-term Care: Meanings and Management of Bodies and Spaces', *Health and Place* 11: 173–85.

Edginton, B. (1997), 'Moral Architecture: The Influence of the York Retreat on Asylum Design', *Health and Place* 3: 91–9.

Elliot, E., Williams, G. and Rolfe, B. (2004), 'The role of lay knowledge in HIA', in J. Kemm, J. Parry and S. Palmer (eds) *Health Impact Assessment* (Oxford: Oxford University Press) pp. 81–90.

Foucault, M. (1965), *Madness and Civilization: A History of Madness in the Age of Reason* (New York: Random House).

Foucault, M. (1979), *Discipline and Punish: The Birth of the Prison* (London: Penguin).

Foucault, M. (1991), *The Birth of the Clinic* (London: Routledge).

Francis, S. and Glanville, R. (2001), *Building a 2020 Vision: Future Health Care Environments* (London: Nuffield Trust).

Francis, S., Glanville, R., Noble, A. and Scher, P. (1999), *50 years of Ideas in Health-Care Buildings* (London: Nuffield Trust).

Gastaldo, D. (1997), 'Is Health Education Good for You? Re-thinking Health Education Through the Concept of Bio-power', in A. Petersen and R. Bunton (eds). *Foucault, Health and Medicine* (London: Routledge) pp. 113–33.

Geores, M. and Gesler, W.M. (1999), 'Compromised Space: Contests Over the Provision of Therapeutic Environments for People with Mental Illness', in A. Williams (ed.) *Therapeutic Landscapes: The Dynamic between Place and Wellness* (Lanham, MD: University Press of America) pp. 99–121.

Gesler, W. (1992), 'Therapeutic Landscapes: Medical Issues in Light of the New Cultural Geography', *Social Science and Medicine* 34: 735–46.

Gesler, W. (2003), *Healing Places* (Lanham, MD: Rowman & Littlefield).

Gesler, W.M., Bell, M., Curtis, S., Hubbard, P. and Francis, S. (2003), 'Therapy by Design: Evaluating the UK Hospital Building Program', *Health and Place* 10: 117–28.

Gesler, W. and Kearns, R. (2002), *Culture/Place/Health* (London: Routledge).

Gillespie, R. (2002), 'Architecture and Power: A Family Planning Clinic as a Case Study', *Health and Place* 8: 211–20.

Holahan, C.J. (1979), 'Environmental Psychology in Psychiatric Hospital Settings', in D. Canter and S. Canter (eds) *Designing for Therapeutic Environments* (New York: Wiley).

Hutton, J.D. and Richardson, D. (1995), 'Healthscapes: The Role of Facility and Physical Environment on Consumer Attitudes, Satisfaction, Quality Assessments, and Behaviors', *Health Care Management Review* 20: 48–61.

Joseph, A.E. and Phillips, D.R. (1984), *Accessibility and Utilization: Geographical Perspectives on Health Care Delivery* (New York: Harper & Row).

Kearns, R. A. (1995), 'Medical Geography: Making Space for Difference', *Progress in Human Geography* 19: 249–57.

Lawson, B. and Phiri, M. (2003), *The Architectural Healthcare Environment and its Effects on Patient Health Outcomes*. A Report on an NHS Estates Funded Research Project (London: The Stationery Office).

Main, T. (1980), 'Some Basic Concepts in Therapeutic Community Work', in T. Main, *The Therapeutic Community* (London: Croom Helm).

Manning, N. (1989), *The Therapeutic Community Movement: Charisma and Routinization* (New York: Routledge).

Moon, G. (2000), 'Risk and Protection: The Discourse of Confinement in Contemporary Mental Health Policy', *Health and Place* 6: 239–50.

Moos, R. (1997), 'The Social Climate of Hospital Programs', in R. Moos, *Evaluating Treatment Environments: The Quality of Psychiatric and Substance Abuse* 2nd edn (New Brunswick, NJ: Transaction Publishers) pp. 23–44.

NHS Estates (2001), *Achieving Excellence Design Evaluation Toolkit*. A Toolkit for Evaluating the Design of Healthcare Buildings from Initial Proposals through to Post Project Evaluation: Instructions. NHS Estates: <http://www.nhsestates.gov.uk/patient_environment/index.asp?submenu_ID=chd.publications_guidance>

NHS Estates (2006), *AEDET Evolution Design Evaluation Toolkit*.

Parr, H. (1999), 'Mental Health and the Therapeutic Geographies of the City: Individual and Collective Negotiations', in A. Williams (ed.) *Therapeutic Landscapes: The Dynamic between Place and Wellness* (Lanham, MD: University Press of America) pp. 123–151.

Parr, H. and Butler, R. (1999), 'New Geographies of Illness, Impairment and Disability', in R. Butler and H. Parr (eds) *Mind and Body Spaces*. (London: Routledge).

Philo, C. (1989), 'Enough to Drive one Mad: the Organization of Space in 19th Century Lunatic Asylums', in J. Wolch and M. Dear (eds) *The Power of Geography* (Boston, MA: Unwin Hyman).

Philo, C. (2004), *A Geographical History of Institutional Provision for the Insane from Medieval times to the 1860s in England and Wales: The Space Reserved for Insanity* (Lewiston, UK: the Edwin Mellen Press).

Philo, C. and Parr, H. (2003), *Muddying the Therapeutic Geographies of Mental Health Care: An example of Facilities in the Scottish Highlands*. Paper prepared for the 10th International Symposium in Medical Geography, Manchester.

Pinfold, V. (2000) '"Building up safe havens ... all around the world": users' experience of living in the community with mental health problems', *Health and Place* 6: 201–12.

Reizenstein, J. (1982), 'Hospital Design and Human Behavior: A Review of the Recent Literature', in A. Baum and J. Singer (eds) *Advances in Environmental Psychology* (Hillsdale, NJ: Erlbaum) pp. 137–70.

Ricketts, T., Savitz, L., Gesler, W. and Osborne, D. (1994), *Geographic Methods for Health Services Research* (Lanham, MD: University Press of America).

Sack, R.D. (1986), *Human Territoriality: Its Theory and History* (Cambridge: Cambridge University Press).

Scull, A. (1979), *Museums of Madness: The Social Organization of Insanity in Nineteenth Century England* (London: Allen Lane).

Smith, C. and Giggs, J. (eds) (1988), *Location and Stigma: Contemporary Perspectives on Mental Health and Mental Health Care* (London: Unwin Hyman).

Smith, C. and Hanham, R. (1981), 'Proximity and the Formation of Public Attitudes Towards Mental Illness', *Environment and Planning A* 13: 147–65.

Smith, R.G. (2000), *Windows in Time: the Spatial Experience of Chronic Illness and Disability in Men from Childhood to Adulthood.* Unpublished PhD thesis. (London: Queen Mary, University of London).

Spencer, C. and Blades, M. (1986), 'Pattern and Process: An Essay on the Relationship between Behavioral Geography and Environmental Psychology', *Progress in Human Geography* 10: 230–48.

Sylvester, J. (1998), 'Attributional Coding', in G. Symon and C. Cassell (eds) *Qualitative Methods and Data Analysis in Organizational Research: A Practical Guide* (London: Sage Publications) pp. 228–41.

Tomlinson, D., Carrier, J. and Oerton, J. (1996), 'The Refuge Function of Psychiatric Hospitals', in D. Tomlinson and J. Carrier (eds) *Asylum in the Community* (London: Routledge) pp. 111–34.

Ulrich, R.S. (1984), 'View through a Window May Influence Recovery from Surgery', *Science* 224: 420–21.

Waller, S. and Finn, H. (2004), *Enhancing the Healing Environment: A guide for NHS Trusts* (London: King's Fund).

Weaver, M. (2001), 'Strange Roads Ahead for Public Architecture', *Guardian*, 19 October.

Chapter 11

The Writing's on the Wall: Decoding the Interior Space of the Hospital Waiting Room

Valorie A. Crooks and Joshua Evans

Introduction

Recently, geographers and others have looked 'inward' to document the ways in which health care settings are designed, organized, and decorated (Gesler et al., 2004). Interior designs have significant implications for how people live out their lives 'inside' (see Gillespie, 2002). Spaces designed for the delivery of health services in particular have received a great deal of attention in this regard, specifically in relation to physical layout, selection of paint colours, and organization of furnishings (Arneill and Devlin, 2002), and, more importantly, how these elements combine to produce a therapeutic milieu (see Gesler, 1999). What has received less attention is the way these interior spaces are further scripted through signage. It is our contention that after initial design choices are made, additional 'layers' of meaning are added to the space of care provision; these layers send messages to users, shaping their experiences within them. We argue here that in the case of hospital waiting rooms, signage placed on the walls (for example, health promotion posters, regulatory postings) is an important material artefact that becomes a layer upon the initial canvas of the health care setting. Signs are significant in the way they send messages to patients and others that, in turn, have the potential to shape their use of the space, their actions during doctor–patient interactions, and even their emotional responses to the space of care provision.

In this chapter we focus on how underlying ideologies of patient regulation and responsibility are communicated through signs that are located in hospital waiting rooms. Thus, we draw on Gesler's (1999) suggestion that language (in this case, written language) is an important part of place-making. We need to develop a better understanding of how written language is employed in the delivery of health services and further explore the intersection(s) between language, health and place (see also Gesler and Kearns, 2002). While the design and content of health promotion posters and other types of formal signage are most often the products of wider bureaucratic or public health campaigns (see, for example, Montazeri and Sajadian, 2004), informal regulatory signs meant to inform patient behaviour in the waiting room and/or space of care provision are often not. These informal layers of signage are place-specific and thus reflect social dynamics inherent to one place but not necessarily another.

We weave these two types of postings together to explore this textual layer on the waiting room environment.

Utilizing observations undertaken in twelve waiting room settings in three hospitals in the city of Hamilton, Ontario, we seek to unpack the concept of the 'therapeutic landscape' by examining the literal messages such spaces send to users – this being something that can shape the 'healing affect of landscapes' (Williams, 1999, vii) – at the micro-scale. Our purpose in undertaking this analysis is to suggest that we, as geographers, must be careful in applying the therapeutic landscapes concept at too large a scale to specific places because, in doing so, micro-elements that can potentially send contradictory messages (those not in keeping with the planned therapeutic design or healing elements of the environment) may be overlooked. In addressing this, we respond to recent calls from geographers to think critically about the application of the therapeutic landscapes concept to 'real' sites and at different scales (Andrews, 2004; Conradson, 2003; 2005; Wilton and DeVerteuil, 2006; Wilson, 2003).

We have divided this chapter into five main sections. In the first section we provide a brief overview of therapeutic geographies with specific reference to the design and layout of hospital spaces. We then move to a discussion of the waiting room which provides an important contextual background for the study. After this an overview of the methods and techniques is provided. Following this is the analytic discussion of our findings which is organized according to two explanatory themes. Finally, in our concluding discussion, we summarize the main findings and revisit the stated purpose of our chapter by thinking about the implications for how we apply the therapeutic landscapes concept to health care spaces in particular.

Therapeutic Geographies and the Hospital

Research examining therapeutic environments is a growing area of study within health geography (Parr, 2003; Smyth, 2005). Existing studies have largely focused on the interplay between practices of care, experiences of healing/well-being, and the social construction of place. In doing so, particular attention has been paid to mediating influences such as human–environment relations, social structure, and cultural values and beliefs (Gesler and Kearns, 2002). Following a recent review by Smyth (2005), therapeutic geographies can be conceptualized in three ways, as: (1) places with reputations for healing; (2) healing spaces within the health care system; and (3) networks of care support in the community. Geographers have contributed to understandings of therapeutic geographies through studies of natural environments such as mountain environments in the state of Alaska, USA (Palka, 1999); spaces of care provision such as hospital wards (Kearns and Barnett, 1997; 1999; 2000; Gesler et al. 2004) and family practice clinics (Gillespie, 2002); and informal care networks supporting the elderly in the home (Milligan et al., 2004) and others in the community (Parr, 1999; Wiles and Rosenberg, 2001).

The 'therapeutic landscape' concept is a common thread which connects many of the disparate studies on therapeutic geographies. This concept serves as a framework for understanding how certain environments contribute to the development and maintenance of health and well-being (Gesler and Kearns, 2002; Kearns and

Gesler, 1998; Williams, 1999). The therapeutic landscape framework emphasizes holistic definitions of health that move beyond biomedical understandings of bodily processes to include physical, mental and spiritual healing (Williams, 1999). The novelty of the concept is derived from its connection to landscape theory developed within cultural geography. Geographers seeking to shift the study of health from a focus on disease mapping to a more humanistic understanding of well-being have found the reformulated landscape concept as a useful heuristic for thinking about health–place linkages.

Spaces within the health care system which are designed to promote healing represent an important but somewhat understudied therapeutic environment that geographers have only recently begun to explore (see, for example, Kearns and Barnett, 1997; 1999; 2000). Following Gesler et al. (2004), the health care system can be thought of as a therapeutic landscape consisting of an array of health care settings such as hospitals and clinics. Furthermore, following Kearns et al. (2003, 2303), the hospital can be treated as a 'text in itself, as well as a producer of texts'. Clinical environments have increasingly taken on symbolic value as social forces have reshaped the health care system. In this regard, health geographers have traced the growing influence of 'consumerist' approaches to health care provision on health care settings (Kearns and Barnett, 2000; Kearns et al., 2003; Wiles and Rosenberg, 2001). As consumerist ideologies have proliferated, hospital environments have been reinvented, evolving from bland, impersonal spaces to appealing spaces not unlike sites of consumption like the shopping mall (Sloane, 1994). These place-selling motivations reflect larger ideological shifts in the organization of health care delivery that have reshaped the symbolic character of the health care landscape.

In modern hospital systems much attention is paid to the architectural design and quality of interior spaces and their potential therapeutic value. Most of this literature has been produced outside of geography (within environmental psychology and architecture specifically) and has paid attention to 'ambient effects of cleanliness, spaciousness, lighting, colour, noise, and ventilation on patient recovery times' (Gesler et al., 2004, 119). Modern clinical designs must balance variations in patient needs and experiences with diverse cultural interpretations of elements that make up a therapeutic health care setting (Gesler et al., 2004). These designs, often imbued with power, are informed by different stakeholders including patients, staff, administrators, architects, policy makers and the general public (Gesler et al., 2004) and have the potential to shape interactions within such spaces (Adams and Schlich, 2006). As a result, researchers have begun to critically examine the 'emplacement' of power relations within health care settings such as hospitals and clinics (Poland et al., 2005). Following Foucault's work on space and disciplinary power (1975), researchers have explored how the organization of specific medical spaces and practices undertaken within facilitate the control and regulation of behaviours, including those which are health-related. Gillespie (2002, 218), for example, traced how the organization of space within a family planning clinic functioned to reproduce a series of unequal power relations surrounding reproductive health, stating that: 'the architectural layout … represented a metaphor for individuals' private, intimate sexual lives, opened up to the medical gaze, and disciplinary power over the human body.'

Conceptualizing the Hospital Waiting Room

There is an underlying desire to facilitate patient flow in the design and operation of hospitals and health clinics. Patient flow is, however, affected by a number of factors including the availability of health practitioners and 'gatekeeper' activities such as record keeping (Côté, 1999). As a result, patients and others can spend significant periods of time in waiting rooms. These spaces provide the first impression of clinics (Sutherland, 2002) and communicate to others about the characteristics of health practitioners who practise within them (Arneill and Devlin, 2002). For the most part, such spaces often contain many of the same elements that are present in home living rooms: lighting and accessories (for example, plants, floral arrangements) that assist in setting the mood, furnishings, and a layout that facilitates traffic flow and circulation (MacLean, 2006). These elements also send signals about the quality of care delivered within. For example, Arneill and Devlin (2002) found that waiting rooms which are nicely furnished, light in colour, contain artwork, and are warm in temperature bring about perceptions both of comfort and of good quality care. Sutherland (2002) adds that recent magazines, quiet music, play areas for children, and fish tanks can also enhance how the waiting room is read by its temporary occupants. As such, micro-elements of the clinic waiting room such as the colour of the walls, amount of light, or arrangement of furniture can affect how one experiences and interprets this space (Gesler, 1999).

Waiting rooms in clinics can serve multiple purposes in addition to acting as a repository for patients and visitors. Kearns (1991), for example, has shown that spaces within the clinic where patients gather and wait serve as sites of socialization for important community events and activities. Tanner (2002) further suggests that waiting rooms are transitional spaces that are intended to prepare people and their bodies for the medical gaze applied in the practice room. In addition, waiting rooms are increasingly being viewed as ideal health promotion spaces within which information can be transferred to patients (Ward and Hawthorne, 1994). More specifically, health promotion and informational pamphlets and posters have become regular elements of waiting rooms, turning this space into a classroom (McGrath and Tempier, 2003). These materials have been shown to be an effective medium for health education within this space (Craven et al., 2005; Montazeri and Sajadian, 2004). Other types of postings are also commonplace within the waiting room environment, including decorative posters and art (Arneill and Devlin, 2002), regulatory signage such as mission statements, and fee and service listings (Sutherland, 2002). These postings on the walls may act as a reminder of the treatment(s) and care provided within the clinic (Radley and Taylor, 2003a) or even facilitate the discourse of medicine or the clinical gaze through their content (Radley and Taylor, 2003b; Rapport et al., 2006). As such, waiting rooms are complex spaces that can have multiple functions and meanings based on how the microenvironment is read, interpreted, and understood.

In a short commentary piece appearing in the *BMJ* addressed at health practitioners, Sutherland (2002, 1443) posits:

> … I have visited more than 100 general practitioners and have been surprised by the effect the time spent in waiting rooms can have on my mood, ranging from very positive to

> very negative feelings. If someone who is well can be so affected then what impact does a waiting room have on patients' wellbeing? Could the time spent in your waiting room affect your patient consultations?

Here she contends that the waiting room environment has the potential to shape the senses of well-being experienced by those who spend time within this setting. More broadly, Blumberg and Devlin (2006) have found that the design and layout of hospital spaces can play an important role in establishing and/or maintaining well-being. However, the relationship between environmental design and the practice of medicine, including the role(s) that specific health care settings and micro-spaces play in the process of healing, has been under-conceptualized (Arneill and Devlin, 2002). More specifically, there is a lack of knowledge about the roles that postings and signs within the waiting room play in informing experiences within it, despite this being an important material layer of the clinic environment.

The hospital itself is a metaphorical text that we can read in order to understand its ideological underpinnings (Kearns et al., 2003). In the case of the micro-space of the hospital clinic waiting room, the postings and pamphlets that adorn its walls create another literal layer of text for users to interpret their behaviours and health. This interpretation takes place both within and beyond this material environment, and could potentially have a role in shaping the well-being of those who occupy it.

Methods

In the summer of 2006 we undertook formal observations of twelve waiting room spaces in three hospitals located in the city of Hamilton, Ontario, Canada. We first walked around the open/public areas (typically the first floor of the site and exterior) of seven hospitals either together or individually in order to identify differences and commonalities between potential sites, and also to share our common casual observations of waiting room spaces that we had viewed. Upon completing this preliminary work, we developed a synthesis document of our preliminary field notes and began to discuss which hospital sites we wanted to seek formal approval[1] from in order to undertake detailed observations of the waiting rooms within.

It was ultimately determined that we wanted to undertake formal observations of waiting rooms in three hospital sites which all had different design characteristics and foci. One site was a new hospital which exclusively provided cancer care and featured very modern design elements and a large collection of original art pieces. The second site was located closer to the city's downtown area and was established in the late nineteenth century, with its current structure having been built in the 1950s and added to by a large addition undertaken only a few years ago; it provided a range of care in addition to being a regional mental health centre. Our third site was built early in the twentieth century, was in need of renovation, and provided a

1 Because this work did not involve human participants, we were not required to receive ethics approval from our institutions. However, we did have to get formal permission from the hospitals in order to take photos during our observations. This permission came with the agreed upon condition that we would not include any people in our photos.

range of general health services and orthopaedic specialties. We selected these three sites because they offered a range of services and featured elements that spanned a century of hospital design and a combination of older and newer signs, posters/ postings, and artwork.

In order to gain access to the waiting rooms of our three selected hospitals, we contacted their public relations (PR) offices. We requested to undertake formal observations in waiting room spaces which would involve note-taking within them and also taking extensive photos of the postings, posters and other layers on the environment we found to be noteworthy. The two PR offices we dealt with (in that two hospitals shared one PR office) had different responses to our request. In one hospital we were given a tour of the waiting rooms by a PR representative during which notes were taken but photography was not allowed. For the other two hospitals we were asked to hire a PR intern who would supervise the photography while escorting us through the sites, after receiving official permission to undertake data collection from both PR and the site directors. In this instance we were given unrestricted access to all waiting rooms in the two sites for a one-week period during which we undertook our observations and made notes. After visiting these sites and taking notes within them, we then hired an intern for a day so photos could be taken of features that we had identified as important during our observations.

Upon completion of data collection, we had visited a total of twelve waiting rooms within our selected sites. Our extensive field notes were compiled in a word processing document and the accompanying photographs were embedded within. We each reviewed the notes and photographs separately and then came together to discuss emergent themes contained within the dataset, a process known as thematic analysis (see Aronson, 1994). Our discussion was also informed by a literature review of waiting room spaces that we had undertaken prior to data collection. Of course, our own cumulative experiences as patients/others within medical spaces also inevitably shaped how we read these rooms and our discussion of themes. We agreed that two broad explanatory concepts characterized most of what we had observed in terms of layers of meaning within the waiting room space, they were: (1) control and regulation; and (2) marketing the hospital. In the remainder of the chapter we draw on our observations to explain these concepts, their importance to the waiting room space, and how and where they were conveyed by textual layers added onto this fixed micro-space within the hospital.

Analytic Discussion

By collecting data in multiple waiting rooms at three different sites, we had the opportunity to observe many different room styles and designs. Before we delve into a deeper discussion of our analytic findings, we think it important to first share a rudimentary typology of the waiting rooms encountered. Two main distinctions between different waiting room spaces were noted: (1) in layouts; and (2) in intended user(s). With regard to the layouts of waiting rooms, it was observed that while some were fully enclosed, others were open areas in hallways or situated at the end of a corridor as shown in Figure 11.1. Waiting rooms were also designed with different

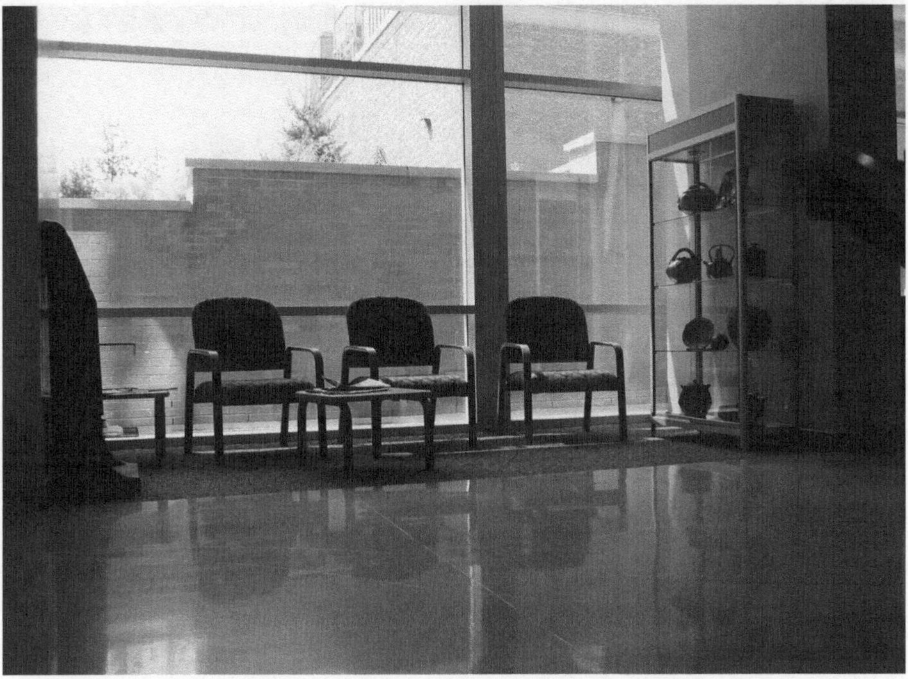

Figure 11.1 The 'waiting area' is a form of hospital waiting room

users in mind. Some rooms served as a waiting space for patients, while others were holding rooms for family members and others. It became apparent to us that the layouts and intended users of waiting rooms were interrelated factors in shaping the design and layout of these spaces. For example, waiting rooms designed for patients were more likely to contain a reception desk and one or more staff persons, have an entrance point off of a hallway, and may or may not have been completely contained. Oppositely, waiting rooms designed for someone other than the patient were never staffed, were less likely to contain clocks, and were more likely to be completely enclosed. Distinctions aside, we found remarkable similarities between waiting rooms, and even across hospital sites, in terms of what was posted on the walls within them and the messages that such postings sent to their occupants. In this section of the chapter we provide an analytic discussion of these similarities, organized by the two explanatory concepts previously introduced.

Sites/Sights of Control and Regulation in the Waiting Room

The entrance and exit (door)ways to and from waiting rooms were frequently used as places to post signage which conveyed regulatory messages to occupants regarding expected behaviours and actions. These areas introduced people to the waiting room

Figure 11.2 An example of informal entrance way signage

and also acceptable practices within it. They served as 'liminal'[2] micro-spaces in that they were a demarcated site of transition between citizen and patient, or person and visitor, and thus were likely viewed as an appropriate place for information transfer. In this way they were literal – both real and in place – regulatory gates. Signage on or beside doors and in other types of entrance ways (such as a bulletin board beside a waiting area) regularly featured posted information regarding expected (and regulated) rules and practices such as: (1) the prohibition of cellular phones and latex items such as balloons; and (2) having clean hands, being scent free, and not smoking. Such regulations were mandated by the hospitals; however, while postings at entrance ways to hospitals and wards were often formal in nature (on plates, printed on glossy posters), at entrance ways to waiting rooms we observed more informal postings as shown in Figure 11.2. To illustrate these formal postings, one found a large brass plate at a main entrance way to a hospital of which read: *Latex Allergy Alert – Out of consideration for the health of our patients and staff with latex allergies, please do not bring latex balloons or other unauthorized products into the hospital*. Similarly, at entrance ways to various wards a formal blue poster read: *LATEX RESTRICTED AREA – No Novelty Latex Balloons No Latex Gloves*. Meanwhile, the entrance ways to waiting rooms did not feature these formal printed and posted signs regarding this particular regulation. If

2 Liminal spaces are sites of transition, transformation and 'in betweenness' that may correspond with changes in identity or even larger passages through life stages (see Teather, 1999 for further discussion).

it was noted at all it was done on an informal sign made by a staff person. At the same time, it was observed that, with few exceptions, informal signs in waiting rooms were more likely to explain why certain regulations were in place while formal ones did not allocate space for such a purpose.

Figure 11.3 The check-in counter is an important site of information dissemination

While entrance and exit ways to waiting rooms conveyed messages regarding rules and regulations, areas around staff persons' work spaces, and specifically check-in counters, enforced controlling messages of access and admittance. More specifically, signage regarding who could and could not have access to the clinic space and practice rooms was posted at this site which, in turn, made it a figurative gate, and the enforcers, namely administrative staff, literal gatekeepers. One such sign read: *Attention – Radiation Treatment Area. Please Do Not Enter Unless Escorted By A Therapist. Thank You.* In this instance, one was made to understand how the space beyond the waiting room was for staff and patients only and not family and/or other caregivers present. While such signs enforced regulations regarding the use and segmentation of specific hospital spaces, signage around the check-in counter sent other messages regarding control and specifically criteria for access to clinic services to be enforced by the staff gatekeepers. Postings regarding fees, bookings and health cards were abundant in this particular area of the waiting room. Signs noting things such as *Check In – Please present your health card* made it very clear, at all clinic waiting rooms visited, that services would not be provided without showing a health card.[3] Informal signs enforcing this regulation were highly visible at check-in counters, in that single counters often featured multiple informal (and sometimes formal) signs enforcing this as a measure of controlling who did and did not ultimately gain access to the hospital clinic and its services. Check-in areas as a whole, such as the one depicted in Figure 11.3, were important and busy sites of information exchange and often housed signs of different types, both formal and informal, which sent a variety of controlling, regulatory and health promotion messages.

Waiting rooms that did not contain staff or check-in counters were, however, not devoid of the controlling postings discussed above. Informal postings regarding visiting hours were visible in such spaces, and specifically at empty 'volunteer' desks. Signs also informed these visitors of their expected roles and how they should present themselves within the clinic space. More specifically, such visitors to the hospital were expected to be the embodiment of a 'healthy individual' as communicated by signs which contained messages such as *Do not visit patients if you have any cold or flu-like symptoms thank you.* With regard to the roles of such visitors, signage indicated that they were expected to perform certain duties despite not being formal staff. One such sign told these individuals to *Please use telephone before visiting patients in the Intensive Care Unit* while another asked that *Families/ Visitor Please answer phone in absence of volunteer*. While telephones were present in many such spaces, their use was controlled by the hospitals and enforced by the messages posted on the walls.

Private Money and Public Health: Marketing Hamilton's Hospitals in the Waiting Room

A second type of formal posting campaign came to our attention while undertaking field work: signs that marketed the hospital's charitable foundation to waiting room occupants. In 1996, five hospitals in the city of Hamilton, two of which were our study sites, were combined to form Hamilton Health Sciences (HHS). This integration served as a response to increasing fiscal pressure on provinces and reflected a desire to protect existing services. A commensurate response was the creation of the 'Cornerstone of Care' campaign, a philanthropic initiative to acquire private donations for investment into HHS. The campaign had a visible presence throughout the city, with prominent billboards reading *Help make something great even greater* at busy intersections, on the exterior walls of hospitals, and in the waiting rooms of our study sites. Together these symbols marketed HHS as an important fixture of excellence in a city often portrayed as a place in decline (Wakefield and McMullan, 2005).

We observed many permanent Cornerstone of Care installations within waiting rooms, along with posters and pamphlets inviting visitors to make donations to the HHS. For example, Figure 11.4 shows a sign inviting the patient to make a donation as it may directly enhance the quality of care he or she would receive. The slogan

3 Although hospitals in Ontario (and Canada) are publicly funded and therefore can be used by all, health cards are given to citizens of the province which must be shown at hospitals, family doctors and other sites as they demonstrate that individuals are entitled to access services.

Figure 11.4 Calling on the patient to become a donor

at the bottom reads *Please support the HHS Grateful Patient Program*. Signs on vending machines in waiting rooms were observed which read: *Your purchase makes a difference – All profits donated to enhance patient care*. Another sign at an entrance way stated that *Whether buying a cup of coffee or a bouquet of flowers you make a difference with every purchase. On Behalf of the patients and families of Hamilton Health Sciences, we thank you for your support!* While such signage was not directly linked to the campaign, it enforced the notion that HHS hospitals were interested in investing in and enhancing patient care through whatever means available. The marketing materials remained consistent across waiting rooms and hospitals, effectively extending the philanthropic request throughout the city.

Concluding Discussion

Waiting Rooms as Regulated and Controlled Transitional Micro-Spaces

Waiting rooms are intermediate spaces within hospitals (Tanner, 2002). Generally speaking, they are the products of expert design discourse emphasizing efficiency and functionality. Such spaces also act as a transitional site between the public space of the hospital or clinic and the private space of the examination room. Thus they serve as a spatial boundary which distinguishes between the 'lay' public and 'expert' physician. Our data reveal that the check-in area, in particular, acts as a gate for this boundary and the signage posted there serves as a reminder that there are indeed rules controlling who does and does not have access to the space(s) beyond. Yet one cannot get to the check-in without first having gone through the entrance way to the waiting room. This micro-space was observed to house a great deal of informal regulatory signage which enforced the hospital-wide expectations of patients and visitors. Signage in waiting rooms for visitors also called them to action in order to facilitate the flow of information from the clinic via telephone to waiting room occupants and within and outside the hospital.

As noted earlier, Gesler (1999) argues that language shapes the larger process of place-making. Findings show how the built environment is not a neutral backdrop to health care practices but rather is directly involved in the control of hospital patients and space. Following Gillespie (2002), the interior spaces of waiting rooms are regularly written upon to invoke underlying ideologies of patient regulation. We can also see from the above examples that written language can make and demarcate place. Sites that housed most of the informal and formal regulatory postings were important transitional micro-spaces in the waiting room, and the texts layered upon them enforced this: the entrance way was a type of 'liminal space' where an individual transitioned from person to patient (in waiting), the empty desk in the visitor waiting room became the site at which the occupant became part of the chain of health-care communication despite not being a paid employee, and the check-in counter served as a policed point of access. We can understand these transitional micro-spaces to be important sites of not only regulation and control but also power within the waiting room which reflects the larger notion that the built environment has implications for the practice of health care within (Gillespie, 2002).

Waiting Rooms as Promotional Spaces

McGrath and Tempier (2003) tell us that the presence of health promotion posters in waiting rooms make them a form of classroom. Our observations reveal that waiting rooms are many other things: they are regulated spaces; they are liminal spaces; and they are also promotional spaces. Our finding that the waiting room was used as a site for hospital promotion demonstrates the notion that underlying ideologies of consumerism are also being literally inscribed into the material landscape of the hospital (Kearns and Barnett, 1997; 2000; Gesler et al., 2004).

Corporate and philanthropic place advertisements are significant for two reasons. First, placing these objects within the waiting room makes them material expressions of 'consumer-orientated' spaces. In this sense, the meaning of consumer could be expanded beyond someone who purchased or bought a product to describe an elevated level of involvement or stake in a setting or process (namely the provision of quality care). We find this subject-position to be at odds with ones projected through formal and informal regulatory signage that explicitly or tacitly communicated the regulated and regulatory nature of the space. In contrast, the waiting room visitor is invited to contribute to the system and services they are there to consume. This points to a larger tension between docile, passive patients and active citizens, both of which are encoded into the visual environments of hospital spaces and specifically the waiting room.

Second, we see significance in the way that place-advertisements scattered throughout waiting room environments and hospitals thread together an extended sense of place. More specifically, the presence of place-based advertisements with consistent messages and images spread throughout the waiting room, the hospital grounds and the city provided a scalar continuity which unites these spaces. Our observations support Hanlon's (2001) claim of health services providing an 'extended sense of place', specifically with regard to what we refer to here as the scalar continuity of the HHS marketing campaign and visitors being called upon to contribute to the greater public good through making a donation. Such a finding also extends the work of Kearns and Barnett (1999; 2000; 2003) who have argued that contemporary hospital environments are marked by the ideological fingerprints of a neo-liberalizing health care system, resulting in a burgeoning array of consumer-orientated spaces of care. Taking its philanthropic and corporatist messages into consideration, one can imagine how the Cornerstone of Care campaign messages and their continuous presence at multiple scales combined to recast Hamilton's hospital system as a landscape of consumption, mixing private money with public health.

Our exploratory analysis has revealed much about the waiting room space and textual messages within it, and we encourage further work to be undertaken in order to develop a more nuanced understanding of micro-spaces within hospital and other clinical environments.

Revisiting the Therapeutic Landscape Concept

It was stated earlier that our purpose in undertaking this analysis is to suggest that we, as geographers, must be careful in applying the therapeutic landscapes concept at too large a scale to specific places because, in doing so, micro-elements that can potentially send contradictory messages (those not in keeping with the planned therapeutic design or healing elements of the environment) may be overlooked. While, as outlined earlier, the hospital may be designed with therapy and healing in mind, we have shown here that the textual layers added onto the built environment send strong messages to occupants which enforce practices of control and regulation and attempt to compel the visitor into becoming a donor, all of which are powerful and not necessarily an enhancement to the potentially therapeutic properties of the hospital space as a whole. However, our work in the waiting room was undertaken in order to make a larger point, which is that we must read the multiple layers (textual and otherwise) present in spaces and consider their implications for practices undertaken at a particular site before focusing exclusively on the healing properties of a particular place or proclaiming a space to be a therapeutic landscape. With regard to built environments in particular, we must look beyond architecture and design in our consideration of how place-bound practices contribute to the interior characteristics of health care settings in ways that may negatively affect emotions and, in turn, processes of healing.

Acknowledgments

We must first thank the PR staff who allowed us to gain access to these waiting room spaces. Valorie Crooks was funded by a Canadian Institutes of Health Research (CIHR)/Canadian Health Services Research Foundation Postdoctoral Fellowship in Health Services Research and a CIHR Postdoctoral Fellowship in Health Care, Technology, and Place (HCTP) during data collection and analysis. Joshua Evans is currently funded by a CIHR HCTP Doctoral Fellowship.

References

Adams, A. and Schlich, T. (2006), 'Design for Control: Surgery, Science, and Space at the Royal Victoria Hospital, Montreal, 1893–1956', *Medical History* 50: 303–24.

Andrews, G. (2004), '(Re)thinking the dynamics between healthcare and place: therapeutic geographies in treatment and care practices', *Area* 36: 307–18.

Arneill, A.B. and Devlin, A.S. (2002), 'Perceived Quality of Care: the influence of the waiting room environment', *Journal of Environmental Psychology* 22: 345–60.

Aronson, J. (1994), 'A pragmatic view of thematic analysis', *The Qualitative Report*, 2 (1). Online journal. Available: <http://www.nova.edu/ssss/QR/BackIssues/QR2-1/aronson.html>, accessed 26 June 2006.

Blumberg, R. and Devlin, A.S. (2006), 'Design Issues in Hospitals – The Adolescent Client', *Environment and Behaviour* 38: 293–317.

Conradson, D. (2003), 'Geographies of care: spaces, practices and experiences', *Social and Cultural Geography* 4: 450–54.

Conradson, D. (2005), 'Landscape, care and the relational self: Therapeutic encounters in rural England', *Health and Place* 11: 337–48.

Côté, M.J. (1999), 'Patient flow and resource utilization in an outpatient clinic', *Socio-Economic Planning Sciences* 33: 231–45.

Craven M.A., Nicolaou, L., Allen, C.J., Crustolo, A.M. and Kates, N. (2005), 'Patient education materials for mental health problems in family practice: does location matter?' *Patient Education and Counseling* 56: 192–6.

Foucault, M. (1975), *Discipline and Punish: the Birth of the Prison* (New York: Random House).

Gesler, W.M. (1999), 'Words in wards: language, health and place', *Health and Place* 5: 13–25.

Gesler, W., Bell, M., Curtis, S., Hubbard, P. and Francis S. (2004), 'Therapy by design: evaluating the UK hospital building program', *Health and Place* 10: 117–28.

Gesler, W. and Kearns, R. (2002), *Culture/Place/Health* (New York: Routledge).

Gillespie, R. (2002), 'Architecture and power: a family planning clinic as case study', *Health and Place* 8: 211–20.

Hanlon, N. (2001), 'Sense of place, organizational context and the strategic management of publicly funded hospitals', *Health Policy*, 58: 151–73.

Kearns, R.A. (1991), 'The Place of Health in the Health of Place: the case of the Hokianga special medical area', *Social Science and Medicine* 33: 519–30.

Kearns, R.A. and Barnett, J.R. (1997), 'Consumerist ideology and symbolic landscapes of private medicine', *Health and Place* 3: 171–80.

Kearns, R.A. and Barnett, J.R. (1999), 'To boldly go? Place, metaphor and the marketing of Auckland's Starship hospital', *Environment and Planning D* 17: 201–26.

Kearns, R.A. and Barnett, J.R. (2000), '"Happy Meals" in the Starship Enterprise: interpreting a moral geography of health care consumption', *Health and Place* 6: 81–93.

Kearns, R.A., Barnett, J.R., and Newman, D. (2003), 'Placing private healthcare: reading Ascot hospital in the landscape of contemporary Auckland', *Social Science and Medicine* 56: 2303–15.

Kearns, R.A. and Gesler, W.M. (1998), *Putting Health into Place: Landscape, Identity and Well-Being* (Syracuse: Syracuse University Press).

MacLean, M. (2006), 'Lessons from the waiting room – Doctors' office creates a sense of comfort with design ideas you can use at home', *The Edmonton Journal* Thursday 23 February 2006.

McGrath, B.M. and Tempier, R.T. (2003), 'Is the Waiting Room a Classroom? (Letter to the Editor)', *Psychiatric Services* 54: 1043.

Milligan, C., Gatrell, A. and Bingley, A. (2004), "Cultivating Health': therapeutic landscapes and older people in Northern England', *Social Science and Medicine* 58: 1781–93.

Montazeri, A. and Sajadian, A. (2004), 'Do women read poster displays on breast cancer in waiting rooms?' *Journal of Public Health* 26: 355–8.

Palka, E. (1999), 'Accessible Wilderness as a Therapeutic Landscape: Experiencing the Nature of Denali National Park, Alaska', in A. Williams (ed.) *Therapeutic Landscapes: The Dynamic Between Place and Wellness* (Lanham, MD: University Press of America) pp. 29–51.

Parr, H. (1999), 'Mental health and the therapeutic geographies of the city: individual and collective negotiations', in A. Williams (ed.) *Therapeutic Landscapes: The Dynamic Between Place and Wellness* (Lanham, MD: University Press of America) pp. 123–51.

Parr, H. (2003), 'Medical Geography: Care and Caring', *Progress in Human Geography* 27: 212–21.

Poland, B., Lehoux, P., Holmes, D. and Andrews, G. (2005), 'How place matters: unpacking technology and power in health and social care', *Health and Social Care in the Community* 13: 170–80.

Radley, A. and Taylor, D. (2003a), 'Remembering one's stay in hospital: a study in photography, recovery and forgetting', *Health: An Interdisciplinary Journal for the Social Study of Health, Illness and Medicine* 7: 129–59.

Radley, A. and Taylor, D. (2003b), 'Images of Recovery: a photo-elicitation study on the hospital ward', *Qualitative Health Research* 13: 77–99.

Rapport, F., Doel, M.A., Greaves, D. and Elwyn, G. (2006), 'From manila to monitor: biographies of general practitioner workspaces', *Health: An Interdisciplinary Journal for the Social Study of Health, Illness and Medicine* 10: 233–51.

Sloane, D.C. (1994), 'Scientific Paragon to Hospital Mall: The Evolving Design of the Hospital, 1885–1994', *Journal of Architectural Education* 48: 82–98.

Smyth, F. (2005), 'Medical geography: therapeutic places, spaces and networks', *Progress in Human Geography* 29: 488–95.

Sutherland, M. (2002), 'Rating waiting', *Biomedical Journal* 324: 1443.

Tanner, L. (2002), 'Bodies in Waiting: Representations of Medical Waiting Rooms in Contemporary American Fiction', *American Literary History* 14: 115–30.

Teather, E.K. (ed.) (1999), *Embodied Geographies – Spaces, Bodies and Rites of Passage* (New York: Routledge).

Wakefield, S. and McMullan, C. (2005), 'Healing in places of decline: (re)imagining everyday landscapes in Hamilton, Ontario', *Health and Place* 11: 299–312.

Ward, K. and Hawthorne, K. (1994), 'Do patients read health promotion posters in the waiting room? A study in one general practice', *British Journal of General Practice* 44: 583–5.

Wiles, J. and Rosenberg, M. (2001), 'Gentle caring experience: Seeking alternative health care in Canada', *Health and Place* 7: 209–21.

Williams, A. (ed.) (1999), *Therapeutic Landscapes: The Dynamic Between Place and Wellness* (Lanham, MD: University Press of America).

Wilson, K. (2003), 'Therapeutic landscapes and First Nations people: an exploration of culture, health and place', *Health and Place* 9: 83–93.

Wilton, R. and DeVerteuil, G. (2006), 'Spaces of sobriety/sites of power: examining social model alcohol recovery programs as therapeutic landscapes', *Social Science and Medicine* 63: 649–61.

Chapter 12

Therapeutic Landscapes for Older People: Care with Commodification, Liminality, and Ambiguity

Malcolm P. Cutchin

Aging and place, a relatively obscure dimension of geographical inquiry, is having new life breathed into it through a variety of scholars (for example, see Andrews and Phillips, 2005). The innovative work is arguably leading to a new 'geographical gerontology' (Andrews et al., 2007), and the concept of therapeutic landscapes has been suggested as a potentially important component in such work (Kearns and Andrews, 2005; Milligan et al., 2004; Cutchin, 2005). At this point, however, there is still much room for exploration of the utility and meaning of therapeutic landscapes and older adults, and the need for research is pressing because of the growth in older age groups. In particular, there is little if any in-depth analysis of residential care settings as therapeutic landscapes. Moreover, there is a need to extend any such analysis with a critical understanding of the unhealthy dimensions of such landscapes that make them more complex than heretofore realized and less therapeutic than often depicted.

In this chapter, I will address the important issue of how the concept of therapeutic landscapes is useful in the analysis of residential care environments for older people. Beyond that analysis, however, I provide a critical examination of how such seemingly therapeutic landscapes also are problematic. This extension of analysis into the unhealthy, problematic side of residential care places complements other recent analyses that use the therapeutic landscape concept to assess potentially negative dimensions of health contexts (Wakefield and McMullan, 2005). I provide an understanding of how places created as therapeutic for older adults in the United States become inherently unhealthy in a number of ways that diminish the intended, and often real, therapeutic benefit. In essence, this chapter describes and analyzes the conundrum of a type of therapeutic landscape that is being replicated for millions of older adults across the world.

I begin the analysis with an introduction to the phenomenon of residential care (also known as assisted living). While the concept of therapeutic landscapes has not been readily applied to residential care settings, the evidence that I will present suggests that the concept would be a useful addition to the current state of gerontological knowledge. The argument proceeds by offering three additional ways of conceptualizing processes at the heart of assisted living. In addition to applying the concept of therapeutic landscapes to this residential context for older adults, the concepts of commodified landscapes, liminality, and ambiguity are used to portray unhealthy dimensions of residential care. I conclude the chapter with an assessment of implications for residential care/assisted living, as well as how the concept of therapeutic landscapes is enhanced by the inclusion of such critical analysis of places of care.

Assisted Living, Transitions, and Well-being

'Residential care settings' is a generic phrase that may be used to describe a range of senior housing and long-term care options in the United States.[1] States have used different terminology for larger group residential care settings, including: board and care homes, rest homes, adult care homes, domiciliary care homes, personal care homes, community-based residential facilities, and assisted living (Mollica and Johnson-Lamarche, 2005). A continuum of these settings begins with independent living sites where few care services are offered and ends with nursing homes where 24-hour supervision and health care are provided (Leith, 2006). In the middle zone of this spectrum is assisted living, the term used most commonly now to refer to all types of group residential care.[2] The common use of the term assisted living, along with variation in state definitions of such settings, tend to cause disagreement about what assisted living is, what it should be, and how it should be regulated (Mollica and Johnson-Lamarche, 2005).

Assisted living offers a middle option between independent living arrangements and nursing homes (in other words, long-term care) by including key aspects of both types of places. Those aspects include the provision of congregate housing – single or shared apartment units – and the availability of supportive services for functional, health, and social needs, such as common rooms and services such as dining, exercise classes, and social activities. The common principles of assisted living include: (a) the provision of a homelike environment; (b) the maximization of dignity, independence, autonomy, and privacy; (c) accommodation of residents' changing needs and preferences; and (d) minimizing the required moves when needs increase (Wright, 2004). The interpretation of these principles and their implementation varies from state to state and site to site in: housing structure, resident payment and admission, resident retention, and staffing. Regardless, assisted living is a popular and increasingly widespread form of housing and care for older persons in the US (Chapin and Dobbs-Kepper, 2001; Fonda et al., 2002; Maddox, 2001; Zimmerman et al., 2001). In 2005, there were over 935,000 licensed assisted living apartments in the United States (Mollica, 2006). While the growth rate of assisted living units has slowed in recent years, there was a 97 percent increase between 1990 and 2002 (Harrington et al., 2005).

---

1 I restrict my specific analysis to the United States although many of the descriptions and analyses would pertain in a greater or lesser degree to other contexts as well.

2 An exception to this is a relatively new form of residential care setting in the United States, the continuing care retirement community, where the full range of care types, including assisted living, are combined on one campus.

Assisted living is, theoretically, a type of setting that offers a therapeutic option for more frail older adults because of the social interaction created inside and outside the setting by way of informal and formal activities and outings (Cannuscio et al., 2003). Adjustment to assisted living can be a difficult process, however; the process may entail risks to health and well-being for the older adult that are not dissimilar to the health risks of adjustment to other types of facilities (Carp, 1974; Hays, 2002). Health outcomes research on assisted living is mixed, suggesting that it is associated with improved health and well-being (Fonda et al. 2002), that is has limited and idiosyncratic effects (Zimmerman et al., 2005), or that it has no effect on health outcomes compared to nursing homes (Frytak et al., 2001; Pruchno and Rose, 2000). Other research has identified various social and environmental components of assisted living as being associated with improvement in resident satisfaction and self-reported quality of life (Ball et al., 2000; Brandi et al., 2004; Mitchell and Kemp, 2000; Sikorska, 1999).

Several researchers have moved beyond the outcomes of satisfaction and quality of life to assess key processes in the assisted living experience of older adults. Indeed, perspectives oriented toward the social construction of well-being (as opposed to more individualistic perspectives) argue that the transition to residential care settings is most importantly influenced by social environments and social interactions (Hochschild, 1973; Kontos, 1998). For example, Eckert et al. (2001) discovered the theme of connectedness, developed from qualitative data collected by observers in assisted living residences. Those authors posit connectedness as a central but multifaceted theme because it describes important relationships between residents and contextual aspects of assisted living: staff, other residents, the facility, and the community. While social relationships are essential to connectedness, the physical features and design of assisted living environments can enhance those relations and provide a sense of home. Connectedness is deemed important because it 'begins to illustrate the nature and diversity of connections between residents, their care providers, and the places and locations where they live' (Eckert et al., 2001, 309). Other studies have investigated the characteristics of being at home in settings similar to assisted living, including a congregate housing community (Young, 1998) and life care and long-term care facilities (Hammer, 1999). These two inquiries suggest some universal aspects of the experience of feeling at home in senior housing, such as autonomy, self-identity, close social relationships, mutual respect, comfort, involvement, and security.

With colleagues, I have investigated assisted living using different research approaches. From an in-depth qualitative study of two assisted living settings based primarily on observations and in-depth interviews with residents, family, and staff, I produced a model of 'mediated aging-in-place' (Cutchin, 2003). I proposed a complex interaction among the wider community context, uses of space and place in assisted living settings, place-centered activities, and resident–staff conflict. One result of this process is an 'approximation of home and community' that assists the resident in becoming integrated into their new residence. For a significant number of residents, however, the new residential experience only partially replicates life previous to assisted living (Cutchin, 2003). Using quantitative data collected from several hundred residents at 25 residences in four states, colleagues and I provided some corroboration of the model (Cutchin et al., 2003). In that analysis, findings from a structural equation analysis indicated that whether or not older adults in assisted living consider their residence as 'home' was influenced most significantly (as determined by a sum of direct and indirect path coefficients) by 'non-family social involvement' and to a lesser but significant extent by 'place attachment' and 'place valuation.' In the study, non-family social involvement denoted engagement in a range of social transactions with people inside and outside the assisted living residence. Place attachment was measured as affective bonds with town or community, and place valuation was composed of resident ratings of various place qualities. The findings of the analysis may be seen as evidence that assisted living can provide a valuable type of aging-in-place experience for many residents. An additional analysis of both the quantitative and qualitative data from the questionnaire discovered that, in addition to the previous findings, the social and personal experience of place inside an assisted living setting is structured to an important degree by a range of purposeful activities in which staff often play a central role (Cutchin et al., 2005). A negative finding of that analysis was that a majority of survey respondents reported being less involved in community affairs after their move to assisted living.

The research reviewed collectively implies that assisted living settings are complex places that entail a correspondingly complex experience for the older adults who move into them. Ecological theories of aging (for example, Lawton, 1982; Kahana, 1982; Nahemow, 2000) provide much of the current basis for thinking about how assisted living creates an environment to enhance the well-being of frail older persons. Central to these theories is the equilibrium ('fit' or 'congruence') of person and environment. The concept of *place*, although not typical of ecological perspectives on aging, adds an important dimension to gerontological theories of person–environment relationships. The literature on place that pertains to assisted living (social gerontology and residential experience) is diverse. Place theory varies in philosophical orientation, but core place concepts include the following: sense of place (for example, Tuan, 1975; Rowles, 1983), place identity (for example, Cuba and Hummon, 1991; Proshansky et al., 1983), place attachment (for example, Rubinstein and Parmalee, 1992), and place integration (Cutchin, 2001). These concepts' utility is to describe and explain dimensions of the complex human relationship with meaningful environments. Such place concepts align better with constructivist positions of older person–environment relations that emphasize sociality, meanings, and context (Gubrium and Holstein, 1999). Indeed, other research to date has begun to address the complexity and nuances of the relationship to place and the social dynamics in settings that combine housing and care (Abbott et al., 2000; Biggs et al., 2000; Bland, 1999; Eales et al., 2001; Kontos, 1998; Andrews and Phillips, 2002). Much of this work is based on qualitative methodologies, and the emerging understanding is fertile and critical in that it portrays the positive and negative dimensions of such places, develops an understanding of the intricate processes involved, and suggests that such accommodation can be better than it currently is. There is much left to do in this particular research area, and there will be continued need for inquiry as residential care continues to evolve and grow.

The so-called post-medical, or cultural, turn in health geography during the last decade (for substantive reviews and examples, see Kearns and Gesler, 1998;

Gesler and Kearns, 2002; Kearns and Moon, 2002) has opened up analytical avenues that are useful for exploring place issues in older people's health in the context of assisted living. While the concept of place remains an important pillar of social and geographical gerontology, the concept of therapeutic landscapes has been rarely employed but is of potentially important consequence (Kearns and Andrews, 2005). In the next section I will take up the therapeutic landscape idea to examine assisted living places and the positive experiences they engender for older adults. That discussion sets the stage for a more critical perspective on assisted living.

Theorizing Assisted Living as Both Therapeutic and Unhealthy Landscapes

Because assisted living residences have grown in number (and therefore importance) in the United States only recently, a relative paucity of theoretical research on the topic currently exists. The concomitant cultural turn in health geography, along with an even more recent re-flourishing of geographical gerontology, make this an appropriate opportunity to theorize assisted living. Moreover, assisted living settings are unique because of their positioning between home and institution – they are places of particular interest for the theoretician. It is not simply a matter of good (home-like) and bad (institution-like) aspects of these places that make them intriguing. As various scholars have stressed, the home can itself become a site of contested space and negative implications (Wiles, 2005; Williams, 2002; Dyck et al., 2005). My primary (but not sole) focus below is a set of dynamics that come together in assisted living to produce places and thereby affect place experiences and health outcomes; and these processes and effects are evaluated as to their therapeutic and their unhealthy nature.

Assisted Living as Therapeutic Landscapes

The concept of 'therapeutic landscapes' (Gesler, 1991; 1992; Williams, 1999) translates well for inquiry about older persons and residential care places. Although used for inquiry into a broad set of health and care issues as well as to denote places, spaces, and networks (Smyth, 2005), therapeutic landscapes – the idea of places imbued with meaning and the relationship people have with those places as having healing effects – has significant potential for further development in the context of places designed specifically for older persons. As Andrews (2004, 308) has written, the concept of therapeutic landscapes allows researchers to 'embrace more cultured perspectives that view health places as complex and symbolic constructions.' Inquiry into the therapeutic and unhealthy dimensions of care is of particular relevance because there is much at stake regarding the well-being of millions of older adults and their relatives.

While the deepening literature on therapeutic landscapes offers various ways of understanding assisted living, Conradson's (2005) view of the relational experience of person and place (landscape) that leads to the therapeutic 'outcome' is of particular utility. Similar to my own arguments about the primacy of the relational (transactional) in considering person, place, and well-being (Cutchin, 1997; 2001;

2003; 2007), this view prioritizes the person–place 'encounter' over the therapeutic landscape as a thing. There are various ways in which assisted living is the context for person–landscape encounters and experiences that tend to be therapeutic. One common thread that connects these therapeutic dimensions is the manner in which they can be seen as potential moderators of psychological and physiological stress – an overlooked factor in the person–place–well-being nexus for older adults (Cutchin, 2005).

Although assisted living residences vary in their size, ownership status (for-profit or non-profit), amenities, and overall quality, there are a number of aspects of life in such residences that seniors value and/or that are therapeutic in nature. One of these is the security that such residences provide compared to independent living in the community. Security is a complex and multifaceted dimension of the older person and assisted living relationship. One dimension of security is the dependability that such managed housing provides. Residents do not have to make repairs or worry about housekeeping as they would in an independent living situation. In addition, most assisted living residences have multiple forms of security against potential outside threats (for example, burglars, scams). The security structures (monitored doors, alarms) and personnel provide a sense of ease that lessens the stress that the older adult may have experienced in their previous home. The congregate nature of the assisted living setting also provides security by making the older resident aware that if they are in need, they have multiple staff and co-residents that are available to help them. A specific need of many frail older adults is eating well. Assisted living provides regular, balanced meals that relieve both the burden of cooking from many residents, and their family members from worrying whether or not their loved one is obtaining sufficient nutrition.

As our research has previously discovered, the social and active aspects of assisted living also are important and therapeutic. Assisted living residences usually have social and activity directors in place that are given the task of creating healthy activities and interactions for residents. While some residents view structured activities as important to their well-being, we have observed and heard from residents that some such activities and interactions can become contrived, boring, and under-stimulating. More positive activities are varied, ranging from parties and entertainment of different sorts, to poetry clubs and exercise classes. Some activities take place outside the assisted living setting in contexts such as museums, malls, and public lectures. The fact that these are place-based both in the assisted living residence and the surrounding community means that the therapeutic landscape encounter is larger than the housing complex itself. This translates into a more stimulative set of transactions, with older adults feeling more a part of the world than they would otherwise. Activities most commonly done with others build social ties and bring pleasure. More solitary activities, such as reading, knitting, and so on also are encouraged and made available by the space and resources in many assisted living sites; however, the social activities appear to be the most therapeutic. This conclusion is echoed in the findings of Bryant et al. (2001) and Milligan et al. (2004) with regards to older people and the therapeutic landscape effects of communal gardening. Belonging, doing, and achieving in multiple ways is a positive outcome of the senior and assisted living therapeutic landscape encounter (Cutchin, 2003).

One of the most important therapeutic outcomes of assisted living is the formation of a sense of home and community. Home is the locus of significant attachment and meanings (Rowles, 1987). Because older adults who had been living independently have experienced either a decline in their sense of place or have been extracted out of emplaced meaning of their former home and community, the formation of new meanings through a sense of home and community in assisted living is important and therapeutic. One important way that the assisted living–older person transaction works to foster home and community is through the social activity discussed above. Often overlooked, home has a significant social dimension for older adults (Rubinstein, 1989), but rarely before moving into a congregate living situation is the communal experience of shared dining, shared spaces, and interactions so closely tied with home. While the challenges and problems of merging the communal and home in one micro-scale place are discussed below, the need for community is noteworthy. Because assisted living is, to some degree, housing that is segregated from the rest of society, the formation of communal relationships among residents, staff, family, and other visitors is necessary to maintain some continuity with life processes as they were before assisted living. Of course, this varies by person; some older adults have shared how difficult the communal aspects of assisted living are for them to grow accustomed to. The sharing, trust, and many other components of communal life that can be re-created in assisted living appear to be a therapeutic outcome for most residents, however.

There are certainly other dimensions of assisted living transactions that we could suggest as therapeutic, but the role of security, social activity, and the re-creation of home and community are especially important. These exemplary outcomes are difficult to achieve for frail older persons outside the assisted living context. Indeed, assisted living has promised, and to some degree achieved, a therapeutic landscape for older persons who need such outcomes. But as a number of more recent analyses of therapeutic landscapes have suggested (Martin et al., 2005; Wakefield and McMullan, 2005; Wilson, 2001), landscapes pertaining to residence and care may be more or less therapeutic – and the less therapeutic have been underemphasized when utilizing the therapeutic landscape concept. The following sections illustrate dynamics in assisted living that are distinctly unhealthy (anti-therapeutic) in nature.

Assisted Living as Commodified Landscapes

Alongside the development of the more positive concept of therapeutic landscapes, the idea of 'landscapes of consumption' was developed to describe how consumerist ideology and the commodification of places of care has significant effects for the experience of that care (see Gesler and Kearns, 2002 for an overview). Kearns and Barnett (1997) illustrate how therapeutic and commodified landscapes of care sometimes coexist and merge in market-based care systems. The central argument in such cases is that the insertion of agendas of consumption into care places is detrimental to the good and therapeutic dimensions those places are intended to present. In essence, the implementation of consumerist ideology creates commodified places that demean and diminish the care ethic and the quality of care.

In an assisted living market where private, for-profit entities are large players, the creation of commodified landscapes should not come as a surprise. The research on consumerist ideology and places for older adults has focused on retirement communities, the residential precursor to assisted living for many older adults. Lucas (2004, 458) clearly describes the logic underlying the commodification of senior communities:

> The positive images used to describe the retirement community lifestyle, therefore, have a distinct purpose. The geographic experience of aging they describe was created not necessarily for the benefit of the residents, but to promote a specific view of later life that itself had been constructed in order to encourage continued consumption.

Lucas explains that it is not just lifestyle that is being commodified, but the spaces and places of retirement are being sold as part and parcel of the identity of successful aging. Lucas' argument is extended by McHugh and Larson-Keagy (2005) using Del Webb's master planned retirement communities in the Phoenix, Arizona area as a good example. Such communities are promoted through 'plasticized' images of utopian places and ageless residents who are overly active (McHugh, 2000). Such critical views of aging and place owe much to the pioneering research of Glenda Laws (1993; 1995) who carefully examined how ageism and commodification have long been intertwined in the creation and portrayal of built environments for older people.

Assisted living settings are different than retirement communities in their geographic scale and their scope of services, but they incorporate consumption landscapes nonetheless. An analysis of printed and web-based promotional matter for assisted living firms provides good evidence. These materials use imagery to project the therapeutic qualities of assisted living:

> Imagine living in a world where your health, safety and security are top priorities. One where you don't really need to bother much about everyday chores like cleaning. It's a place, too, where you can have some real quality time to meet with family and friends, exercise, read, take an art class, play cards and enjoy a great meal. … it's where you'll find this kind of world and much more in affordable and elegant surroundings (Senior Care Communities, 2004).

This imagery promotes healthy, successful aging as noted by Lucas (2004), even though the population served is on average more frail than the population that moves to retirement communities; that is, the assisted living population is less likely to be able to take advantage of all the positive attributes of assisted living. Some promotional materials own up to the greater needs of the assisted living population but also commodify through idealistic imagery. The text that accompanies a photograph of a sun-drenched, home-like façade of an assisted living brochure[3] provides one example:

3 Other research was conducted at this New England site and confidentiality requirements prevent me from naming the source of the materials. Similar sales pitches can be found in most assisted living promotional materials.

> The best of both worlds – independence and care when you need it. … Sunlight streams into the library, casting shadows of trees across the rich carpeting. Two women chat by the fire, one knitting, the other playing with her pet.

Older adults, and their families who often help them choose an assisted living residence, are provided with many choices of place and experience in which to consume. Aspects of comfort, security, ease, activities, fun, and elegance are the commodities in these examples. But similar to other types of landscapes where symbolism can be read and cultural meaning interpreted (Lewis, 1979), both promotional images and actual assisted living landscapes can be read. While the exterior (outdoor) landscape is important in projecting the commodities on offer by an assisted living residence, the indoor aspects of assisted living environments become a more important commodity, compared to places such as retirement communities, because assisted living populations are frailer and likely to spend more time indoors. A single image from one promotional brochure can illustrate this adequately. In it, the resident (older man) sits in a library at a table interacting with a younger man and woman (presumably the older man's adult children or an adult child and an in-law) and a child (presumably the older man's grandchild). They are engaged in working a jigsaw puzzle, and they look happy to be doing so. A globe, numerous newspapers ready at hand, shelves full of elegant volumes and artifacts denote sophistication and connections with the world. The image symbolizes and projects a future for a potential resident that entails family, home-like surroundings, intellectual activity, fun, and general well-being. Moreover, the image presents commodities of the landscape that are of particular interest to persons of higher socioeconomic class with particular cultural backgrounds and tastes.

The consumption landscape of assisted living echoes that of retirement communities, but it offers a somewhat distinct set of commodities to a different set of older persons. Much of the logic and techniques of the creation of consumption landscapes are the same across both types of settings, but assisted living emphasizes the interior landscape that is likely to be more important to a frailer population. The projection of these landscapes is antithetical to the therapeutic landscape ideal because they can never be realized in practice – they are to a lesser or greater degree misleading – and they divert attention from the care that residents will need. Such consumerist text and images emphasize the positive and eliminate the negative, or the potentially challenging aspects of assisted living that all potential residents (and their families) should be aware of. These techniques sell a landscape rich with promise and something for everyone – everyone that can afford such a landscape and culturally fit into it.

Assisted Living as Liminal Places

As I alluded above, the move by older persons to an assisted living residence requires the leaving of one place and the re-creation and integration with another. Because assisted living is a residential environment that is qualitatively different than that of the typical pre-move environment, the transition from one place to another is significant and, for some, difficult. Frank (2002) took a distinctively critical view of the relationship between assisted living residents and their place, and she emphasized the transitional challenge. From analysis of in-depth qualitative data collected in two assisted living residences, Frank suggested that residents did not form sufficient social bonds with other residents or staff. Frank argued that new residents typically experienced a state of 'limbo' and a 'heightened sense of liminality,' in other words, a 'suspended state' of transition, that inhibited them from becoming 'at home' in assisted living (Frank, 2002, 17). The implications, as Frank (and I) see it, are negative for residents' well-being.

Frank's conceptualization of liminality follows the anthropological literature's focus on the concept as it pertains to rites of passage. As a result, the concept as she uses it for assisted living is non-geographical. While the early work of Van Gennep, (1960; cited in Andrews, 2002) is still important in understanding liminality, it can be expanded. Van Gennep argued that most rites of passage are composed of three fundamental stages: separation, transition, and incorporation. The stages pertain to passage between social statuses, and when one is in transition between statuses, liminality defines one's situation. Shields (2003, cited in Madge and O'Connor, 2005) extends this view by suggesting 'liminal zones' where liminality and movement to a new social status happens. This view suggests that in addition to the liminality of some residents who do not readily integrate into assisted living, we should try to understand the place of assisted living as a central part of that liminal experience.

There are various manifestations of the concept of liminality in assisted living. Frank argues that residents remain in that suspended state, part of which is lacking at-homeness. I have also suggested (Cutchin, 2003) that the development of home and community was at best an approximation of the previous state. Indeed, some older persons (and even their family members) view assisted living as a 'way-station' on the route to a more institutional setting (Cutchin, 2004). This somewhat pessimistic view is supported by evidence that suggests the average tenure in assisted living is only about two years (Chapin and Dobbs-Kepper, 2001). Moreover, that view also aligns with the common idea that assisted living is in the middle of a continuum of care and that older persons with more complex health problems should be cared for in skilled nursing facilities.

The liminality of assisted living also applies in a different sense to certain subgroups in particular. Because the vast majority of assisted living residents in the United States are non-Hispanic whites, other racial or ethnic groups are *de facto* exceptional persons in these settings. I never witnessed any overt racism or cultural bias in assisted living, but I was told stories by members of minority groups that suggested they viewed assisted living as comfortable but liminal – and liminal in the sense that their concerns about subtle racism or cultural bias were not understood by the white (majority) older adults. In brief, there was a lack of connection between minority members of the community and those of the majority that prevented them from full status in that community. This state of affairs was viewed by minority residents as a continuation of racism experienced through their life course in American society. While these residents chose to remain in place in assisted living, it was acceptance of a liminal experience that was countered by other benefits.

While this understanding of assisted living as a liminal place for some residents must be viewed as suggestive (not conclusive) of an important dynamic, it appears to be related to the process of 'dislocation' of older adults (Milligan, 2003). Discussing a key feature of dislocation, Milligan (2003, 462) states:

> As the embodiment of identity, the anthropological place of the home further places limits on the extent to which an individual can be objectified and depersonalized – stripped of their history and identity – to become anonymized within a collective (institutional) regime.

Liminality based primarily in the difference and newness of place – something we all experience during a significant move – surely occurs to all older persons who enter assisted living. The expectation is that most older people will pass through the liminal phase quickly and develop a therapeutic relationship with a new place. The evidence from assisted living (Frank, 2002; Cutchin, 2003; 2004) indicates that liminality probably takes different forms and lasts longer for many residents than we might expect, with the result probably being increased stress and decreased well-being.

Assisted Living as Ambiguous Places

The liminal experience of being between statuses, identities, and/or places implies a form of ambiguity and confusion for older adults. The concept of ambiguity is more than a corollary of liminality in assisted living, however. Because assisted living holds the middle ground in the housing and care continuum, it inherently creates much of the ambiguity in question. This type of residential care attempts to merge attributes from each side of the spectrum with the result of ambiguities about what it is, who it is for, and how people are to negotiate it. The result is considerable uncertainty for older adults and their families (Cutchin, 2004; Martin et al., 2005).

One cause of the ambiguity in assisted living comes from the so-called 'blurring of boundaries' between home and institutional care, as well as home and institutional spaces (Dyck et al., 2005; Milligan, 2006). For instance, rules and regulations are necessary to the functioning of a managed, communal living context like assisted living, but those rules and regulations run counter to the autonomy associated with home. This situation creates ambiguity about who is in control of the place. While residents have a good deal to say about how policy should be set, in reality, management holds the majority of power in any such dialogue. Similar to this form of ambiguity, the blurring of private and public space is of concern. The result is a 'semi-public' place (Milligan, 2006) where rules of private and public behavior and control are confounded. A good example is the widespread permeability of public space and private rooms, where staff members have the right to enter a resident's room under particular circumstances. The rationale for such rules are clearly to protect residents and enable their care, but the ambiguity of what is public and what is private remains.

Several other dimensions of ambiguity are present in assisted living. One is the ambiguity of independence versus dependence. Assisted living strives to provide a mix of both independence and care determined by individual circumstances. The optimal goal is independence which is likely associated with better health and quality of life. But the provision of services to dozens of seniors at the same time means that habits and standardization will cause dependent care for some who do not need it. For example, a person may be wheeled to dinner – to get them there on time – when they could walk and obtain some exercise. What mix of independence and care is best for the older person is frequently not clear.

In a similar fashion, assisted living is ambiguously a place for both the psychologically healthy as well as those with dementia. I have listened to residents tell me how this ambiguity causes them concern. Are they living in a place for the demented or for the well? Are they in turn to be viewed as one or the other? This impact on place identity is theoretically a negative one that continues to plague assisted living. Likewise, the ambiguity of assisted living as a place for the living and the dying is noteworthy. While the active, healthy life entailed in the consumerist images of assisted living occurs for some places and residents, there are an increasing number of residents who die in assisted living (Sloane et al., 2003). By pointing this out, I am not suggesting that the dying should be segregated from the living in residential care; the issue is that the inclusion of a significant number of the dying in a place advertised primarily as for a more able and healthy population adds additional ambiguity to the meaning of assisted living as a place, especially when compared to other places for older adults toward either end of the care spectrum.

The 'best of both worlds' is not necessarily a good thing in assisted living. People of all ages need to understand their places and their role in them. Assisted living as an experience and a place is rife with ambiguities that cloud such understanding. That situation can only enhance liminality and contradict the purely positive messages of assisted living marketers. These combined features of the place and experience contribute to stress for older adults who move to assisted living and, as I have indicated above, well-being is put at risk.

Conclusions

Therapeutic landscapes is a concept now being used more critically and usefully to understand an increasingly wide array of places. Very little work in this vein has been focused on older people and the quasi-institutional places, such as assisted living, through which they live and receive care. In this chapter, I have considered residential care (assisted living in particular) in an effort to address this gap and, importantly, to attempt to understand how place of residence and care can be understood as both therapeutic and unhealthy at the same time. While assisted living for older people can easily be understood as therapeutic, I have argued that problematical dimensions – commodified landscapes, liminal places, and ambiguous places – are also prevalent and should be considered. There are perils in places that we think are therapeutic, and those perils are potentially expanded by ignoring their anti-therapeutic aspects.

What are the implications? For aging studies it means we need to examine the unhealthy dimensions of assisted living places and their effects on the well-being of residents and their families and/or caregivers. There has been too much focus on the assumed positive aspects of assisted living. For aging policy, it means that we need to grapple with how to maximize residential care options that mix lower and higher levels of care by minimizing the stressful, negative effects of the unhealthy aspects discussed here. In my view, a more fully democratic organizational and management system could be implemented in assisted living to give residents and their family greater involvement and power in how that living is to be carried out. Much of what might be called alienating effects of commodification, liminality, and ambiguity arise from an imbalance of control, power, and decision-making in the hands of management who do not live in assisted living or have their loved ones there. To address this imbalance would mean developing models of care antithetical to the for-profit, corporate model that is currently dominant in assisted living in the United States. This is not an indictment of all assisted living; even residences with quality management could benefit from recognition of these problems and act to increase the therapeutic dimensions of their environments and, by so doing, diminish unhealthy dimensions, such as commodification, liminality, and ambiguity.

So what of the concept of therapeutic landscapes? The problem I have presented illustrates the need to understand health and place vis-à-vis therapeutic landscapes as multifaceted, complex, and full of relative goods and bads. Theorists are needed to generate and utilize fertile concepts such as therapeutic landscapes to make sense of the ever-changing terrain of care and well-being. Therapeutic landscapes, such as assisted living, almost always include unhealthy dimensions; given this, they can be seen as moral landscapes in need of imaginative action and enhancement – they demand that we ask the question of what they ought to become. A critical social science must therefore see the bad as well as the good, and it should help envision a better future. To that end, the metaphor of the Roman god Janus, suggested by Gunnar Olsson (1991, 13), is worth considering:

> Janus is a pivoting symbol of gatekeeping, whose major characteristic is not that he can see in opposite directions as the same time, but that he is able to merge seemingly contradictory categories into a meaningful whole. … Janus' concern is with creativity.

But, of course, translation of the meaningful whole also is necessary. Citizens who partake in social life and social care systems should be encouraged to understand what works and what does not work in those systems and to change them for the better. Seeing the whole of the therapeutic landscape concept in a creative way through conceptualizing the unhealthy and the therapeutic together is a necessary first step to making places better and healthier.

Acknowledgments

Funding for the author's research relevant to this chapter was provided by the National Institute on Aging (grant R15 AG17028).

References

Abbott, S., Fisk, M. and Forward, L. (2000), 'Social and Democratic Participation in Residential Settings for Older People: Realities and Aspirations', *Ageing and Society* 20: 327–40.

Andrews, G.J. (2002), 'Towards a More Place-sensitive Nursing Research: An Invitation to Medical and Health Geography,' *Nursing Inquiry* 9: 221–38.

Andrews, G.J. (2004), '(Re)thinking the Dynamics Between Healthcare and Place: Therapeutic Geographies in Treatment Care Practices', *Area* 36: 307–18.

Andrews, G.J., Cutchin, M., McCracken, K., Phillips D.R. and Wiles, J. (2007), 'Geographical Gerontology: The Constitution of a Discipline', *Social Science and Medicine* 65: 151–168.

Andrews, G.J. and Phillips, D.R. (2002), 'Changing Local Geographies of Private Residential Care for Older People 1983–1999: Lessons for Social Policy in England and Wales', *Social Science and Medicine* 55: 63–78.

Andrews, G.J. and Phillips, D.R. (2005) *Ageing and Place: Perspectives, Policy, Practice* (London: Routledge).

Ball, M.B., Whittington, F.J., Perkins, M.M., Patterson, V.L., Hollingsworth, C., King S.V. and Combs, B.L. (2000), 'Quality of Life in Assisted Living Facilities: Viewpoints of Residents', *Journal of Applied Gerontology* 19: 304–25.

Biggs, S., Bernard, M., Kingston, P. and Nettleton, H. (2000), 'Lifestyles of Belief: Narrative and Culture in a Retirement Community', *Ageing and Society* 20: 649–72.

Bland, R. (1999), 'Independence, Privacy and Risk: Two Contrasting Approaches to Residential Care for Older People', *Ageing and Society* 19: 539–60.

Brandi, J.M., Kelley-Gillespie, N., Liese, L.H. and Farley, W. (2004), 'Nursing Home vs. Assisted Living: The Environmental Effect on Quality of Life', *Journal of Housing for the Elderly* 18: 73–88.

Bryant, L.L., Corbett, K.K. and Kutner, J.S. (2001), 'In Their Own Words: A Model of Healthy Aging', *Social Science and Medicine* 53: 927–41.

Cannuscio, C., Block, J. and Kawachi, I. (2003), 'Social Capital and Successful Aging: The Role of Senior Housing', *Annals of Internal Medicine* 139: 395–99.

Carp, F.M. (1974), 'Short-term and Long-term Prediction of Adjustment to a New Environment', *Journal of Gerontology* 29: 444–53.

Chapin, R. and Dobbs-Kepper, D. (2001), 'Aging in Place in Assisted Living: Philosophy versus Policy', *The Gerontologist* 41: 43–50.

Conradson, D. (2005), 'Landscape, Care and the Relational Self: Therapeutic Encounters in Rural England', *Health and Place* 11: 337–48.

Cuba, L. and Hummon, D.M. (1991), 'A Place to Call Home: Identification with Dwelling, Community, and Region', *Sociological Quarterly* 34: 111–31.

Cutchin, M.P. (1997), 'Physician Retention in Rural Communities: The Perspective of Experiential Place Integration', *Health and Place* 3: 25–41.

Cutchin, M.P. (2001), 'Deweyan Integration: Moving Beyond Place Attachment in Elderly Migration Theory', *International Journal of Aging and Human Development* 52: 29–44.

Cutchin, M.P. (2003), 'The Process of Mediated Aging-in-Place: A Theoretically and Empirically-based Model', *Social Science and Medicine* 57: 1077–90.

Cutchin, M.P. (2004), 'A Deweyan case for the study of uncertainty in health geography', *Health and Place* 10: 203–13.

Cutchin, M.P. (2005), 'Spaces for Inquiry into the Role of Place for Older People's Care', *International Journal of Older People Nursing* in association with *Journal of Clinical Nursing* 14 (8b): 121–9.

Cutchin, M.P. (2007), 'John Dewey's Metaphysical Ground-Map and Its Implications for Geographical Inquiry', *Geoforum* DOI:10.1016/j.geoforum.2007.01.014.

Cutchin, M.P., Chang, P.F. and Owen, S.V. (2005), 'Expanding Our Understanding of the Assisted Living Experience', *Journal of Housing for the Elderly* 19: 5–22.

Cutchin, M.P., Owen, S.V. and Chang, P.F. (2003), 'Becoming "At-home" in Assisted Living Residences: Exploring Place Integration Processes', *Journal of Gerontology: Social Sciences* 58B: S234–43.

Dyck, I., Kontos, P., Angus, J. and McKeever, P. (2005), 'The Home as a Site for Long-term Care: Meanings and Management of Bodies and Spaces', *Health and Place* 11: 173–85.

Eales, J., Keating, N. and Damsma, A. (2001), 'Seniors' Experiences of Client-centered Residential Care', *Ageing and Society* 21: 279–96.

Eckert, J.K., Zimmerman, S. and Morgan, L.A. (2001), 'Connectedness in Residential Care: A Qualitative Perspective', in S. Zimmerman, P.D. Sloane and J.K. Eckert, (eds) *Assisted Living: Needs, Practices, and Policies in Residential Care for the Elderly* (Baltimore: The Johns Hopkins University Press) pp. 292–313.

Fonda, S.J., Clipp, E.C. and Maddox, G.L. (2002), 'Patterns in Functioning Among Residents of an Affordable Assisted Living Housing Facility', *The Gerontologist* 42: 178–87.

Frank, J.B. (2002), *The Paradox of Aging in Place in Assisted Living* (Westport, CT: Bergin & Garvey).

Frytak, J.R., Kane, R.A., Finch, M.D., Kane R.L. and Maude-Griffin, R. (2001), 'Outcome Trajectories for Assisted Living and Nursing Facility Residents in Oregon', *Health Services Research* 36: 91–111.

Gesler, W.M. (1991), *The Cultural Geography of Health Care* (Pittsburgh: University of Pittsburgh Press).

Gesler, W.M. (1992), 'Therapeutic Landscapes: Medical Geographic Research in Light of the New Cultural Geography', *Social Science and Medicine* 34: 735–46.

Gesler, W.M. and Kearns, R.A. (2002), *Culture/Place/Health* (London: Routledge).

Gubrium, J.F. and Holstein, J.A. (1999), 'Constructionist Perspectives on Aging', in V. Bengtson and K.W. Schaie (eds) *Handbook of Theories on Aging* (New York: Springer Publishing Co.) pp. 287–305.

Hammer, R.M. (1999), 'The Lived Experience of Being at Home: A Phenomenological Investigation', *Journal of Gerontological Nursing* 25: 10–18.

Harrington, C., Chapman, S., Miller, E., Miller, N. and Newcomer, R. (2005), 'Trends in the Supply of Long-term-care Facilities and Beds in the United States', *Journal of Applied Gerontology* 24: 265–82.

Hays, J.C. (2002), 'Living Arrangements and Health Status in Later Life: A Review of Recent Literature', *Public Health Nursing* 19: 136–51.

Hochschild, A.R. (1973), *The Unexpected Community* (Berkeley: University of California Press).

Kahana, E. (1982), 'A Congruence Model of Person-environment Interaction', in M.P. Lawton, P.G. Windley and T.O. Byerts (eds) *Aging and the Environment: Theoretical Approaches* (New York: Springer) pp. 97–121.

Kearns, R.A. and Andrews, G.J. (2005), 'Placing Ageing: Positionings in the Study of Older People', in G.J. Andrews and D.R. Phillips (eds) *Ageing and Place: Perspectives, Policy, Practice* (London: Routledge) pp. 13–23.

Kearns, R.A. and Barnett, J.R. (1997), 'Consumerist Ideology and Symbolic Landscapes of Private Medicine', *Health and Place* 3: 171–80.

Kearns, R.A. and Gesler, W.M. (eds) (1998), *Putting Health into Place: Landscape, Identity, and Well-being* (Syracuse, NY: Syracuse University Press).

Kearns, R. and Moon, G. (2002), 'From Medical to Health Geography: Novelty, Place and Theory After a Decade of Change', *Progress in Human Geography* 26: 605–25.

Kontos, P.C. (1998), 'Resisting Institutionalization: Constructing Old Age and Negotiating Home', *Journal of Aging Studies* 12: 167–84.

Laws, G. (1993), "The Land of Old Age": Society's Changing Attitudes toward Urban Built Environments for Elderly People', *Annals of the Association of American Geographers* 83: 672–93.

Laws, G. (1995), 'Embodiment and Emplacement: Identities, Representation and Landscape in Sun City Retirement Communities', *International Journal of Aging and Human Development* 40: 253–80.

Lawton, M.P. (1982). 'Competence, environmental press, and the adaptation of older people', in M.P. Lawton, P.G. Windley and T.O. Byerts (eds) *Aging and the Environment: Theoretical Approaches* (New York: Springer Publishing) pp. 33–59.

Leith, K.H. (2006), "Home is Where the Heart is … or is it?" A Phenomenological Exploration of the Meaning of Home for Older Women in Congregate Housing', *Journal of Aging Studies* 20: 317–33.

Lewis, P. (1979), 'Axioms for Reading the Landscape', in D.W. Meinig (ed.) *The Interpretation of Ordinary Landscapes*. (New York: Oxford University Press) pp. 11–32.

Lucas, S. (2004), 'The Images Used to "Sell" and Represent Retirement Communities', *The Professional Geographer* 56: 449–59.

Maddox, G.L. (2001), 'Housing and living arrangements: A transactional perspective', in R.H. Binstock and L.K. George (eds) *Handbook of aging and the social sciences* (San Diego, CA: Academic Press) pp. 426–43.

Madge, C. and O'Connor, H. (2005), 'Mothers in the Making? Exploring Liminality in Cyber/space', *Transactions of the Institute of British Geographers* NS 30: 83–97.

Martin, G.P., Nancarrow, S.A., Parker, H., Phelps, K. and Regen, E.L. (2005), 'Place, Policy and Practitioners: On Rehabilitation, Independence and the Therapeutic

Landscape in the Changing Geography of Care Provision to Older People in the UK', *Social Science and Medicine* 61: 1893–1904.

McHugh, K.E. (2000), 'The "Ageless Self"? Emplacement of Identities in Sun Belt Retirement Communities', *Journal of Aging Studies* 14: 103–15.

McHugh, K.E. and Larson-Keagy, E.M. (2005), 'These White Walls: The Dialectic of Retirement Communities', *Journal of Aging Studies* 19: 241–56.

Milligan, C. (2003), 'Location or dis-location? Towards a Conceptualization of People and Place in the Care-giving Experience', *Social and Cultural Geography* 4: 455–70.

Milligan, C. (2006), 'Caring for Older People in the 21st Century: "Notes from a Small Island"', *Health and Place* 12: 320-31.

Milligan, C., Gatrell, A. and Bingley, A. (2004), "Cultivating Health": Therapeutic Landscapes and Older People in Northern England', *Social Science and Medicine* 58: 1781–93.

Mitchell, J.M. and Kemp, B.J. (2000), 'Quality of Life in Assisted Living Homes: A Multidimensional Analysis', *Journal of Gerontology: Psychological Sciences* 55B: P117–P127.

Mollica, R.L. (2006), 'Residential Care and Assisted Living: State Oversight Practices and State Information Available to Consumers', AHRQ Publication No. 06-M051-EF (Rockville, MD: Agency for Healthcare Research and Quality).

Mollica, R. and Johnson-Lamarche, H. (2005), *State Residential Care and Assisted Living Policy: 2004* (Washington, DC: US Department of Health and Human Services).

Nahemow, L. (2000), 'The Ecological Theory of Aging: Powell Lawton's Legacy', in R.L. Rubinstein, M. Moss and M.H. Kleban (eds) *The Many Dimensions of Aging* (New York: Springer) pp. 22–40.

Olsson, G. (1991), *Lines of Power/Limits of Language* (Minneapolis: University of Minnesota Press).

Proshansky, H.M., Fabian, A.K. and Kaminoff, R. (1983), 'Place-identity: Physical World Socialization of the Self', *Journal of Environmental Psychology* 3: 57–83.

Pruchno, R.A. and Rose, M.S. (2000), 'The Effect of Long-term Care Environments on Health Outcomes', *The Gerontologist* 40: 422–8.

Rowles, G.D. (1983), 'Place and Personal Identity in Old Age: Observations from Appalachia', *Journal of Environmental Psychology* 3: 299–313.

Rowles, G.D. (1987), 'A Place to Call Home', in L.L. Carstensen and B.A. Edelstein (eds) *Handbook of Clinical Gerontology* (New York: Pergamon) pp. 335–53.

Rubinstein, R.L. (1989), 'The Home Environment of Older People: A Description of the Psychosocial Processes Linking Person to Place', *Journal of Gerontology: Social Sciences* 44: S45–53.

Rubinstein, R.L. and Parmalee, P.A. (1992), 'Attachment to Place and the Representation of the Life Course by the Elderly', in I. Altman and S. Low (eds) *Place Attachment* (New York: Plenum Press) pp. 139–63.

Senior Care Communities (2004), <http://www.seniorcarecom.com/AssistedLiving/AssistedLiving.html>, accessed 6 March 2004.

Shields, R. (2003), *The Virtual* (London: Routledge).

Sikorska, E. (1999), 'Organizational Determinants of Resident Satisfaction with Assisted Living', *The Gerontologist* 39: 450–56.

Sloane, P.D., Zimmerman, S., Hanson, L., Mitchell, C.M., Riedel-Leo, C and Custis-Buie, V. (2003), 'End-of-life Care in Assisted Living and Related Residential Care Settings: Comparison with Nursing Homes', *Journal of the American Geriatrics Society* 51: 1587–94.

Smyth, F. (2005), 'Medical Geography: Therapeutic Places, Spaces, and Networks', *Progress in Human Geography* 29: 488–95.

Tuan, Y.F. (1975), 'Place: An Experiential Perspective', *Geographical Review* 65: 151–65.

Van Gennep, A. (1960), *The Rites of Passage* (Chicago: University of Chicago Press) originally published 1909.

Wakefield, S. and McMullan, C. (2005), 'Health in Places of Decline: (Re)imagining Everyday Landscapes in Hamilton, Ontario', *Health and Place* 11: 299–312.

Wiles, J. (2005), 'Home as a New Site of Care Provision and Consumption', in G.J. Andrews and D.R Phillips (eds) *Ageing and Place: Perspectives, Policy, Practice* (London: Routledge) pp. 79–97.

Wilson, K. (2001), 'Therapeutic Landscapes: The Dynamic Between Place and Wellness', (book review) *Health and Place* 7: 346–8.

Williams, A. (1999) *Therapeutic Landscapes: The Dynamic Between Place and Wellness* (Lanham, MD: University Press of America).

Williams, A. (2002), 'Changing Geographies of Care: Employing the Concept of Therapeutic Landscapes as a Framework in Examining Home Spaces', *Social Science and Medicine* 55: 141–54

Wright, B. (2004), *An Overview of Assisted Living: 2004* (Washington, DC: AARP Public Policy Institute).

Young, H.M. (1998), 'Moving to Congregate Housing: The Last Chosen Home', *Journal of Aging Studies* 12: 149–65.

Zimmerman, S., Sloane, P.D. and Eckert, J.K. (eds) (2001), *Assisted Living: Needs, Practices, and Policies in Residential Care for the Elderly* (Baltimore: The Johns Hopkins University Press).

Zimmerman, S., Sloane, P.D., Eckert, J.K., Gruber-Baldini, A.L., Morgan, L.A., Hebel, J.R., Magaziner, J., Stearns, S.C. and Chen, C.K. (2005), 'How Good is Assisted Living? Findings and Implications from an Outcomes Study', *Journal of Gerontology: Social Sciences* 60: S195–S204.

Chapter 13

Home as Therapeutic Landscape: Family Caregivers Providing Palliative Care at Home

Rhonda Donovan and Allison Williams

The context and provision of health care services have undergone rapid change as a result of technological advancement, system restructuring and decreased investment, such that the delivery of care is increasingly taking place in non-traditional settings, such as the home. The trend towards home-care[1] is pervasive; in Canada, for example, it accounts for the largest growing sector in the health care system (Coyte and Young, 1997). This shift has increased expectations for family members[2] to actively participate in the care of their relatives (Fast and Keating, 2000). The type of care and the degree to which it is provided continues to expand (Shamian, 2006) and will not likely abate, as demand on services will increase as the population ages. There is a much greater understanding of the impact of caregiving on family caregivers (Haley, 2003; Hauser and Kramer, 2004), yet little is known about how medical care in the home affects the physical and social environments for those who live there (Williams, 2002). There is considerable evidence to suggest that the home operates as a therapeutic environment for the ill and the dying; it is not known if this exists for family caregivers, given that their experience of home is altered by both the medicalization of the home as well as the caregiving process.

In this chapter, we build on the work of Williams (2002; 2003; 2004) by applying the concept of therapeutic landscapes to examine how providing palliative and end-of-life (P/EOL) care[3] in the home affects caregivers' experience and meaning of home. As the project's principal investigator, Williams led and implemented the research. Using the therapeutic landscapes concept as a framework, Donovan carried out the analysis of the data. We have used a political economy approach

1 The terms community care, home care and community-based services are often used interchangeably to refer to the range of services (social support, health promotion and home care) delivered outside of traditional institutions. It is used here to refer to traditional home care programmes, community-based supportive services and family caregiving.

2 Family members may be those persons related by blood or legal relationships as well as through affiliations based on mutual commitment and obligation.

3 Palliative care is intended to relieve suffering and meet the physical, psychological and spiritual needs of the patient and family to improve the quality of life for the dying (Carstairs and Beaudoin, 2000). Also referred to as hospice care, it is delivered by a multi-disciplinary team, including various health care practitioners, social workers and spiritual advisors.

(see Williams, 2002 for an extended discussion) to examine how family caregivers, impelled into the delivery of health care through system restructuring, are impacted by this experience. Such knowledge has implications for policy reform and service improvement (Williams, 2002).

The goals of palliative care are strongly linked to the right to die with dignity (Carstairs, 2005; Carstairs and Beaudoin, 2000) and, while it can be delivered anywhere, it is primarily concerned with supporting people to die in their homes (Stajduhar and Davies, 1998). Research suggests that patients and families prefer the home setting for care and death because the circumstances are more intimate, peaceful and less distressing (Kinsella et al., 2000); however, it appears that various factors contribute to caregivers' ability to provide and sustain care, including their ability to access appropriate home and palliative care services. In Canada, these services are fragmented, with each province providing various degrees of public and private services (Kirby and LeBreton, 2001; Romanow, 2004; Shapiro, 2002), creating gaps in the necessary support structures for family caregivers to sustain care in the home.

Therapeutic Landscapes

Therapeutic landscapes have traditionally referred to places with enduring reputations for healing, but can also include places associated with health promotion and maintenance (Williams, 1999; 2002). Recent developments have moved the concept beyond specific idealized sites reputed for health and healing, visited only temporarily by those who are able; therapeutic landscapes now encompass commonplace, ordinary geographies – such as the home (Williams, 2002; Milligan et al., 2004). Further, scholars have recognized that places are not always, nor consistently, healthful for all who frequent them; particular places deemed healthful may also present any number of health risks (Wakefield and McMullan, 2005; Andrews and Holmes, this volume). In addition, there has been a call for the recognition of therapeutic landscapes to be context-dependent; that is, affected by local social and economic conditions and changes (Conradson, 2005). These developments have allowed for and provided a discriminating view of ordinary, everyday landscapes such as the home, while opening up different understandings of how these landscapes can contribute to the maintenance of well-being, such as in the case of family caregivers.

As the core setting for everyday life, home is often considered *the* place of most personal significance and meaning; this significance, as Relph (1976) suggests, is far deeper than the explicit functions it has for us. While a diverse scholarship on the meaning of home exists (see Despres (1991), Somerville (1997) and Mallett (2004) for extensive reviews) we understand home to be a multidimensional concept 'that is physically, psychologically and socially constructed in both real and ideal forms' (Somerville, 1997, 226). The linkages between health status and 'being at home' are widely acknowledged and any disruption of these links – tangible or existential – can result in changes to well-being (Chaudhury and Rowles, 2005). Factors in

Table 13.1 Demographic information and family formation history (adapted from Williams et al., 2003, 12)

Case	Pseudonym a) Caregiver b) Patient	Relationship	Age	Occupation/ Status	Residence/ Tenure/ Length of Time	a) Children b) Grandchildren	Dependants in Home	Number of Interviews
1	a) Sheila	Daughter	<40	Fitness Instructor/ Part-time	Apartment/ Rent/Less than one year	a) 0 b) 0	0	2 – during Chronic Living Dying stage
	b) Susan	Mother	67	Nurse/Retired	Apartment/ Rent/Less than one year	a) 2 (adult) b) 2	0	–
2	a) Anne	Wife	62	Nurse/Retired	Home/Owner/ Five years	a) 2 (adult) b) 6	0	1 – during Chronic Living dying stage 2 – during Bereavement
	b) Andy	Husband	70	Pharmacist/ Retired	Home/Owner/ Five years	a) 2 (adult) b) 6	0	–
3	a) Bill	Husband	56	Carpenter/Self-employed	Home/Owner/ Less than two years	a) 2 (adult) b) 3	0	1 – during Chronic Living dying stage 2 – during Bereavement
	b) Betty	Wife	55	Real Estate Agent/Retired	Home/Owner/ Less than two years	a) 2 (adult) b) 3	0	–

the home that impact caregiving can be enhanced or controlled for to improve the experience and preserve the health and well-being of family caregivers.

As a setting for health care, the home has yet to be fully interrogated. Feminist geographers have challenged traditional ideas on home space, explicating how the relationships one has with home are influenced by prevailing social norms (see Dyck, 2001; Dyck et al., 2001; Williams, 2002). They suggest that the home becomes a site of labour for both paid health service providers and unpaid family caregivers as medical care enters the home. This can alter or disturb the activities and meaning of the lived environment of the home (Dyck et al., 2001). Additionally, caregivers must renegotiate their identities and activities in relation to place as the nature of caregiving changes (Wiles, 2003). Brown (2004) demonstrates how home hospice (or end-of-life care) produces a tradition of home death that masks inequalities around the labour of caregiving behind the freedom one has to die there.

Methods and Analysis

The overall objective of this research was to determine if and how the home can operate as a therapeutic environment for family caregivers during the palliative process. Using a case study design, two methods (semi-structured interviews and videography) were used to collect data with participants throughout the caregiving and bereavement process, as discussed at length in Williams (2004).

Interviews were transcribed verbatim and data analysed using content analysis (Hsieh and Shannon, 2005). The interview schedules provided a framework for analysis, and through immersion in the data, themes and patterns were identified. The qualitative research data software package, NUD*IST, was used to manage the large amount of data. Consistent with case study research, analysis occurred first within cases and ultimately across cases (see Merriam, 1998).

Case Summaries

Three in-depth case studies were conducted and are summarized in Table 13.1.

Case One – Sheila and Susan

The first case included a caregiver daughter, Sheila, and her palliative mother, Susan. Due to Sheila experiencing 'burnout' and being no longer able to care for her mother in the home, Susan was moved first to a hospital and then a long-term care home where she died. Sheila was designated as primary caregiver by the family. There were minimal resources for informal support and clear financial constraints. Formal publicly provided services included a one-hour weekly visit from a palliative nurse. Sheila and Susan lived in a one-bedroom rental apartment, chosen for its view of the river, accessibility (elevator) and proximity to amenities in the city. The small size of the apartment was seen as a disadvantage, as it was cramped and the narrow doors rendered wheelchair access problematic. Sheila was very dissatisfied with the outcome and was particularly disappointed with the quality and quantity of care available from the public health care system.

Case Two – Anne and Andy

The second case was made up of a caregiver spouse, Anne, and her palliative husband, Andy. Anne was the primary caregiver, although she had a great deal of informal support from their grown children, her brother (who visited at least every second day), and the church community. The retired couple lived in a custom-built home on a rural acreage about 20 minutes outside of the city. It was a large bi-level with the main floor operating as a complete home and the lower level as an apartment/workspace. Wide door frames, access ramps and various other home modifications were put in place at the time of construction. Formal publicly provided services were limited given that the residence was rural, but included a one-hour weekly visit from a regular home care nurse; when Andy reached the terminal stage, a specialist palliative care nurse was involved. The patient died at home and the caregiver was pleased with the outcome.

Case Three – Bill and Betty

The third case consisted of a caregiver spouse, Bill, and his palliative wife, Betty. Bill was the primary caregiver, with limited support from the couple's two grown children who lived nearby. Formal support included twice-daily visits from a home care practitioner and weekly visits from a palliative care nurse. The couple lived in their two-bedroom home in an inner-city neighbourhood. Betty died at home and Bill was pleased with the outcome.

Results

The experience of home for family caregivers in this study was significantly impacted by the caregiving experience. The findings point to the home as both a therapeutic (health-promoting) and non-therapeutic (health-detracting) environment, as demonstrated through a discussion of both the health-promoting and health-detracting properties of home along a number of themes: the essence of home, changing social relations, transforming home space, fulfilling the caregiver role, and the financial impact of caregiving. The dichotomy of the health-promoting/detracting properties of home is informed by the degree to which caregivers attempted to preserve certain properties of the home or had to adapt in response to the change in properties produced by the caregiving situation.

The Essence of Home

The home was the preferred site for care and the decision for care in the home was mutually decided upon by both the patient and family caregiver. Central to this desire were caregivers' beliefs that patients would be more comfortable and receive better care at home. In all three cases the meaning and value of home were significant, suggesting that the essence of home provided a health-promoting element:

> I don't really have a value on our home. It's just our place, just what we have. I don't know what you mean by value, but it's the only place to be. It's the best place to be.
> (Bill)

The ability to derive meaning from the home was essential to normalize the process, and maintaining an emotionally safe environment for the patient was important. For caregivers, this included adapting their own behaviour to meet the needs of the patient through, for example, speaking openly about dying or by avoiding the topic altogether. Caregivers often took their cue from the patient, even when doing so was harmful to their own emotional well-being, as Bill noted:

> She spoke to everybody else about their deaths and everything but she would never speak about her own. Her whole attitude was that she was getting better. Everyday she would say I'm getting better.

After her passing, it was revealed that he found avoiding the topic distressing:

> I just wished we could have talked about that. It never happened.
> (Bill)

The sanctity of home space was often challenged by well-meaning friends and family members who, at times, presented behaviour that was health-detracting. The number of visitors, at times, became overwhelming, resulting in the loss of personal space and time, while creating extra work to keep the home clean and entertain:

> It's a little much. We have a lot of good friends and they're all concerned but it's a little much. It's a little taxing once in a while … lots of nights you sit there and you want to be by yourself and you're not.
> (Bill)

Conversely, the intensity of visitors during caregiving was contrasted during the bereavement period, where:

> You're very encumbered with a lot of people for a particular short period of time and all of a sudden there's nobody. There's still nobody. Well, there's people but not like it was … the phone doesn't ring as much or doorbell doesn't ring as much, things like that.
> (Bill)

Interestingly, it was not just the frequency, but the nature of the visits that could be distressing; thus it became necessary to delicately negotiate boundaries between family and friends' needs and those of the patients:

> But we do have one [rule], the only law we have in our house is when guests come they cannot cry with my mother because there is one set of guests that come, they cry for three hours … Mom is like, they have got to stop, they just have to stop.
> (Sheila)

The meaning of home for caregivers did not change significantly over the caregiving process or following the death of their family member in the home, although for both spousal couples, it became very lonely. One caregiver temporarily avoided the living room where the death occurred:

> I've basically kind of rejected the area to a certain extent I suppose. Oh, I'll sit down and watch a movie or something and fall asleep, but normally I'll just go to bed.
> (Bill)

Many compelling factors are evident in the decision for caregivers to care for family members at home, as reflected in the theme: the essence of home. The determination of the caregivers in this study to preserve the essence of the home, albeit numerous detractors, is indicative of the potential of the home to operate as a therapeutic environment for both patients and caregivers alike.

Changing Social Relations

Numerous types of relationships existed for caregivers, including those with the patient, with family and friends, and with the health care team. The relationships between caregivers and patients proved to be health-enhancing for caregivers. These relationships were strong prior to the patient's illness and, although changed in terms of the activities once shared, were generally strengthened throughout the dying trajectory.

> We probably got closer. Yes, probably far closer.
> (Bill)

Such continuous interaction could be difficult, however, thereby challenging the relationship:

> It just drives me crazy because I am not a person to always be with someone. It drives me crazy, but that is my problem. But I go out and do different things. I know it has changed my Mom because she doesn't like to be dependent on someone and that drives her crazy … but our relationship is still very strong in terms of being best friends and stuff.
> (Sheila)

Interaction with others, either in person or by phone, helped to alleviate some of this tension. In fact, although well supported by families and friends, the caregivers in this study were reluctant to seek informal assistance for help with tasks directly related to caregiving, instead relying on family and friends for emotional and spiritual support:

> … sometimes I do [ask for help], but not all that often because I do what I can. We do what we can without leaning too heavily on people …
> (Anne)

Not only was there reluctance to burden family members with physical tasks, but caregivers were cognizant of the limitations that family members had as a result of their own employment and family obligations. This reluctance also reflected the desire to avoid unwanted, but well-intentioned, advice and help. For example, caregivers sometimes only wanted someone to listen to and not necessarily 'fix' any particular problem (Anne).

Establishing and maintaining relationships with the health care team was essential for successful home care and proved to be health-enhancing for caregivers. Professionals with significant experience in palliative care were considered most valuable as sources of information and support because they were both knowledgeable and industrious:

> Oh she [nurse 1] was very knowledgeable. She had instant rapport with Andy. She had her way around the health care system with great ease. If she thought Andy needed something she would just get on the phone and talk to the doctor and order it and look after it. She also had what she called her bag of tricks. She had all sorts of stuff in her bag. Try this drink, try this, do this ... so palliative care nurses, I would say, are the cream of the crop when it comes to caring for someone who is back at home. The other nurses, like [nurse 2] was an excellent homecare nurse, but she didn't have quite the experience of [nurse 1] and [nurse 1] was an old war-horse. She was in her sixties and she'd been around. She just knows.
> (Anne)

Accessing these resources was possible when there was some degree of continuity and dependability in terms of home support, and when professionals were respectful of the patient and family roles. For example, when speaking of 'nurse 1', as noted above:

> ... she [nurse 1] was totally great. She was always here, you know, Monday morning, eleven o'clock; I knew she would be here. No problem. I knew she would be here. You know, she would say your feet are swollen: have you tried doing this ... try this medication ... And then she would also talk to me, but the best thing is that she also talked to my mother. The other ones just talked to me, they didn't talk to my mother and I thought that is wrong. I've had 'how is your breathing?' and they look at me ... no, talk to my mom, you know? That is the thing that I didn't really like ... [nurse 1] treated me like a regular person, very nice, what do you want, but the other ones treated me as just the daughter. I really didn't know too much. I told one [visiting nurse] her [Susan's] feet are swelling, oh really, they could have been swelling for days. No, I noticed this from yesterday, give me some credence, you know, I am keeping a mental track, I am not stupid.
> (Sheila)

Having access to appropriate support, from family, friends, and health care professionals, was crucial to sustaining caregivers at home. Respect for the patient and the role that caregivers played in their care, along with knowledge that family support was available *if needed*, was often adequate.

Transforming Home Space

The home, as a physical entity, was generally not conducive to caregiving. Home modifications and the reconfiguration of space and objects within that space allowed the home environment to become more conducive to caregiving, although limited residential space – as in the apartment and inner-city house – proved to be problematic. Numerous strategies were incorporated to permit patient functioning, prevent accidents, and minimize caregiver fatigue and ill health. The living room or kitchen often became the focal point of care as the patient's condition deteriorated. Not only was the care easier to provide in a central location, but doing so allowed for greater interaction when family and friends were visiting.

In addition to altering the functional use of rooms, furniture was often rearranged with patients' comfort in mind. Mobility within the home was restricted as patients' functional status deteriorated, requiring modifications and assistive devices. The most significant changes occurred in the bathrooms, for ease in the toileting and bathing needs of the patient. As can be seen in Figure 13.1, these modifications included the addition of support rails, and shower and toilet seats. In the apartment, a hydraulic bath lift was added, by specific request of the caregiver. The therapeutic value of these home modifications and assistive devices varied by caregiver; their use often was dependent upon the availability of space within the home.

Figure 13.1 Modifications made to the bathroom

Other restrictions included the patients' inability to manoeuvre on and off furniture, leaving them immobile or requiring caregivers to lift and assist with transfers. The provision of Saskapoles, floor discs, hoyer lifts, along with suggestions to raise furniture onto blocks, as seen in Figure 13.2, were helpful to promote independence and reduce caregiver strain.

The use of the hoyer lift in the apartment was burdensome due to the limited space available and therefore not used. Where space was not an issue, as in the rural acreage, the hoyer lift was recognized as a remarkable asset:

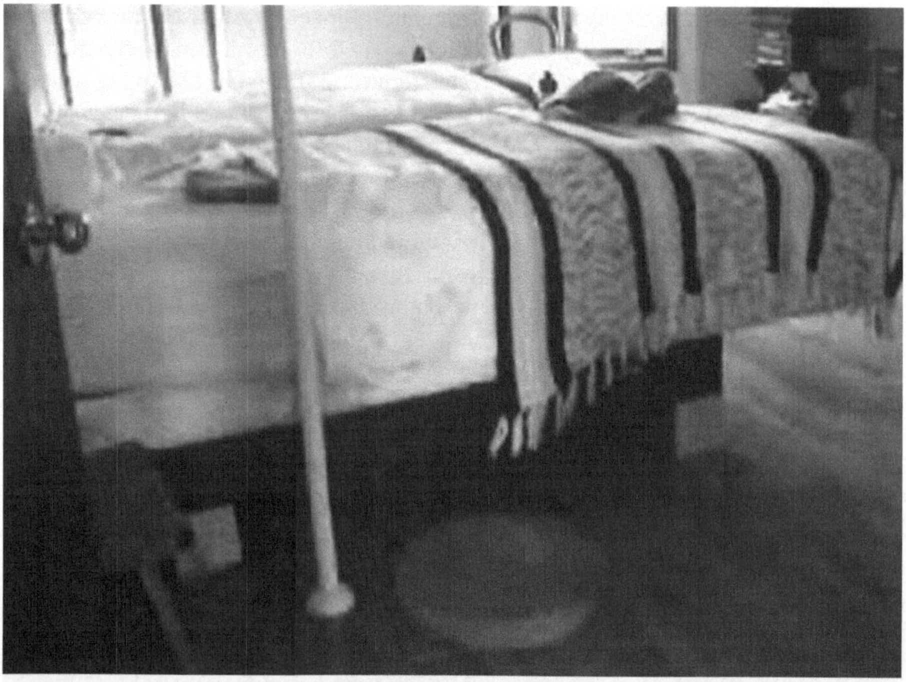

Figure 13.2 Modifications made in the bedroom

> I couldn't always manage without it. Actually the last two weeks I don't think Andy moved without me using that lift, so it was a great help. I just couldn't have handled him … he lost … strength … he just lost it. The only thing he could move was his left arm.
> (Anne)

In addition to limited space for equipment, narrow doorways and stairways made navigating walkers and wheelchairs difficult. Anne and Andy had a wheelchair ramp and wider-than-average doorways built into the home when it was constructed, as depicted in Figure 13.3.

When these modifications were not available, patients were confined to areas of the home which did not require stairs. Further, caregivers found it physically challenging to get patients outside of the home and into vehicles to run errands, attend appointments or just take a break from the confines of home. For example, in the house in the inner-city:

> I have to get her out of the house and I have no way of doing that at the present time. We need to get a wheelchair lift in … I have to get her outside and I'm not going to walk her down those stairs or take the wheelchair down the stairs and see her fall.
> (Bill)

The use of home space was altered for caregivers as well, as mentioned, due to the reconfiguration of various objects and rooms. The consequences could be health-detracting when they were unable to engage in regular activities and hobbies that could

Figure 13.3 Wheelchair ramp connecting house to garage

be viewed as coping strategies. For example, Anne, an avid seamstress, was unable to keep up with her sewing given that her sewing room was located in the basement of their rural acreage. Andy's condition required her to be nearby at all times; spending extended periods of time in the basement was not possible.

While much of the medical equipment was generally available, access was sometimes problematic and inconsistent. There were different channels to go through to access equipment, and not all health care professionals involved in the care were aware of the different items available; it was often up to the caregiver to reach out to their caseworkers when in need. Additionally, caregivers could not always rely on the equipment being delivered when needed. Anne, for example, who lived on a rural acreage, was advised by the occupational therapist:

> I have these at my office but I'm not in your area for several days, so if you want to pick it up you can, or you'll have to wait for it.
> (Anne)

Additionally, not all families were aware of these services and either purchased items on their own or did without. Some items were not provided nor recommended in a timely manner:

> … teaching me that I can also get clothes with no back in them. I didn't know that. Nobody told me this … nobody. You could have told me this two months ago and I would have no problem putting clothes on and off. Last week one of the aides [during respite] said why don't you buy clothes with no backs? You can do this? There is a special company, so I got the catalogue. Tomorrow I am making some orders.
> (Sheila)

Following the two home deaths, the removal of medical equipment and supplies from the home was swift, as indicated by Bill:

> I got rid of everything that immediate day.

The functional use of rooms was restored immediately; however, in some cases, retaining modifications was necessary:

> Well, the hand bars and stuff are still in there [the bathroom] and there's no point in taking them out because you will just have holes in the wall, so I will just leave them alone. I bought them anyway so I will just leave them there.
> (Bill)

The changes made to the home also brought comfort to caregivers in bereavement. When asked about the kitchen, where the patient and caregiver spent so much time when Andy was alive, Anne replied:

> Oh I love it. I didn't like to spend as much time in the kitchen as he did. Now I am finding that I moved this chair into the kitchen to watch TV!
> (Anne)

With the desire to maintain care and have the death occur in the home, significant modifications to the home environment were necessary. When equipment and home modifications were provided – if available, accessible and fitting (depending upon availability of residential space) – these changes enhanced the caregiving experience by reducing caregiver strain and increasing both patient and caregiver mobility.

Fulfilling the Caregiver Role

Fulfilling the caregiver role was, overall, a health-detracting element. Caregivers in this study performed many tasks in relation to caregiving with very little formal or informal support. This included administering and monitoring medications; treating wounds (for example, bed sores); personal care such as basic hygiene, feeding, bathing and toileting; lifting, moving and transferring the patient; and, massaging and exercising them when bed-bound. They also entertained visitors, scheduled and attended various medical appointments, and communicated with health service providers. Along with these tasks, caregivers were responsible for regular household chores, including meal preparation, housekeeping, grocery shopping, laundry, managing finances and general home maintenance. Two of the caregivers in this study were also employed outside the home.

Balancing these many tasks with continuous caregiving, coupled with disruptions to sleep, left caregivers exhausted. They felt it necessary to have breaks from the caregiving situation, and sought stress relief through employment, fitness programmes, relaxation techniques and hobbies. However, as the patients' condition deteriorated and caregiving became more intense, fewer opportunities for self-care were available and restrictions were placed on activities and other routines. A lack of time and the inability to concentrate or derive pleasure prevented caregivers from engaging in these activities. Additionally, normal standards of cleanliness in the home declined as caregivers dedicated their time and energy to the needs of the patient.

Some activities were easier to forego than others. For example, employment was necessary both financially and as a coping strategy, offering a break from caregiving responsibilities and the confines of home:

> I need to get back to work. It's not just financial. I need to get back to work … I need to get away from this.
> (Bill)

It was much easier to balance work, caregiving and household responsibilities with the patient at home compared with the patient being institutionalized. This was discovered by Sheila following an unfortunate incident during respite care in a long-term care facility. When asked whether the caregiver had received respite, she replied:

> No, no respite at all. It has been more instead if Mom was here. I have had less time this week than I had when she was at home. At least when I was at home I could stay, vacuum, clean the house, do the dishes, come and go. Because Mom has been in respite, go to work, go to the home. Come home, stay there three or four hours. I can't do anything at the house when I am over there, so by the time I do all that I come home like … I am tired, so I would sit down on the couch, feel a little bit sorry for myself and start trying to do the other things.
> (Sheila)

Learning new skills was necessary and caregivers felt competent in their abilities, suggesting that fulfilling the role was health-promoting:

> I'm quite comfortable with the way we are doing things. I don't think that I'm hurting [Betty's] situation. I don't think that anybody else could come in here and do it any better. As a matter of fact, I know they can't. I would think that my wife would say the same thing.
> (Bill)

Caregivers felt there was little that could be done to prepare them in advance for this role, suggesting that it was 'something that was in you or not' (Sheila). They required assistance to learn specific caregiving tasks, but this assistance also took the form of moral support:

> … for the most part, I'm not sure, I think I really need her [nurse] more for the support than for anything else because when she came in I was doing everything right, everything that she could suggest I had been doing. But it's a bit reassuring, shall I say, to have her come in.
> (Anne)

Despite the need to get away to run errands, work or take a break, caregivers experienced high levels of anxiety when they had to leave, worried at what might happen in their absence:

> I worry about him. I worry about him. That's when I might be weepy.
> (Anne)

Not surprisingly, caregivers felt relief upon return:

> I am just so happy that things are okay when I walk in the door and when he says hello … I am just glad to be home.
> (Anne)

Anxiety over leaving could also provoke feelings of resentment over the loss of personal control and was therefore health-detracting:

> Grocery shopping was hurry up, go, go, go, go. If there was a line-up, my God, you know. All this pressure being home within the hour. In, out, here you go. Always constant pressure being home, being home, being home. I didn't like that.
> (Sheila)

In addition to exhaustion, caregiving responsibilities took their toll on the physical health of the caregivers, aggravating previous injuries and pre-existing conditions. While most ailments were widely ignored, it was necessary for the caregiver daughter (Sheila) to seek chiropractic treatment on a regular basis as a result of lifting and transferring her partially paralyzed mother. In some cases, negative health conditions appeared during bereavement as the intensity of caregiving subsided:

> It's [my health] has gotten worse actually. I suppose it's just caught up with me … I have Chrone's, I've got high blood pressure … The last year has probably caught up with all my circumstances and now I've got this back and sciatic nerve thing; I don't know what it's all about or why it is and I'm trying to treat it the best way I can so I can walk and work, but it's difficult.
> (Bill)

Although feeling confident about their caregiving role, the tasks involved in caregiving were onerous, leaving caregivers exhausted and exacerbating pre-existing health conditions. Although self-care activities such as catching up on sleep, getting away from the caregiving situation for short periods of time, and engaging in hobbies, would alleviate some of this stress, reluctance to leave the patient, coupled with fatigue, prevented caregivers from doing so.

The Financial Impact of Caregiving

Providing care to a family member in the home was not without financial consequence to the caregivers and overall was health-detracting. This took the form of costs associated with the care as well as reduced and/or lost wages as a result of providing care. As previously noted, working outside the home operated as a coping mechanism, but was also necessary to ease the financial burden. For example, Bill, who was off work for almost two months following his wife's diagnosis, felt it necessary to return to work as their savings were dwindling, stating that:

> Financially, our burden is coming up ...
> (Bill)

Due to being self-employed, he had some autonomy in terms of his work and was able to return to work with reduced hours. Home ownership and dual retirement pensions made it possible for Anne and Andy to sustain care with little financial impact. For Sheila, the combination of her wages from her part-time job coupled with her mother's pension was sufficient only because they maintained a simple lifestyle:

> A little extra would be nice but we are not desperate or anything like that. Between mom and me, we are bringing in enough because we both don't live extravagant lifestyles.
> (Sheila)

Outside of institutions and prior to being on the palliative care caseload, many items necessary for care are assumed by the family unit. For example, caregivers in this study were responsible for certain medications (not covered by the palliative programme), dressings, antibiotic creams, support stockings, oxygen tank rentals and certain equipment (not covered by the palliative programme). In fact, it was as a result of purchasing expensive medications for Betty's treatment that Bill became aware of home-care services at all, despite her condition:

> I wouldn't have known about it unless somebody told me … and she [the pharmacist] was the first one to tell me about it. The reason she told me is that we do not have Blue Cross [private health insurance]; we do not have a prescription plan and the drugs are very expensive and that in circumstances like that [I] should be eligible for this care.
> (Bill)

Without financial security, these expenses could be difficult, and the strain therefore health-detracting for caregivers. Being assigned to the palliative caseload was health-promoting as many of the drug expenses were immediately covered.

Caregivers were able to access the support of home care practitioners and, although this was not free, it was available at a reduced rate. For Bill, accessing this support enabled him to return to work, as Betty was unable to move around on her own:

> She can't get up and walk around or anything like that so they help her up and they take her to the bathroom, they bring her back. They come in at 10:00 and 3:00, which allows me to take care of Betty in the morning and leave by 9:00 so I can go to work, and I don't have to home until at least five or six.
> (Bill)

Paradoxically, the one service that caregivers paid for – home care/support (homemaking and personal care) – was the least predictable; there was no continuity in terms of who would show up, and the type and quality of services provided was limited. For Betty, at least four different home care practitioners were involved in her care, above and beyond the palliative care nurse. As Betty's conditioned worsened, the home care practitioners were unable to help anymore and it was necessary for Bill to suspend working until after Betty's passing.

> No, I actually cancelled home care because they couldn't do anything for her. They couldn't move her or help her, so I just cancelled them all.
> (Bill)

Purchasing this service was not as successful for Sheila and following one 'fiasco', was not used again:

> When I got home the cat was in the hallway, the door was wide open. Mom was propped up on the couch … she couldn't reach her water or her pills ... there was a lot of water all over the bath and it took me two hours to clean up the mess that she made.
> (Sheila)

Having to work, coupled with a lack of support due to mistrust with the health care system, was health-detracting and greatly contributed to Sheila's inability to sustain care in the home. When faced with being unable to do both adequately any longer and feeling that her mother's care and her own health were being compromised as a result, it became necessary to have Stella institutionalized.

While many of the services were provided at a minimal charge, this was not well-articulated to the caregivers in this study. A lack of familiarity with the system, in terms of eligibility and for availability of services, left caregivers expecting to receive bills for this assistance. For example:

> I don't know whether or not there will be a bill from homecare. I'm a little bit fuzzy on when the nurse comes in whether we pay. I have the idea that we have to pay for her but they haven't sent us a bill yet so I don't know.
> (Anne)

In general, the financial impact of caring for a dying family member at home was a burden and clearly a health-detracting element. This is most notable with the issue of balancing caregiving with the need to earn an income on which to survive. While employment offered a means of coping with the caregiving situation, the dual roles brought many tensions which were, in effect, health-detracting. Expenses were reduced somewhat by having access to community-level palliative care services. Access to such services provided health-promoting properties for caregivers in the home.

Limitations

Although this study provides an in-depth look at the impact of caregiving over time, findings cannot be generalized to all caregiving situations. First, it is common in case study research to recruit a small sample size in order to attain depth as opposed to breadth, and this study did consist of a small number of participants. Second, the difficulty in recruiting participants is well known in palliative care research (Miller et al., 2003) and the sample may be biased towards those coping best with the situation. For example, other participants who expressed interest in participating may not have had the time to do so, given that they were overwhelmed with the dying and caregiving process. Third, the caregivers in this sample were not caring for other dependants in the home; thus any effect that additional dependants may have had on the caregiving experience was not explored.

Conclusions

The purpose of this study was to determine if and how the home can operate as a therapeutic environment for family caregivers providing palliative care, given their increased role in the provision of care as a result of health care restructuring. The case studies presented here illustrate that caregiving significantly alters family caregivers' experience of home but, with adequate support (physical, emotional and financial) and appropriate modifications to a spacious home environment, the potential for the home to be therapeutic indeed exists. Both therapeutic (health-promoting) and non-therapeutic (health-detracting) properties were found along a number of themes: the essence of home, changing social relations, transforming home space, fulfilling the caregiver role, and the financial impact of caregiving. In addition to the policy implications discussed in full elsewhere (Williams, 2002; 2004), this research has allowed us to better recognize the properties of home that are both health-promoting and health-detracting. In so doing, we have been able to uncover a number of factors that can be either enhanced or controlled for during caregiving, in order to better preserve caregiver health and well-being.

Providing care in the home was important to the caregivers in this study, both for their own well-being and for that of their dying family member. We have discussed various measures that family caregivers undertook to maintain an emotionally and physically safe home environment. Although such dynamics are quite difficult to control for, advising family caregivers of the potential to feel overwhelmed by visiting family and friends, while encouraging and assisting them in setting boundaries to manage this, can help prevent unnecessary tensions throughout the caregiving process.

The critical role of caregiver relationships with health care providers was revealed. Certainly increasing the number of face-to-face visits with specialized palliative care nurses would enhance the support felt by caregivers. Available, continued contact with these health professionals via other modalities – whether phone or e-mail, for example – would enhance this therapeutic aspect of caregiving at home. Managing the tasks associated with caregiving was overwhelming for caregivers, suggesting that it is necessary to enhance this assistive and supportive aspect of the caregiving experience.

This research demonstrates that the addition of medical equipment and the reorganization of space in the home were health-promoting for caregivers; thus, providing access to a wide variety of supportive devices and equipment, along with making suggestions for modifications, have proven beneficial. An inventory of the home should be taken as early as possible in the caregiving process to determine their current and anticipated needs, as well as the suitability of the home environment for such equipment.

The caregivers in this study demonstrated that they require further assistance and more opportunities to escape the caregiving situation and to avoid feeling isolated and overwhelmed. In addition to accessing respite services, family caregivers should be encouraged to take breaks from caregiving through hobbies, social activities or even employment, as the family caregivers in this study demonstrated.

Financial assistance for various costs associated with palliative caregiving in the home was revealed as overall health-detracting, reflecting the importance of the context of the home as a therapeutic environment; as a site of care, it is affected by both social and economic conditions in the larger society in which it is located. The financial aspects of caregiving can be controlled for through continued and increased investment in resources to support caregivers in the home. This should include access to adequate support so that family caregivers can continue working if necessary.

Reflecting the developments in the therapeutic landscape concept, which have recognized that therapeutic environments can be ordinary and commonplace while simultaneously healthful and hurtful, this research has shown that the home provides both health-enhancing and health-detracting elements for caregivers. As a site of care, we must remember that the home is not an extension of the hospital. When care enters the home, the home's normal functioning becomes lost in the day-to-day grind of caregiving. Preserving the sanctity of the home and maintaining respect for patients, caregivers and family members is essential. Although the caregivers in this study would not change their decision to care for their dying family members at home, it was not without consequences to their health and well-being.

Acknowledgments

Funding for the research and publication of this study was received by Allison Williams from the Social Sciences and Humanities Research Council of Canada (SSHRC) (#410-99-0627). Great appreciation is felt to the participant families and primary caregivers to whom this work is dedicated.

References

Andrews, G.J. and Holmes, D. (2007), 'Gay Bathhouses: The Transgression of Health in Therapeutic Place' in A. Williams (ed.), *Therapeutic Landscapes* (Aldershot: Ashgate).

Brown, M. (2004), 'Between Neoliberalism and Cultural Conservatism: spatial divisions and multiplications of hospice labor in the United States', *Gender, Place and Culture* 11 (1): 67–82.

Carstairs, S. (2005), 'Still Not There, Quality End-of-Life Care: A Progress Report', <http://sen.parl.gc.ca/scarstairs/PalliativeCare/PalliativeCare_e.asp>, accessed September 2005.

Carstairs, S. and Beaudoin, G.A. (2000), 'Quality End-of-Life Care: The Right of Every Canadian', Subcommittee to update 'Of Life and Death' of the Standing Committee of Social Affairs, Science and Technology.

Chaudhury, H. and Rowles, G.D. (2005), 'Coming Home', in G.D. Rowles and H. Chaudhury (eds) *Home and Identity in Late Life: International Perspectives* (New York: Springer Publishing Company).

Conradson, D. (2005), 'Landscape, care and the relational self: Therapeutic encounters in Rural England', *Health and Place* 11: 337–48.

Coyte, P.C. and Young, W. (1997), 'Applied home care research', *International Journal of Health Care Quality Assurance: Incorporating Leadership in Health Services* 10 (1): 1–4.

Despres, C. (1991), 'The Meaning of Home: A literature review and directions for future research and theoretical development', *Journal of Architectural and Planning Research* 8 (2): 96–115.

Dyck, I. (2001), 'Centring the Home in Research', *Canadian Journal of Nursing Research* 33 (2): 83–86.

Dyck, I., Kontos, P., Angus, J. and McKeever, P. (2001), 'The home as a site for long-term care: meanings and management of bodies and spaces', *Health and Place* 11: 173–85.

Fast, J.E. and Keating, N.C. (2000), 'Family Caregiving and Consequences for Carers: Toward a Policy Research Agenda', CPRN Discussion Paper.

Haley, W.E. (2003), 'The costs of family caregiving: implications for geriatric oncology', *Critical Review in Oncology/Hematology* 48: 151–8.

Hauser, J.M. and Kramer, B.J. (2004), 'Family Caregivers in palliative care', *Clinics in Geriatric Medicine* 20: 671–88.

Hsieh, H. and Shannon S.E. (2005), 'Three Approaches to Qualitative Content Analysis', *Qualitative Health Research* 15 (9): 1277–88.

Kinsella, G., Cooper, B., Picton, C. and Murtagh, D. (2000), 'Factors influencing outcomes for family caregivers of persons receiving palliative care: toward an integrated model', *Journal of Palliative Care* 16 (3): 46–54.

Kirby, M.J.L. and LeBreton, M. (2001). 'The Health of Canadians – Interim Report; Volume One: The Story So Far', The Standing Senate Committee on Social Affairs, Science & Technology, Parliament of Canada.

Mallett, S. (2004), 'Understanding home: a critical review of the literature', *Sociological Review* 52 (1): 62–89.

Merriam, S.B. (1998), *Qualitative Research and Case Study Applications in Education* (San Francisco, CA: John Wiley & Sons, Inc.).

Miller, K.L., McKeever, P. and Coyte, P.C. (2003), 'Recruitment issues in healthcare research: The situation in home care, *Health and Social Care in the Community* 11 (2): 111–23.

Milligan, C., Gatrell, A. and Bingley, A. (2004), 'Cultivating Health: Therapeutic Landscapes and Older People in Northern England', *Social Science and Medicine* 58: 1781–93.

Relph, E. (1976), *Place and Placelessness* (London: Pion Limited).

Romanow, R. (2004), 'Health Care in Canada; Sustaining Medicare: The Commission on the Future of Health Care in Canada', in: F. Baylis, J. Downie, B. Hoffmaster and S. Sherwin (eds) *Health Care Ethics in Canada,* 2nd edn (Toronto: Nelson).

Shamian, J. (2006), 'New strategies for health care in the home', Waterloo Smarter Health Seminar Series, <www.infranet.uwaterloo.ca.>

Shapiro, E. (2002), 'Home care. The Health transition fund synthesis series', Minister of Public Works and Government Services Canada. Catalogue #H13-6/2002-2.

Somerville, P. (1997), 'The Social Construction of Home', *Journal of Architectural and Planning Research* 14 (3): 226–45.

Stajduhar, K.I. and Davies, B. (1998), 'Death at home: challenges for families and directions for the future', *Journal of Palliative Care* 14 (3): 8–15.

Wakefield, S. and McMullan, C. (2005), 'Healing in Places of Decline: (Re)imagining everyday landscapes in Hamilton, Ontario', *Health and Place* 11: 299–312.

Wiles, J. (2003). 'Daily geographies of caregivers: mobility, routine, scale', *Social Science and Medicine* 57: 1307–25.

Williams, A.M. (ed) (1999), *Therapeutic Landscapes: the Dynamic Between Place and Wellness* (Lanham, MD: University Press of America).

Williams, A.M. (2002), 'Changing geographies of care: employing the concept of therapeutic landscapes as a framework in examining home space', *Social Science and Medicine* 55: 141–154.

Williams, A.M., Chessie, K. and Marche, A. (2003), 'The impact of palliation on familial space: Examining home space from the perspectives of families who are living (and caring) for their dying loved one at home', Unpublished Research Report.

Williams, A.M. (2004). 'Shaping the Practice of Home Care: Critical Case Studies of the Significance of the Meaning of Home', *International Journal of Palliative Nursing* 10 (7): 333–42.

PART 4
Contesting Landscapes as Therapeutic: Contemporary Advances

Chapter 14

Gay Bathhouses: The Transgression of Health in Therapeutic Places

Gavin J. Andrews and Dave Holmes

Introduction

A significant progression in recent therapeutic landscape research has been that geographers have used the concept far less as an explanatory 'bumper sticker' for labeling and describing empirical phenomena, and have started to actively investigate and (re)inform the concept itself by thinking far more carefully and directly about the constitution and nature of therapeutic and its relationships to place (Smyth, 2005). This greater reflection has involved a fundamental recognition that therapeutic places and effects are relative to the many different groups involved in their production, consumption and occupation (Wilson, 2003; Baer and Gesler, 2004; Andrews, 2004; Gastaldo et al., 2004; Wakefield and McMullan, 2005; Conradson, 2005, Johnsen et al., 2005; Andrews et al., 2005). Indeed, there has been a recognition, for example, that no universally experienced therapeutic place exists, and what might be an extremely therapeutic place experience for one person, might be a diametrically opposite 'anti-therapeutic' experience for another, and indeed any number of perceptions, feelings and possibilities might exist for any person and place (Smyth, 2005).

Continuing this critical line of inquiry, based on a recent ethnographic study of three gay bathhouses in a large Canadian city, this chapter considers how these establishments – whilst possessing a therapeutic name, long heritage and hosting many facilities within them traditionally recognized as therapeutic – for users, are attractive and welcoming also for non-traditional reasons related to their sub-culture and sexuality. Moreover, this contribution considers how they host sexual behaviors which are potentially extremely hazardous to health and the target of public health measures. Although by comparison to more socially mainstream bathing spaces (see Gesler, 1998; 2003; Andrews and Kearns, 2005), the gay bathhouse might be regarded as a rather extreme or infrequent phenomenon, we argue that the insights it provides, into both the generation of alternative versions of therapeutic and the potential misalignment of therapeutic experiences and health impact, provides useful lessons for public health. Indeed, public health messages can fail to achieve their intended goals if they fail to connect with their target audience. Here then, the specialist place of the bathhouse teaches us how potentially difficult and hard fought understanding and connecting to people, their motivations and behaviors potentially is, and how this understanding can be determined through detailed qualitative investigation. We start the chapter by providing a brief history of the bathhouse, and specifically the gay bathhouse, and move on to describe one of our three study sites in detail and some key informants' views.

From Bathhouses to Gay Bathhouses

Communal public bathing has a long history with varied roots in the very earliest of civilizations and societies. Historical records tell us that communal bathing was commonplace in Greece from the sixth century onwards, albeit the preserve of the rich and powerful and those who participated publicly in athletic competition. As is evidenced in archaeological remains and documentation, throughout their former empire, the Romans thereafter adopted the Greek idea of bathing and conviviality and made communal bathing far more commonplace and accessible, now part of a famous cultural legacy and heritage that remains and is revisited to this day (Gesler, 1998; 2003). Although not as commonplace, communal bathing can be also traced in many forms throughout medieval Europe and later surfaces as having important cultural and religious significance in such times and places as the sixteenth century Ottoman Empire and, as Sento, part of daily life in Japan from the mid nineteenth century onwards. In more modern times, communal bathing has been less about symbolic or religious meaning and is more utilitarian. In Britain, for example, as part of the orthodox medical beliefs of the time, the Victorians sea-bathed in order to cure themselves of various illnesses (Andrews and Kearns, 2005), and by the early late nineteenth and early twentieth century, public baths in many UK cities and towns combined affordable washing facilities for the urban working class, whose houses often lacked them, and basic swimming pools for their recreation. Here the communal element was the swimming pool whilst, signifying a lasting legacy of Victorian modesty, the baths themselves were more typically individual and segregated by cubicles. These public baths can still be found in contemporary British urban landscapes but now more in name than anything else. The local 'City Baths' or 'Swimming Baths', for example, typically refers to a building that contains a modern rectangular competition-rated swimming pool, as the actual baths for washing having long since been replaced by sports-related facilities such as squash courts and gyms. Moreover, as local governments gradually sell off or contract out sports services to private companies, even these traditional names are disappearing from local urban landscapes, replaced by more marketable postmodern titles, often that reflect times and places elsewhere (Bale, 2003) (for example – 'The Pyramids Pools'; 'Fifthdimension'; 'South Pacific Pool').

In the late nineteenth and early twentieth centuries, bathhouses in North America emerged as facilities where men had sex with men. These were not typically heterosexual facilities that had suddenly transformed into gay spaces, but for the most part were new facilities making use of other types of commercial buildings, very often in central city neighborhoods. At the time, gay sex had been made illegal, creating sexual outlaws where they had not previously existed, and forced gay men to 'become experts at stealing moments of privacy and at finding cracks in society where they could meet and not get caught' (Bérubé, 2003, 34). As an initial response, societal spatial 'cracks' expanded with time to include places such as public parks, washrooms, storehouses, alleys, and railway cars, to name a few. These, however, were all highly dangerous places to frequent because if the gay men were discovered, they could be arrested, blackmailed, beaten, robbed, or killed. As a result, gay men began to regard bathhouses as places of relative safety and privacy. Moreover, because bathhouses existed as places of refuge from an unwelcoming heterosexist society, they provided a degree of camaraderie and belonging through participation in a gay culture, as well as erotic facilities (Bérubé, 2003). Meanwhile, in Europe itself, things developed slightly differently. Bathhouses certainly emerged, but even more secretively and were less recognizable, both individually and as a social phenomenon.

For 40 years, gay bathhouses have been an important element of the gay male community (Tattelman, 1999). A typical North American city with a thriving gay culture might have dozens of these establishments, particularly if a definable gay village exists as an open planning space where authorities 'tolerate' their existence. Otherwise, in less gay accepting cities and towns they might be far fewer in number and very carefully disguised. Bathhouses have provided a 'space to define, support, or flaunt sexual interests' (Tattelman, 1999: 71). Indeed, over time, the gay bathhouse has evolved into a safe space for both sex and the development of male gay pride and gay self-acceptance.

Welcome to Gay Bathhouses

Three bathhouses were selected as study sites based on their local reputation as popular gay spaces and the willingness of the owner managers to participate in the study. In total 147 hours of research time was spent collectively at these establishments, whilst data collection techniques within the ethnography included interviews, observation of design features (and interactions related to them), and observation of social behavior. In order to capture changes throughout the day, for each location, the researcher was regularly present in the daytime, early and late evening. Our data below is extracted from observation and key interviews with clientele conducted in all three study sites. The four themes that structure the findings below – participation and community, social interaction, design and interaction, and safe and unsafe sex – formed the basis of many questions asked, but sub-themes more strongly emerged as categories during the analysis. Aided by computer software, the analytical approach was focused and systematic, certainly being far more structured and linear than grounded theory, for example, and its derivatives. Because this is a qualitative study, the generalizability of the findings is not a concern. However, when reflecting on the potential transferability of the findings, it should be noted that the study is based on North American sites for gay male activity. Consequently, situations in rural areas, Europe and elsewhere, and in lesbian cultures, might differ considerably. Indeed, there is no universal form of gay bathhouse and gay bathhouses are certainly not part of all non-heterosexual cultures.

The architecture of bathhouses studied creates an environment of multiple public, semi-public, and private spaces where men can sexually encounter each other. In terms of an introduction, please step with us into one of our bathhouse study sites. Apart from certain differences that are noted in the following discussion, physically, it is fairly typical of the three bathhouses studied, and indeed of North American gay bathhouses in general. Holmes et al. (2007, 276–77), comment:

> The bathhouse was discreetly located with no signage for identification on a downtown side street in a major Canadian city. It had three floors, but because the basement was employed as a laundry and storage area, only the main floor and the second storey were designed for use by clients. Upon entering the bathhouse, the researchers noted that the inside walls and ceiling were painted in flat black and the lighting was low, thereby significantly decreasing visibility. Trance, techno, and house music played alternately throughout the entire establishment.
>
> Because the hallways were narrow, when two males passed one another, each had to turn partially sideways or brush against the other. These hallways were circular, maze-like, and seemed connected to each other without a discernable pattern. Mirrors were positioned sporadically on the walls, and there was one television set projecting gay leather pornography. This provided the only light present within the backmost hallway. In that hallway, there were functional chains and shackles fastened to the walls; these were not used during our visits, but the manager informed us that patrons do use them regularly.
>
> There were twenty-two rooms, located on the two accessible levels. All rooms were equipped with a locker with lock, a small side table and a small bed with black sheets and pillowcases. All but one room had at least one large mirror on one of its walls; the mirror in one room was a one-way viewing window, allowing the neighbouring room full visual access. Other room-to-room access points occurred in the two rooms fitted with *glory holes*. These are small, oval holes cut into a wall to form connections at waist height between adjacent rooms. The longer side of the oval aligns with the vertical axis thus allowing for the introduction of a male's genitals into an adjoining room without the rest of the body being seen by the invisible individual on the other side of the wall. Both rooms with glory holes had three holes per room, allowing genital access to three adjacent rooms. In addition, two additional rooms were designated as sling rooms – rooms within which there was a leather seat suspended from the ceiling at waist height by four chains. In addition, there were twenty-five lockers with locks available along the wall of one hallway for those not interested in renting a room.
>
> As well as the private spaces described above, other, semi-private, spaces were freely available to all bathhouse patrons. These semi-private spaces included two bathrooms, one located on each floor, the dry sauna, the steam room, and the rectangular common shower room with its eight fixed showerheads spraying into an area with one central drain. Annexed to the shower room was another space, which had a *douche* nozzle positioned at waist height and designed to have a rubber hose attached to it for rectal douching. This douche station had a drain located directly below the nozzle attachment to prevent the water in this area from running into the centre of the shower room. Located at the end of a hallway, in the backmost corner of the bathhouse, was an open space measuring approximately four square metres, which was designated as a dark room. This area had no enclosing walls, and was a room by designation only. Within this space, it was impossible to see one's hand in front of one's face. The final semi-private space was the snack bar lounge, which was furnished with a leather couch. Here, the clients could purchase snack food, coffee, paper towels, cigarettes, and sex toys such as leather paraphernalia, Crisco, rubber or latex gloves, dildos, lubricants, and cock rings. The space was used for relaxing, reading newspapers, smoking, and conversation with the manager and bathhouse employees.

The price of admission varied; eight hour rentals cost between $18.00 and $25.00 depending upon the size, the location, and the amenities (windows, glory holes, slings, etc.) of the room being rented, and the time of day. During the 'lunch-time special,' between 11:00 and 13:00 room prices were reduced to a range from $10.00 to $15.00. Regardless of time of day, however, a locker cost the standard price of $5.00. The length of time that clients could stay in this bathhouse was limited; they must leave the premises after 16 consecutive hours (two eight hour blocks) and could not rent either the same room or a different one immediately.

The bathhouse was very crowded during the evenings. Men were everywhere; some wandering around, circling, some standing against the walls watching the spectacle. Men gathered in private rooms searching for direct sexual contact – some of the doors to these were open revealing men sitting or lying on their beds in a dim light, or standing in the dark. The men occupying the rooms set the stage for encounters to occur, or not. When a man approached a room, the occupant might, or might not, accept him. It was obvious that all communication was through eye or body signals.

Participation and Community

During interviews, a specific line of questioning was focused on the pull and attraction of bathhouses, and the identity and attachment that clients might develop with them. The data suggests clearly that bathhouses provide not only sex related facilities and opportunities, but a community, a place where individuals can socialize and feel safe and at ease. In this particular sense, contemporary gay bathhouses are perhaps not far removed from their predecessors in ancient Greece or Rome, that similarly provided social as well as physical warmth and conviviality. The key difference being, however, that because the gay community is often marginalized, clients of gay bathhouses are certainly not as high profiled and elite as communal bathers in these former societies.

When asked why he attended a bathhouse, the following client responded:

> Well, you can meet friends, you can go to like a hot tub and sauna and steam room and stuff like that. Its safe, our place.

Indeed, in previous therapeutic landscape research, social features have been highlighted as important to the overall therapeutic experience (Andrews et al., 2005; Gesler, 1998; 2003). In terms of motivation. however, in many cases, a sense of community was a co-equal pull factor for attendance, along with basic sexual pleasure. Clients mentioned both in equal measure, often in the same sentence. For example, when asked specifically about sexual motivations, one client commented:

> Well, it depends. Like if the attraction is there, I'll go for it, but mostly I just chat and hang out with friends.

As a result of the community features, attachments to particular bathhouses were formed. This was reflected in that many clients have their favorite establishments and the ones that they liked far less. Indeed, a client might talk warmly, for example, about 'their' bathhouse, and very negatively about 'other' places. Bathhouses were certainly far from being a home-from-home, but clients identified with them much in the same way as they might a favorite restaurant or night spot, the people and activities found in-situ being as important to them as the physical features and facilities.

In terms of meeting others, however, beyond sex and friendship, for the most part, the clients we spoke to were not looking for long-term loving relationships in bathhouses. Nevertheless, one client highlighted some exceptions:

> Some people they do, especially the younger they do: oh maybe I'll find a boyfriend or a sugar daddy or whatever in here. But for me, no, you can't find a boyfriend in a bathhouse.

Reflecting many other situations in society where people socialize (such as nightclubs and bars), bathhouses were not considered to be the right place to meet 'the one'. However, in contrast to bars and nightclubs, bathhouses were considered to be far more efficient and trouble-free venues for finding sex, primarily because it was understood that, in contrast to patrons of bars and nightclubs, most clients of bathhouses are there specifically for this reason. In this regard, we asked one client to compare a bar and a bathhouse as a meeting place for sex:

> You hang out at the bar or at the club and, if you go there more than two years, you are sick and tired of it because it's the same crowd, same music, same place. You get bored because every weekend is the same thing, you know. And you're spending a lot of money, paying for entrance and sometimes you won't have fun either. But in here, it's different. At least you will have fun or meet new friends. At least you will get something in here. You know what I'm trying to say?

Because the bathhouses were located in known gay city areas with high proportions of gay people, these establishments were visited relatively openly. For most, getting in and out unseen was not a problem or consideration. In short, they were openly and obviously used as a regular local resource.

Social and Sexual Interactions

Another line of inquiry was directed at the nature of social interaction within bathhouses and specifically the norms, codes and rituals of these establishments. One particular finding was that the bathhouse changed significantly during the day in terms of clients and their activity. In particular, where openness in use (as noted above) changed slightly was in the case of married bisexual men; they typically spent brief visits during their working day. In contrast, openly gay men typically visited for much longer periods during the evenings. These two user groups used the space for different reasons and each thereby created a different atmosphere in the bathhouse. A client explained:

> We have two types, two different timings of people. Like afternoon timing is mostly bisexual or married people that come here. They just want to get off and leave. Nighttime and weekend is different type of people. You know.

The changing social composition of therapeutic places through time has been noted in previous therapeutic landscape research (Smyth, 2005). Indeed, different types of people can create different types of environment, with different qualities and attractions (Andrews et al., 2005).

Movement through spaces in the bathhouse was an important feature of their experience. In the internal search for sex, rituals were common, yet a great deal of variability exists in terms of encounters and negotiations. A client commented:

> You hang out or wander. If you see someone, well, you have to have an attraction. Like what his type is. And if he's attractive then, you talk to him, you end up having sex.

Another client commented on the different roles of speaking and body language in sexual negotiations:

> Well it depends on the person. Some people don't really like to talk, you know. If he's hot, why not just have sex, you know. But some people like to talk, and chat about it. They like to talk about what you like to do in bed – like stuff like that – get to know the person better. And communicate with the person.

Moreover, once any negotiation has taken place on whether or not to have sex, a great deal of variability exists with regard to where sex takes place in the bathhouse. Much depends, however, on individuals, their moods and preferences. A client commented on the negotiations and decisions that occur on where, whilst another commented on luring as a tactic:

> It depends on your mood and the other person. Some people are shy and they stay in their rooms and some people like to go walk around and check it out – they're curious.

> Sometimes they turn the light down – you can't even see the person so you have to wait until they come out or you go in closely and look inside at the person there.

On occasion, sex occurred in the public spaces in bathhouses, mainly hallways. This was not common but, as display and a communal event, held its own set of sexual attractions. A client commented:

> During the weekend, they are wild, and they're a little drunk or high, and they will do it [in the hallway].

The number of sexual contacts for each person per visit varied considerably, rather depending on the other clientele and the occurrence of mutual attractions. Only very occasionally did men arrive with preconceived ideas about exactly how many times they would have sex that evening and how. Nevertheless, the number of sexual partners could be high. A client commented on this volume issue:

> How many all depends honestly. Sometimes you get none, sometimes you get seven. It depends on the night. It depends on what is coming into the bathhouse. You know sometimes they are so dead or sometimes there are so many people who are not your type and you can't do anything. It all depends on that day and what kind of people coming in. Is it busy or not?

Given comments such as these, what should be punctuated is that, contrary to heterosexual stereotypes of bathhouses and indiscriminate 'knocking shops', the clientele of bathhouses can be choosy just like any individual; feeling sexual one day and not the next.

Interaction by Design

Mirroring recent therapeutic landscape research into more conventional settings (Andrews, 2004; Gesler et al., 2004), another line of inquiry focused on the role of design features in the overall place experience. The design and layout features described above in our initial description were very important in facilitating sexual activity and in creating an appropriate setting. Take, for example, the fact that whilst that particular bathhouse was maze-like, others were more regular and therefore more straightforward to navigate. Clients held firm opinions on which types of layout they preferred to navigate and why:

> Someone will walk away and you can't find them, but you have to try and find them. You get to go and look around. But other places are dead end. You see the same thing, and most people stay in their rooms and do nothing. And it's too clean.

Sensory modifications were also perceived to be very important to the overall experience. Whilst one bathhouse purposefully had reduced lighting, another was far more brightly lit. Clients possessed very definite views and preferences for these elements. One specifically commented on the anonymity, ambience and excitement that darkness provided:

> Well, honestly, if you go to the other bathhouse where it's bright and it's, well, for me, I don't like it because it looks like a hospital … It's too bright there. Like, most of the people who come here, they don't want you to recognize their face, they just want to have sex. So that's what a bathhouse is. But, you have to know that mostly at night time they are really busy and they are really drunk or they're really high on drugs and they hate light. So they want to make it a little bit darker, and you know … and darkness is more like adventure.

In addition, music was another purposeful environmental modification, which helped create an appropriate atmosphere (see Andrews and Kearns, 2007). Typically this soundscape was composed of fast synthesized dance music (often techno), though other types of music were used on occasion. A client explained:

> When you have all day at work you want to listen to relaxing music like some piano or classical music to relax you. It's the same thing here, you want to have sex you want to hear something more, what's the word? Music helps a lot to get into the mood. You can't have sex if it's so quiet – too quiet. Especially in the bathhouse.

At another level, the specifically designed theme rooms where favored more or less by clients, for personal reasons. A client commented on a steam room:

> It's fun. You see so many people there – so many hot guys. There's this whole like group sex thing. It's more *exciting*.

Along these lines, in one bathhouse the black room was a space that elicited very different and firm opinions. This room was totally devoid of light (both natural and artificial), where any encounters were purely felt, heard and unseen. For one client, this was very much a no-go area:

> I don't go there because you don't know what's there. Sometimes you step on someone's cum or you can't see the person's face. What is the point of going there?

In terms of favorite spaces, much depended of course on the individual and their preferences. As indicated by the above quotation, not all internal spaces were used universally by all clients.

The Safety of Sexual Practices

Because sex between men occurs frequently in bathhouses, which as venues are not well understood, they have become a focus of recent research inquiries into sexual health. While unsafe sex has been reported since the very beginning of the HIV epidemic, the recent underlying assumption in research has been that most gay and bisexual men now do not seek to intentionally have unprotected anal sex. The findings of recent research, however, suggests that unprotected anal sex among both HIV positive and negative gay men occurs frequently, not as unintentional omission or relapse, but as a calculated conscious activity (Holmes and Warner, 2005). Derived from the expression 'bareback horse riding' (or riding a horse without a saddle), *Bareback Sex* acquired popularity in the mid-1990s (Scarce, 1999). Put simply, it means voluntary unprotected anal intercourse (VUAI). According to both journalistic and academic research, gay and bisexual men practice barebacking for various reasons including to increase their physical sexual pleasure; to gain a feeling of true connection and intimacy with their partner; to transgress recommendations from public health organizations as part of their 'prevention fatigue'; for sexual arousal gained by embracing danger; and also because new treatments in the battle against AIDS mean that HIV infection is now, in most cases, no longer fatal (Holmes and Warner, 2005; Holmes et al., 2007; Scarce, 1999). It comes as no surprise that unsafe anal sex between men occurs in gay bathhouses (Holmes and Warner, 2005). One client commented that he frequently witnessed barebacking in his favorite bathhouse, and that he would be specifically asked to perform it. He also noted that:

> [they ask] particularly if they are drunk or high.

This is not to say, however, that bathhouse owners and managers actively encourage or condone this risky behavior. Indeed, the three bathhouses studied provided free condoms, available at their front receptions (and sometimes throughout the premises) and many clients made good use of them and prepared well. The following client was asked specifically about his safe sex practices:

> Well mostly I won't have sex when the person's in public. You'll mostly go back to the room and your room is already prepared with lube and condoms. You mostly prepare in the room.

Alternative forms of abstention, in terms of semen exchange, mean that unprotected sex is not necessarily bareback sex. These forms include ejaculating outside a partner. Indeed, this practice is also perceived as a pleasurable part of a sexual experience. One client commented:

> I like the feeling and seeing cum ... Umm, mostly face to face. Like cumming on my chest or on someone's chest. Most people like to see cum. That's what sex is about, it sexy. Yeah, sex is about the cum. Sometimes you see cum a lot, sometimes you see it a little. You know. It's just like, you know, you're human being. People are always curious, they just want to see it. Me, I like to see it.

Referring to safety and momentary decision-making during intercourse, a client noted:

> You obviously can't get pregnant, but there is a small chance of catching something. The stakes are really different but is all about being carried away and if either of you can be bothered to pull away.

Indeed, in these practices lies a curious mixture of pleasure and an implicit movement towards – but certainly not guaranteeing – greater personal health and safety. Conventional public health education rather unsupportively terms this 'the withdrawal method', typically in reference to perceived inadequate pregnancy prevention strategies in heterosexual encounters. The withdrawal, however, is evidently a complex practice mixing attraction, and a last chance safety precaution, for all sexual orientations.

Summary

In terms of being therapeutic landscapes, gay bathhouses clearly demonstrate the very different and non-traditional ways in which places can be therapeutic for those people who frequent them. Here, although facilities and experiences that are traditionally recognized as therapeutic – such as saunas, massage tables, showers – exist, the attraction is as much associated with the existence of a safe social space for a sometimes marginalized culture and its marginalized activities (see Andrews et al., 2005). Both of these things contribute complexly to some degree. As settings for potentially unsafe sex, bathhouses also demonstrate quite clearly that all that is perceived to be therapeutic is not, in a public health sense, free from disease potential, and necessarily good for health.

Over the last three decades, modern public health has evolved so that it actively seeks to intentionally modify the behavior of individuals in order to make people lead healthier lives. As is well documented and recognized, two fundamental aspects of this movement's foundation are: (1) health promotion (the view of health as not simply the absence of illness but the attempt to improve health without pathology as the reason for doing so); and (2) illness prevention (behavior modification as a method of reducing future risk of particular illnesses). With an emphasis on individual responsibility, the underlying assumption is that a capacity for self-government enables individuals to take control of their health and constitutes a means to make healthy choices (Lupton, 1998). Because of AIDS and other sexually transmitted diseases, sexuality has emerged as a particular phenomenon that can dramatically impact upon the public's health. The underlying assumption here of public health initiatives is that individuals would want to abstain from sexual practices that may put them at risk of acquiring a sexually transmitted infection. As bathhouses and specifically barebacking demonstrate, however, this might be an incorrect assumption. Under public health policy, certain high risk sexual groups have been isolated as in need of particular interventions in order to enable them to achieve the *normal* status of 'respectably "clean" groups in society' (as can be seen, for example, in current interventions which target *Gay Men's Wellness*, and so on; Lupton, 1998). For public health systems, gay men's social deviance is unique and highly problematic as it has strong social structures that serve to recruit, retain, and train participants in unhealthy behavior (Gauthier and Forsyth, 1999). As our exploration of bathhouses demonstrates, these places and their social structures are reinforced by combinations of therapeutic and sexual experiences and not solely by physical pleasure. Only by recognizing the need not to preach responsibility and to communicate effectively, and to understand the complexity of behavior and how it is socially and culturally reproduced, will successful inroads be made, and health connections forged with the gay community.

Acknowledgments

Dave Holmes would like to thank the Canadian Institutes of Health Research, Institute of Gender and Health for funding.

References

Andrews, G.J. (2004), '(Re)thinking the dynamic between healthcare and place: therapeutic geographies in treatment and care practices', *Area* 36: 307–18.

Andrews, G.J and Kearns R.A. (2005), 'Everyday health histories and the making of place: the case of an English coastal town', *Social Science and Medicine* 60: 2697–713.

Andrews, G.J, Sudwell, M. and Sparks, A. (2005), 'Towards a geography of fitness: an ethnographic case study of the gym in British bodybuilding culture', *Social Science and Medicine* 60: 877–91.

Andrews, G.J. and Kearns, R.A. (2007), 'Connecting pop music and health: from celebrity diplomacy to medicinal melodies', *Complementary Therapies in Clinical Practice* (in press).

Baer, L.D. and Gesler, W.M. (2004), 'Reconsidering the concept of therapeutic landscapes in J.D. Salinger's *The Catcher in the Rye*', *Area* 36: 404–13.

Bale, J. (2003), *Sports Geography* (London: Routldege).

Bérubé, A. (2003), 'The History of Gay Bath Houses', in W. Woods and D. Binson (eds) *Gay Bathhouses and Public Health Policy* (New York: Harrington Park Press).

Conradson, D. (2005) 'Landscape, care and the relational self: Therapeutic encounters in rural England', *Health and Place* 11: 337–48.

Gastaldo, D., Andrews, G.J. and Khanlou, N. (2004), 'Therapeutic landscapes of the mind: theorizing some intersections of health geography, health promotion and immigration studies', *Critical Public Health* 14: 157–76

Gauthier, D.K. and Forsyth, C.J. (1999), 'Bareback Sex, bug chasers and the gift of death', *Deviant Behavior* 20: 85–100.

Gesler, W.M. (1998), 'Bath's reputation as a healing place', in R.A. Kearns and W.M. Gesler (eds) *Putting health into place: landscape, identity and wellbeing* (Syracuse: Syracuse University Press).

Gesler, W.M. (2003), *Healing places* (New York: Rowman and Littlefield).

Gesler, W.M., Bell, M., Curtis, S., Hubbard, P. and Francis, S. (2004), 'Therapy by design: evaluating the UK hospital building program', *Health and Place* 10: 117–28.

Holmes, D., O'Byrne, P. and Gastaldo, D. (2007), 'Setting the space for sex: architecture, desire and health issues in gay bathhouses', *International Journal of Nursing Studies* 44: 273–84.

Holmes, D. and Warner, D. (2005), 'The anatomy of a forbidden desire: men, penetration and semen exchange', *Nursing Inquiry* 12: 10–20.

Johnsen, S., Cloke, P. and May, J. (2005), 'Transitory spaces of care: serving homeless people on the street', *Health and Place* 11: 323–36.

Lupton, D. (1998), *Risk and Sociocultural Theory* (Cambridge: Cambridge University Press).

Scarce, M. (1999), 'A Ride on the Wild Side', *POZ Magazine* (February), 32–40.

Smyth, F. (2005), 'Medical geography: therapeutic places, spaces and networks', *Progress in Human Geography* 29: 488–95

Tattelman, I. (1999), 'Speaking to the gay bath house: Communicating in sexually charged spaces', in L. William (ed.) *Public Sex/Gay Space* (New York: Columbia University Press).

Wakefield, S. and McMullan, C. (2005), 'Healing in places of decline: (re)imagining everyday landscapes in Hamilton, Ontario', *Health and Place* 11:299–312.

Wilson, K. (2003), 'Therapeutic Landscapes and First Nations Peoples: an exploration of culture, health and place', *Health and Place* 9: 83–93.

Chapter 15

The Therapeutic Landscapes of the Kaqchikel of San Lucas Tolimán, Guatemala

Julie M. Sperling and Jody F. Decker

This study of the Kaqchikel people of San Lucas Tolimán, Guatemala, focuses on the therapeutic landscapes of everyday life and the survival of people in a Third World society wrought with poverty and violence. The Kaqchikel are a people who have long suffered discrimination, displacement, violence, and poverty. These factors, interacting with their traditional cultural beliefs and values, have shaped the Kaqchikel's therapeutic landscapes, whose underlying practical and symbolic meanings have necessarily become multilayered and complex.

The aim of our study was to develop a context-specific and gender-aware understanding of how the Kaqchikel construct and use what they identified as therapeutic landscapes. To meet this aim, photovoice was used as a powerful research tool as it provided the opportunity for respondents to capture their own photographic images of their therapeutic landscapes, which became the focus of follow-up face-to-face interviews with the researchers involved.[1]

This chapter begins by first identifying the gaps in the therapeutic landscape literature specific to cultural and gendered understandings. Following a regional overview of Guatemala, and the San Lucas Tolimán municipality specifically, the methodology for data collection and analysis is outlined. Results are then presented in the form of four emergent themes: daily life and survival; development, success and a better future; 'getting out' and the utility of natural beauty; and negative ('nontherapeutic') landscapes. We conclude with a few statements on the various points of intersection from the emergent themes.

Notions of Health, Place and Landscape

As defined by Gesler (1993; 171), therapeutic landscapes are places that have 'an enduring reputation for achieving physical, mental, and spiritual healing.' In a later work, Gesler (2003) argued that a sense of place is multidimensional in character; it emphasizes wholeness, connectedness, and integration; it promotes healing from within; and it is conducive to healing as an ongoing and meaningful process. In

1 This chapter is based on more detailed work done by Sperling (2006).

order to understand the many factors at play in the healing process, it is important to examine four interconnected and overlapping landscape types when dealing with therapeutic landscapes: the natural, built, symbolic, and social environments. Working from a non-Western perspective, Dobbs (1997) adds a fifth landscape type – landscapes of belief and/or spirituality. The relative importance of each of these five landscape types varies according to the individual and his or her place, time, and socio-cultural context. This multifarious concept of sense of place that has informed how landscapes can become therapeutic has since evolved to encompass not only specific places with a reputation for healing, but also everyday landscapes and imaginary landscapes that contribute to the maintenance of health and wellbeing, and their role in holistic medicine (Andrews, 2004; Williams, 1998; 1999; Wilson, 2003).

Studies on the therapeutic nature of everyday landscapes are varied and include such topics as community gardens as sites of emotional, physical, and spiritual renewal for the elderly (Milligan et al., 2004), how negatively perceived unhealthy landscapes impact negatively on health (Wakefield and McMullan, 2005) and First Nations' therapeutic landscapes (Wilson, 2003). Wilson's research points to a lack of work done on non-Western conceptualizations of connections between health and place and suggests the need to explore people's everyday lives within a gendered perspective. Our study fills the gaps she identifies, in that it focuses on Indigenous peoples in a Third World context, highlighting how the Kaqchikel create and utilize therapeutic landscapes in a culture of poverty and violence, as well as the ways in which gender shapes these landscapes.

Parr (2004) suggests that although critical geography has tended to be sensitive to difference, health and medical geographers have generally been slow to contribute to research on gender differences. Gender is important to consider when dealing with therapeutic landscapes because the various social roles of men and women may mean they connect to therapeutic landscapes in different ways (Wilson, 2003). This is certainly true in Kaqchikel society, which tends to have a strong male bias. Men are generally responsible for working outside the home, usually in agricultural activities, while women look after the domestic duties (Fischer and Hendrickson, 2003). High levels of violence in Guatemala also disproportionately impact women more than men, with men enjoying a higher degree of mobility. As this study will demonstrate, Kaqchikel men and women have unique therapeutic landscape experiences.

The Guatemalan Context

Guatemala is a country of extreme contrasts. Its cultural and natural heritages are rich in both beauty and diversity, yet they are set against a backdrop of acute poverty, terrifying levels of violence, and endless discrimination and racism. Among the most prominent contrasts are those between rich and poor, urban and rural, *ladino*[2] and Indigenous,[3] and men and women.

---

2 A *ladino* is a non-Indigenous Guatemalan.

3 The term 'Indigenous' will be used here (as opposed to First Nations, Aboriginal, and so on) because in Guatemala the politically correct term for a non-*ladino* individual is

Guatemala has a long history of political violence and unrest. Civil war broke out in 1960 and ravaged the country for 36 years until the Peace Accords were signed in 1996. The insurgency was one of the longest in Latin America, leaving 200,000 dead and 40,000 'disappeared'[4] (Blum, 2003; Green, 1999). During the most intense period of military repression and violence (1979–82), the military burned over 440 rural Maya villages suspected of collaborating with guerrilla forces (Blum, 2003). Survivors of the war still suffer; they live on the economic and social margins of their communities and continue to experience the trauma to both their bodies and their memories (Green, 1999).

There is also a new wave of violence gripping the country. The most prominent forms of violence are gang violence and general criminal delinquency. Contemporary violence is the product of a number of factors, including the desperation caused by extreme poverty, the residual effects of past violence and militarism, and the inability of the police and military to control the violence (Munaiz, 2005).

Homicide levels today are comparable to those of the most violent periods of the civil war (Seijo, 2005). Violence against women is of particular concern – in 2001 there were 1,116 violent deaths of women reported (SNU, 2003). Violence against women is so grave in Guatemala that in 2002 the country had the fifth-highest level of female homicides in the world (Cereser, 2005). For Indigenous Guatemalans the situation is also worrying, as the violence is more generalized, more intense, and dates back to a much earlier time (SNU, 2003). Racism in Guatemala is still very pervasive and heavily engrained in social institutions and norms.

San Lucas Tolimán is a small, predominantly Indigenous municipality located on the southeast shores of Lake Atitlán in the department of Sololá, which is part of the highland region of the country. The population of the municipality is roughly 22,000 and 89 per cent of the population is of Kaqchikel heritage (Abell, 1997; Blum, 2003). The municipality includes the *cabecera* (head town), also named San Lucas Tolimán, and 15 other small towns and villages, seven of which were home to one or more of our research participants. The economy of San Lucas Tolimán is predominantly agriculturally based, with most people growing coffee to sell and maize and beans (and some other vegetables) for subsistence.

Access to land – or lack thereof, as is more often the case – is a central issue in Guatemala, and the municipality of San Lucas Tolimán is no exception. Guatemala has one of the highest inequalities of land distribution in all of Latin America; 2 per cent of the population controls roughly 70 per cent of the arable land (Elías et al., 1997; Green, 1999; Tiney, 2000). Only 24 per cent of all the agricultural lands in the country are Indigenous-owned, while 70 per cent are owned by male *ladinos* (PNUD, 2002). Rural Guatemalans generally have small plots located on steep hillsides whose marginal production levels, insufficient to meet a family's needs, are significantly less than those of flat, irrigated lands (Elías et al., 1997). Women also tend not to own land (only 6.5 per cent nationally), because land is generally passed from fathers to sons (PNUD, 2002).

---

indígena, which in English is Indigenous.

4 The 'disappeared' (*los desaparecidos*) are those people who were forcibly detained or kidnapped, tortured, and killed by government forces and whose bodies were never found.

In terms of land access and tenure, the people of San Lucas Tolimán are arguably more fortunate than others in the region, given the dedication of the Catholic parish to acquiring and redistributing land. Over the years, the church has bought land as it has become available and redistributed it to approximately 2,000 families through the San Lucas Development Program (Abell, 1997).

Despite the fact that more and more *luqueños*[5] own their own small plot of land, the municipality is part of one of the poorest regions of Guatemala. The highlands are the most socio-economically vulnerable region in the country, and poverty, which has been on the rise in the past 20 years, is so extreme that the highlands are considered a 'region in crisis' (Elías et al., 1997, 193). In the seven communities studied, the average yearly income is Q5,940 and families in six of the seven communities run an average annual deficit of Q3,219 (Municipalidad de San Lucas Tolimán, 2002a–g).

Within the region, the most vulnerable groups are those that have been historically marginalized: women, the elderly, children, and the Indigenous (Elías et al., 1997). In male-headed households, as in society in general, women are marginalized and disproportionately burdened. Though it is heavily engrained in social convention that it is the man's responsibility to provide for his family's needs, it is also his 'right' to set aside part of his income for himself; this money is generally spent on alcohol, other women, and socializing with friends (PNUD, 2002). If the family is short on money, it is generally up to the woman to figure out a way to make ends meet (PNUD, 2002).

Due to rising prices, women on a national scale are becoming increasingly integrated into the formal economy, adding work responsibilities outside the home to their domestic activities and extending their working day substantially. The monthly income of these working women, however, is on average only 59 per cent of what men earn (PNUD, 2002). The men of San Lucas Tolimán typically earn about Q20 per day as agricultural laborers (Municipalidad de San Lucas Tolimán, 2002a–g). Although women tend to be less integrated into the formal economy, many of those interviewed ran small businesses out of their homes to help make ends meet, and others sold their weavings for an average price of Q25 to Q30, each piece taking roughly one month to make (Municipalidad de San Lucas Tolimán, 2002c, for example).

Given the economic pressures facing them, it is not surprising that *luqueños* report food insufficiencies in terms of quantity and quality as one of the major problems they face on a daily basis. The most common illnesses are malnutrition, gastrointestinal problems, respiratory infections, and general feelings of being unwell. Most, if not all, of the reported health concerns are caused by a combination of environmental and socio-economic factors, such as poverty, lack of adequate sewage, garbage disposal, potable water, housing, and food, as well as changes in the weather (Municipalidad de San Lucas Tolimán, 2002a–g).

Less than one half of Guatemalans have access to government health services and less than 1 per cent of the country's GDP is invested in the health sector (Hurtado and Sáenz de Tejada, 2001; Vela et al., 2001). Healthcare in San Lucas Tolimán is a combination of Western medicine and traditional Indigenous medicine. Although

---

5 A *luqueño* is a resident of San Lucas Tolimán.

most Kaqchikel now use the former, home remedies are still used by many to treat simple illnesses because they are less expensive and seen as less intrusive than chemical medicines.

Kaqchikel Culture and Worldview

The Kaqchikel worldview is based on the notion of a sacred covenant between humans and cosmic forces, which is intimately tied to notions of balance, reciprocity, and cosmic cycles (Fischer, 2001). The earth is sacred for the Kaqchikel; it is their mother and their source of life (Ventura, 2000). Agricultural cycles are integral to the Kaqchikel worldview, and maize and the *milpa*[6] play a central role in Kaqchikel life. As one Kaqchikel individual states, 'The *milpa* … is more than just corn; the *milpa* is an entire world, a way of life' (Green, 1999, 17–18). The *milpa*, according to Alcorn and Toledo (1998), is both a social institution and process, with its spatial aspect being secondary, and its religious and social importance often exceeding its nutritional and economic importance.

It is predominantly the men who work in the *milpa*. However, Kaqchikel women also play an important agricultural role by helping during the planting and harvesting phases. Women also maintain household gardens that produce supplementary food, medicinal herbs, plants to be used in religious ceremonies, and ornamental plants and flowers (Keys, 1999).

There are many religious ceremonies dedicated to the *milpa* and agriculture in general. Most Maya have converted to Christianity (predominantly Catholicism) and many say that they no longer celebrate these ceremonies. However, Ventura (2000) argues that a spiritual connection to the earth and the *milpa* still exists, whether it is in the form of traditional Maya ceremonies, Christian prayers and services, or a combination of the two.

In addition to the important and sacred relationships between the Kaqchikel and the earth, interpersonal relationships and community cooperation are key aspects of Kaqchikel society. The extended family is the basic social unit and it is embedded in larger community structures (Green, 1999). In Kaqchikel society, every individual plays an important role in community life. The *ucholaaj'* (order) of community life is based on moral and spiritual values, including respect, equanimity, responsibility, honesty, humility, gratefulness, listening, and serving one's community. Practicing these values translates into equilibrium and harmony and guarantees *utz' k'aslen* (wellbeing) and social cohesion (López Batzin, 1997).

Methodology

Feminist methodology guided the research because of the value it places on local lived experience, situated knowledge, subjectivity, power, and difference (Moss, 2002). We used reflexivity to address the issues of how the 'personal' affects research. We were also cognizant of the methodological issues of working in a foreign language.

---

6 A *milpa* is a small plot of land used to grow maize.

Although the first language of the majority of our participants was Kaqchikel, we chose to work in Spanish rather than deal with the problematic issues of using an interpreter. The direct quotes used in this chapter have been translated from Spanish to English for brevity.

The two main methods we used to collect data for the study were structured open-ended interviews and photovoice (also referred to as photo novella or photo elicitation). Through snowball sampling, a total of 28 participants were recruited – 14 women and 14 men. We selected self-identified Kaqchikel individuals who were willing to participate, regardless of their 'traditional' or 'expert' status (or lack thereof). Once participants had agreed to take part in the study, they were immediately given a disposable camera, instructed on its use, and asked to take roughly 12 pictures of their own therapeutic landscapes or, as it was explained to them, places that made them feel good and contributed to their overall health and/or wellbeing, or places that did not make them feel good.

Participants were generally interviewed twice – once when they were given the camera and again once the photographs had been developed. The first interview gathered information about the participants' daily lives – what a typical day was like for them, what responsibilities they had, and what roles they played in their families and in the community – as well as their views on health. The second interview focused on the participants' own therapeutic landscapes. Participants were asked to explain the significance of each of their photographs and how the place captured in each photograph affected (positively or negatively) their health and wellbeing.

Photovoice entrusts cameras to the hands of participants to enable them to act as expert recorders and represent their communities and their lives from their own perspective (McIntyre, 2003). Like feminist theory, photovoice assumes that the research participants are the authorities on their own lives and that what they have to say is important (Wang and Burris, 1994). Through photography, the previously marginalized and disempowered are given a voice with which they can share their lives. Photovoice also facilitates the collection of data by 'insiders' in situations and settings that otherwise would not be open to the researcher (Wang and Burris, 1997).

The photographs were the central elements of our analysis, with the verbal narratives elicited during the interviews playing a complementary role in helping us better contextualize the participants' photographs. To keep the analysis focused and coherent, each of the over 300 photographs and their accompanying narratives were examined with one question in mind: why is this particular place therapeutic to this participant?

Many of the photographs captured multiple therapeutic aspects, and in our analysis these were not ranked at all because no hierarchy was evident; rather, the themes were interconnected, drawing upon and emerging from one another. The themes in each photograph were given equal importance and were all recorded, along with the gender of the participant and the specific place that was shown in the photograph. Four themes emerged from the data and, within each, gender differences were explored.

The Four Emergent Themes

The four themes that emerged from the data are: daily life and survival; development, success, and a better future; 'getting out' and the utility of natural beauty; and negative or nontherapeutic landscapes. All four themes are not equally prominent in the verbal and visual narratives of the research participants, nor are they discrete entities. Rather, the themes are interwoven to reflect the complex and dynamic nature of the Kaqchikel lived experience. Before discussing these themes in detail, participant health beliefs, as determined via data analysis, are outlined.

The participants operationalized and expressed their health beliefs in the context of their daily lives. Many participants spoke of the importance of health in their daily lives because without it they would not be able to work, to fulfill their responsibilities, or to enjoy life. Antonio explained what health meant to him in this way: 'I live well, my head doesn't hurt and neither do my feet … I eat well, I work well, I sleep well.' Or, as Rafael said, 'Health is necessary in order to work – gathering firewood and walking to the *milpa* are chores that require strength and health.' Being able to work is of utmost importance because of the tenuous nature of Kaqchikel daily life and survival.

Participants initially defined health as a predominantly physical condition, treated first with natural remedies, then chemical medicines as needed. When physically ill, most participants preferred to recuperate at home rather than in a hospital, because it was a familiar, comforting environment and their families could care for them there. According to Tomás, 'The home is a cradle, a refuge.' Mercedes also preferred to be at home because 'in the hospital one is like a prisoner.'

Further discussions of health often revealed a strong belief in a mind-body connection. Many participants remarked that being healthy meant that they were not overly worried or stressed, noting that stress could impact their physical health and vice versa. Ailments such as headaches, stomach problems, and fatigue were often attributed to the stresses of everyday life. Having good relationships with their families and friends, as well as with God, also played an important role in determining their health and happiness. For Rebeca, 'moral health' was an integral part of her overall wellbeing, because it meant that she was getting along with others and did not have any conflicts with her family or friends. Being content with oneself was often the first step in having good relationships with others. As Isabel said: 'If we have peace in our hearts, we'll see everything in a positive light; we'll be happy with our family, our friends, our community.' This theme of emotional and social wellbeing was a dominant and common thread in the overwhelming majority of the therapeutic landscapes identified by participants.

Survival and Daily Life

Not surprisingly, the daily struggle to survive is the most prevalent theme and the strong unifying thread that ties the themes together. Survival for poor Indigenous peasants in Guatemala is precarious and achieved on a day-to-day (or very short-term) basis, thus placing a great deal of stress and worry on individuals. The most prominent characteristic of survival-based therapeutic landscapes is the relation to food, whether it is production, purchase, storage, or preparation.

Some crops, such as maize and beans, are grown for home consumption and are the basic staples of the Kaqchikel diet, while others (for example, tomatoes, coffee, and watercress) are sold for profit as well as consumed in the home. Regardless of whether the crop is eaten or sold, a healthy crop (even if it only consists of a single fruit tree) guarantees short-term survival. Not having to worry about whether they would have enough food to feed their families was, for research participants, therapeutic.

There are marked gender differences in terms of which agricultural landscapes were therapeutic for men and women. Men are responsible for the majority of agricultural activities in Kaqchikel society, and thus it is not surprising that their photographs and narratives focused on their fields, which were either primary or supplementary crops. The crops that were photographed the most included maize, beans, coffee, tomatoes, and watercress.

Women, on the other hand, are responsible mainly for domestic activities, and though many women did take photographs of agricultural plots of land (rarely their own), an additional symbol of food security that was almost exclusively female was the fruit trees growing in the yards of their homes. Not only did these trees provide fruit for the family, but they were also a source of home remedies and a good source of vitamins for growing children.

The sacredness of the *milpa* was acknowledged by a number of participants, who talked about the ceremonies they performed before planting and harvesting.

Figure 15.1 Stored cobs of maize from the harvest (photographed by Felipe)

Photographs of the *milpa* prompted some participants to speak of its role in linking them to their ancestors, since the *milpa* is an important symbol in the Kaqchikel worldview, and present-day Kaqchikel continue to sow, harvest, and eat it as did their ancestors. Such was the explanation offered by Felipe, who photographed cobs of maize that had been harvested and were now being stored (Figure 15.1). Felipe spoke of how his ancestors had also cultivated maize and it therefore served as a link to the past; however, the overarching theme of the photograph was not just the sacred and ancestral aspects of the *milpa*, but rather its importance as a vital food source. As Felipe said, 'Even if there is no food but there is maize, life will go on.'

Another aspect of the daily survival theme was income-generating activities that provided participants with an additional way of making ends meet. Men whose primary source of income was not agriculture usually talked about formal employment, emphasizing the fact that the money they earned was essential in sustaining their families. Informal employment, on the other hand, such as small businesses run out of the home, was almost exclusively the domain of women and was both spoken of and photographed. For example, Diego's wife (see Figure 15.2) buys unripe bananas in bulk, covers them until they ripen, and then sells them for one *centavo*[7] more than she originally paid. The crucial role of these few *centavos* earned by women in their small businesses is evident in Diego's statement that 'this is what we live on' – the money earned by Diego's wife plays an essential role in sustaining their family from day to day.

Figure 15.2 Diego's wife and her small banana business

7 One *centavo* is one one-hundredth of a *quetzal*.

A final important activity for the women we interviewed was the traditional Maya art of weaving. More important than the economic benefits of weaving is the cultural significance of this art form. Nearly all women interviewed discussed the important role that both wearing and weaving *tzyäq* (traditional dress) played in their Kaqchikel identity. María photographed herself wearing her favorite *uq* (skirt) and *po't* (blouse), explaining that she puts on these particular items of clothing when she is feeling sad because the colours and designs lift her spirits. Weaving is a woman's art, and though the economic gains accrued are far from indicative of the labor and time invested, weaving is a source of pride and identity for Kaqchikel women.

Also prevalent within the daily life and survival theme is the fundamental importance of the hydrological cycle, evident in many of the participants' photos. Described as 'a vital liquid' or 'the fountain of life,' the importance of water took many forms in the photographs of male and female participants. Two obvious forms of water shown in the photos were Lake Atitlán and rain, while other participants traced the hydrological cycle back to the forests covering the mountains. Verónica photographed the forest and explained, 'Without the forest there is no rain, and without rain there are no crops, and without crops there is no life.'

The lakeshore was also identified as an important place for growing crops because of its proximity to water, which practically guarantees the success of the crop and means they can be grown year-round. One of Felipe's photographs showed three water-related elements – the lake, the lakeshore, and the coming rain – and while the lake, at first glance, appeared to be the focus of the photo, Felipe's description revealed its profoundly agricultural theme with the approaching rains and the fertile lakeshore being the actual focus.

A number of participants also identified threats to both the lake and the hydrological cycle, citing pollution and deforestation as major negative influences on the health of these resources and processes, and ultimately on their own survival. Isabel stated, 'Now there's a view of the swamp, not the lake,' referring to the weeds, algae, and mud that have taken over the lakeshore due to the pollution running into the lake from the streets, especially during the rainy season.

Daily life for the men and women of San Lucas Tolimán is labor-intensive and exhausting. Technologies that make everyday life easier were identified by both men and women as contributing positively to their overall health and wellbeing. The technologies identified by women tended to be focused on the home, such as running water or the more efficient stoves that are replacing the traditional hearth stones set around a fire on the kitchen floor.

Men tended to identify technologies that were outside the home, such as a pickup truck for easier transport of harvested crops or firewood, or technologies at the community infrastructure level (water tanks or new roads). Though these technologies were utilized by both men and women daily, it was the men who were more often instrumental in their construction and thus identified an element of pride rather than simply convenience in them. For example, several participants photographed a recently constructed road to one of the villages located a fair distance off the main highway. As photographed by Gilberto, it means easier travel since pickup trucks can now reach his village, and it is also a symbol of cooperation and accomplishment for the men. For women, however, the road means a faster connection to the 'outside world' in case of emergency (such as complications during childbirth).

Staying healthy was also particularly important in terms of survival and daily life. Though men often spoke of the importance of hygiene (usually in the home setting), they did not specifically identify hygienic places as therapeutic. Women, on the other hand, both spoke of and photographed 'clean' home spaces, such as the kitchen and yard, most likely because they are charged with the responsibility of keeping those spaces clean. Luz's photo of her stove is not only a photo of technology but an example of good hygiene, with the *nixtamal*[8] and beans out of the reach of dogs and cats. Luz also commented that women were in charge of keeping the house clean, while the men were responsible for keeping the streets clean – a clear demonstration of Kaqchikel gendered space. For both men and women, cleanliness was not only a precursor to good health, it was also a source of dignity. For example, many women planted flowers in their yards in order to, as Ana said, 'brighten the home,' and for some, looking after their flowers was a therapeutic distraction in addition to the happiness their flowers brought them.

In addition to hygiene, women also frequently talked about and photographed medicinal plants that were found in their yards. For example, Mercedes photographed her lime tree, saying that the limes were a good remedy for stomach aches, while Verónica pointed out two medicinal plants in one of her photos. Having medicinal plants on hand also represents one less expense for women. In addition, medicinal plants are often traded between neighbors, thereby fostering stronger community ties. An important symbol in one of Dolores' therapeutic places is the eucalyptus tree behind her house because it represents both personal and community health. Members of her community come to her for leaves to make various medicines and she is happy to give them away (rather than sell them) because she is glad her neighbors are using natural medicines rather than buying chemical ones.

Dolores' photograph of her eucalyptus tree is also an example of the second theme to be discussed – community development, success, and building a better future. In some respects, this second theme can be viewed as a natural extension of the survival theme, as individuals and the community go beyond short-term survival to increase their prosperity and better the quality of their lives both now and in the future, thus ensuring longer term survival.

Community Development and Building a Better Future

For many of the participants, community development was an important aspect of their therapeutic landscapes and encompassed everything from actual infrastructure (schools, water tanks, electricity) to the hard work and cooperation necessary to achieve those more tangible results. Community development as symbolized by infrastructure was located in the male sphere. Men frequently talked about the importance of working together to build physically better communities, as seen in

8 *Nixtamal* is softened grains of maize that have been boiled in lime water and will later be ground into a paste to make tortillas.

Figure 15.3 The community's water tank (photographed by Daniel)

Daniel's photo of the water tank his community fought hard to get built (Figure 15.3), and in Antonio's photo of the installation of electrical lines in the new community of Totolyá.

Men's participation in the public sphere is very well established in San Lucas Tolimán, and when women attempt to challenge this dominance their efforts are greeted with resistance from the men, as shown in this quote by Susana:

> Sometimes our husbands don't let us participate in meetings and have fun. Men have the power because they go to the fields and earn money, so they feel very important at home.

Antonio had a more positive slant on his interpretation of a woman's role in society, saying that Maya women play an integral role in society because they support the men. Though the community may see the work Antonio does, it is really the work of Antonio and his wife, because she has fed, clothed, and looked after him so that he could serve his community, or as Antonio said, 'She serves me so that I may serve (the community).' While men's dominance in the public sphere is obvious from their many photos of infrastructure projects, women are very slowly beginning to make inroads into the public sphere.

Many participants spoke of the importance of working together, both at a family and a community level. Cooperation and serving one's community are basic Kaqchikel values that serve to reinforce social bonds and community harmony. As Antonio put it, 'to work together is to be united as brothers.' The participants' photographs showed many scenes of individuals cooperating for the benefit of their families and the community, with no obvious distinction between photographs taken by men and those taken by women. Dolores' photograph of two hard-working boys carrying water in her community is demonstrative of the cooperation that participants spoke of.

Children were a very important theme in women's photographs. Children are in large part the woman's responsibility, and women were proud and satisfied to be raising happy, healthy, and educated children – a strong symbol of the future of the community. As one woman explained, her husband did not concern himself too much with helping raise the children – instead, he only worked, ate, and slept. Though men did mention their children or grandchildren, it was not with the same frequency that women did, and it was often in the context of working together with them in the fields.

For women, their children were almost always at the center of their lives and many felt that a therapeutic place was anywhere their children were, as is the case of Rebeca, who photographed her son playing happily in the street and explained that 'wherever he is, I'm happy.' Women took great pride in giving their children more opportunities than they themselves had had growing up, especially regarding education.

Raising healthy children can be seen as a personal triumph for parents, which introduces the final point of discussion in this second theme: personal and community triumphs or successes. In terms of personal triumphs, men were again were very focused on agriculture. Diego, for example, took a photo of a tomato field (not his), excitedly saying that it looked like it was going to yield a good crop – a huge triumph because tomatoes are very susceptible to pests and because the seeds are quite expensive, thus if the crop fails everything is lost. An even greater personal triumph for men, above successful crops, was being able to buy a livestock animal such as a cow or goat.

Women's personal triumphs included seeing their children grow up healthy and educated, but they also included accomplishing things that went against traditional societal norms. For example, like Diego, Dolores also photographed tomato plants – not a whole field, but rather just a few near her house. They were a source of income, pride and accomplishment, but most importantly, they were a personal achievement because her husband had not supported her in her effort to grow the tomatoes. Similarly, Sara photographed the local school for two reasons; it was a place (1) where education was opening doors for children, and (2) where she and other women were learning to read and write in what was traditionally an environment dominated by men.

Both men and women expressed an overwhelming satisfaction with respect to having their own house and/or parcel of land. A home of their own symbolized both dignity and stability. With a home and a plot of land, individuals could escape *finca* (plantation) life; they no longer had to constantly worry about being fired or be concerned about uprooting their families in order to move on to the next *finca* in search of work. For instance, Luz photographed her home on the outskirts of her village. Even with the potential threat of a mudslide could destroy her home during the rainy season, Luz was still happy to have a place that was hers and to no longer have to move from *finca* to *finca*. Men expressed great happiness about being their own boss, even though they were still poor and life was still difficult, as they were no longer at the mercy of the *finca* owners.

For Sara, having a house of her own was not only a personal accomplishment (particularly because she was a widow), but it was also a symbol of community development:

> I feel very calm and happy here because before I didn't have a home. Now my family is happier because my children have their own rooms … it's tough living all together in one room all the time. Having my own house also means that the community is growing and that achieving these sorts of things is possible.

In addition to personal success, a number of participants photographed and/or spoke of community or group triumphs that were a sign of progress. Once again, men's community triumphs were related to the construction of infrastructure in their communities. Women, however, mentioned one particular collective triumph numerous times: the establishment of women's groups. Women in Kaqchikel society have traditionally been confined to the home, and women's gathering together was essentially prohibited. According to Sara, men did not allow women to gather formally and have meetings because they were 'merely gossips'; thus it was only men who got involved in the community via groups or participation on committees. Although women are still often seen as being 'out of place' away from the home and in the public sphere, they are being presented with opportunities by local churches and NGOs, for example, to become more involved in the public sphere.

'Escaping' and the Utility of Natural Beauty

'Escaping' or 'getting away' from the stressors of daily survival is the third major theme identified in the participants' photographs and narratives and most like the traditional understanding of therapeutic landscapes. This sense of escape was fundamental to the participants' understanding of what was therapeutic. Women often escaped the pressures of home life by going to: group meetings, informal gatherings of friends to weave (Figure 15.4), the forest with family or friends to gather firewood or edible plants, or a friend's house to visit.

In the past, women would escape to the lakeshore to do their washing. Here, they talked, laughed, and forgot about home for a while. The lakeshore is rarely used by large groups of women any longer as most of them now have running water and do their washing in their *pila*.[9] Further, washing at the lakeshore is discouraged in an attempt to limit the pollution of the lake. Additionally, travelling to and from the lake can be dangerous for women.

Though 'home' was an important place for women because it was where they spent the majority of their time and because it was so closely tied to their identities, women did not talk about the home as being as therapeutic or restorative as men did. Men tended to see the home as a place to rest, eat, sleep, and be with their families, and while women often agreed that their families' presence at home was a very important part of their happiness and wellbeing, the home was also viewed as a place

9 A *pila* is a large basin found outside the home in the yard, and is used for washing.

Figure 15.4 Women weaving together under a tree (photographed by Mercedes)

of work, boredom, worry, and loneliness. This is why women placed such a great deal of emphasis on getting out of the house and being with other women – it was a welcome variation to their regular routine and helped them avoid feeling cooped up and bored.

When men needed to escape, they often went to work in their fields (generally alone); this was true for both farmers and those who worked in an office setting. Many farmers said that being there doing physical labor helped them clear their heads, even though they often escaped to their fields by default, having few other options open to them. Daniel, for example, said that he goes to work in his field when he is frustrated because he has no other option; if he had the money, he would travel and visit other towns. Diego also saw it as simply something he must do – a product of his economic and social situation – saying, 'because we're from the countryside, we go to our fields and work.'

Nature, for both men and women, was a place of escape. Going out into the forest or sitting down by the lakeshore were both important escapes for men and women. Men were more likely to go down to the lakeshore to sit, think, and be alone, taking in the beauty of their surroundings. Women often commented on the beauty of the lakeshore, but rarely used it as a place of escape. Many women were afraid to go down to the lakeshore because it was a place frequented by gangs and drug dealers, and travelling to the lake from other towns was also dangerous because of the possibility of getting robbed or worse.

Violence affects both men and women, but it was the women who talked the most about the effects of violence on their everyday lives. When talking of violence, those few men that did discuss it acknowledged that it was undoubtedly worse for women because of the possibility of rape. None of the male participants let the violence curb their mobility, though they always travelled with care. As Daniel put it, 'A lot of people are being killed right now and it's difficult to travel because there are many assaults, but you have to do it and the only thing you can do is ask God that everything go well and that nothing happen to you.'

Rafael photographed the lake at one of his favorite spots in Pachitulul, a village located a few kilometers west of the *cabecera*. His photograph demonstrates men's more frequent use of the lake as a place of escape, but more importantly it is also a symbol of their greater mobility in the face of violence and social norms. Given their restricted mobility, women's therapeutic places did have a tendency to be located closer to the home than men's. For example, in terms of nature, women photographed their own yards or flower gardens, which were places nearby where they could at least get out of the house for a moment.

The forested mountainsides and volcanoes were another important place of escape. While participants almost always commented on their natural beauty and how it lifted their spirits, there was also a theme of utility. Men enjoyed the walk through the forest or up the mountain on the way to their fields, stopping at times to take in and enjoy the scenery. Women went to the forest to gather firewood or edible plants. Olga described the importance of firewood in this way: 'If there is not firewood we can't do anything … we can't make tortillas, we can't cook, we can't do anything,' thus demonstrating the overarching sense of need and urgency that shapes women's trips to the mountain.

Two other common features of participants' escapes were that these natural landscapes were a source of fresh air and privacy, both of which are difficult to find in the villages. Laura took two photographs that captured the occasional profound need for privacy. One photo shows a man sitting alone down by the lakeshore, deep in thought. The fact that the photograph was taken from afar suggests a respect for and understanding of his need for privacy. In the second photo, the focus is a couple sitting under a tent having what Laura interpreted as a private conversation, away from the prying eyes and ears in town and at home.

A final place that served as an escape for both women and men was the church. It played less of a social role than a personal one. Participants went there to be alone and to thank God and ask Him for help, and there they found a sense of peace and tranquility. Participants also found God in nature, especially in the more striking natural features of the region, such as the volcanoes, mountains, and lake. Though Christianity was the dominant form of spirituality discussed and photographed by participants, a small number of participants photographed a sacred site or a traditional religious ceremony, and some also mentioned Kaqchikel spirituality in reference to the sacredness of the *milpa*. Regardless of whether participants subscribed to Christian or Kaqchikel spiritual beliefs (or some combination of the two), connecting with the spiritual was seen as a way of 'escaping' and of actively doing something (praying) to improve their present situation.

'Nontherapeutic' Landscapes

The final theme deals with negative or 'nontherapeutic' landscapes. Many of the negative places or ideas presented by the participants clearly lacked the important therapeutic aspects already discussed. For example, a large number of participants spoke of the informal garbage dumps (garbage thrown over the side of a hill) as being places that contributed negatively to their health. These dumps and other places where garbage and pollution were apparent did not conform to the participants' valuation of hygiene as: (1) a precursor to physical health, both at the individual and community levels, and (2) as a source of pride and dignity.

Private property was also a recurring theme in the discussion of negative landscapes because it symbolized lost or wasted opportunities. A small plot of land means a great deal in the lives of *luqueños*. Large expanses of private land, therefore, were places where participants could not cultivate crops, collect firewood or edible plants, or simply walk for fear of being apprehended for trespassing. Olga photographed a number of very healthy banana plants located on the property of a *finca* owner and identified them as a negative landscape because the community could not access those resources.

Threats to crops were threats to survival and failing crops were another major 'nontherapeutic' landscape because of the stress they placed on participants. Though agricultural chemicals were seen as somewhat positive because they increased yields, they were more frequently perceived in a negative light because of possible health impacts, as well as the additional financial burden they placed on farmers. Felipe photographed bags of coffee 'pulp' (the outer casings of the coffee bean) at a local *finca* because these were useless to the *finca* owner but were a free source of organic fertilizer for the employees.

Negative landscapes were also any place perceived as being dangerous or violent. Dangerous places were exceedingly more prevalent in women's narratives, although they were understandably rarely photographed. Isolated places where women were alone were seen as dangerous and thus 'nontherapeutic,' which is one reason why women so frequently took a companion (male or female) with them when they, for example, went out into the forest to collect firewood.

Concluding Comments

The therapeutic landscapes of the Kaqchikel of San Lucas Tolimán were inextricably tied to daily life, poverty, and violence. First and foremost, anything that ensured their survival was therapeutic. Bettering one's quality of life, thus ensuring longer term survival, was also therapeutic. Another survival-related therapeutic aspect was 'escaping,' which involved a change of scenery and was also often coupled with work.

The therapeutic landscapes of the Kaqchikel were highly gendered and embedded in the rhythms of daily life. Men's landscapes were overwhelmingly centered on their fields and/or jobs and were a greater distance from home than women's therapeutic landscapes, which were mainly focused on their domestic roles and located close to their homes (though this was also due to violence, not just traditional gender roles). Although the landscapes may have differed between men and women, the underlying themes were consistent between the two.

Photovoice was a particularly effective method in a cross-cultural setting because it gave more authority to the participants and elicited longer, more detailed answers about the photographs. We found, as did Harper (2002), that the photographs were able to evoke more emotional stories about the participants' lives. These stories were richer than those that would have been elicited by purely verbal interaction (Orobitg Canal, 2004). It is undoubtedly a difficult task to describe solely in words a place that plays a special part in one's life, health, and happiness; having the photographs made this task easier for the participants because they had a tangible image to draw on and to refer to when explaining their therapeutic landscapes. The participants' photographs also acted as a bridge between our very different experiences of reality (Pink, 2001). We were, in a way, invited into their therapeutic places through the photographs, which enhanced our understanding of their daily lives. Orobitg Canal (2004) posits that using photographs in interviews is a way of producing data through reflexivity, and we believe that the process of taking the photographs gave participants time to reflect more deeply on their therapeutic landscapes before having to explain them.

This research focused on not one, but an assortment of therapeutic landscapes that exist in the daily lives of participants. It also focused on local rather than universal narratives, and on the experience of the place and not the place itself. Kaqchikel culture and worldview define health beliefs and permeate the understanding of places deemed to be therapeutic and 'nontherapeutic.' How men and women moved through and interacted with the world around them played a critical role in determining their therapeutic landscapes. Photovoice enabled them to capture the unique and complex nature of their therapeutic landscapes.

References

Abell, J.D. (1997), *Peace in Guatemala? The Story of San Lucas Tolimán,* <http://faculty.rmwc.edu./jabell/san_lucas_toliman.htm> accessed 1 December, 2004.

Alcorn, J.B. and Toledo, V.M. (1998), 'Resilient resource management in Mexico's forest ecosystems: The contribution of property rights', in F. Berkes and C. Folke (eds) *Linking social and ecological systems: Management practices and social mechanisms for building resilience* (Cambridge: Cambridge University Press).

Andrews, G.J. (2004), '(Re)thinking the dynamics between healthcare and place: therapeutic geographies in treatment and care practices', *Area* 36: 307–18.

Blum, L. (2003), 'Civic activism, political knowledge and political participation in San Lucas Tolimán, Guatemala', *Latin American Studies Association Annual Meeting*, <http://136.142.158.105/Lasa2003/BlumLeonor.pdf> accessed 1 February 2005.

Cereser, L. (2005), 'Feminicidios, en alza constante desde 2001', *Prensa Libre*, 12 July: 10.

Dobbs, G.R. (1997), 'Interpreting the Navaho sacred geography as a landscape of healing', *Pennsylvania Geographer* 25: 136–150.

Elías, S., Gellerty, G., Pape, E. and Reyes, E. (1997), *Evaluación de la sostenibilidad en Guatemala* (Guatemala: FLACSO).

Fischer, E.F. (2001), *Cultural logics and global economies: Maya identity in thought and practice* (Austin: University of Texas Press).

Fischer, E.F. and Hendrickson, C. (2003), *Tecpán Guatemala: A modern Maya town in global and local context* (Boulder: Westview Press).

Gesler, W.M. (1993), 'Therapeutic landscapes: Theory and a case study of Epidauros, Greece', *Environment and Planning D* 11: 171–89.

Gesler, W.M. (2003), *Healing places* (New York: Rowman and Littlefield Publishers).

Green, L. (1999), *Fear as a way of life: Mayan widows in rural Guatemala* (New York: Columbia University Press).

Harper, D. (2002), 'Talking about pictures: a case for photo elicitation', *Visual Studies* 17: 13–26.

Hurtado, E. and Sáenz de Tejada, E. (2001), 'Relations between government health workers and traditional midwives in Guatemala', in B.R. Huber and A.R. Sandstrom (eds) *Mesoamerican healers* (Austin: University of Texas Press).

Keys, E. (1999), 'Kaqchikel gardens: Women, children, and multiple roles of gardens among the Maya of highland Guatemala', *Conference of Latin American Geographers* 25: 89–100.

López Batzin, M.J. (1997), *Formas propias del poder local: Estudio de antropología política en los municipios Kaqchikeles de Patzún y San Martín Jilotepéquez* (Guatemala: Wajxaqib' B'atz').

McIntyre, A. (2003), 'Through the eyes of women: photovoice and participatory research as tools for reimagining place', *Gender, Place and Culture* 10: 47–66.

Milligan, C., Gatrell, A. and Bingley, A. (2004), '"Cultivating health": Therapeutic landscapes and older people in northern England', *Social Science and Medicine* 58: 1781–93.

Moss, P. (2002), 'Taking on, thinking about, and doing feminist research in geography', in P. Moss (ed.) *Feminist geography in practice* (Malden: Blackwell Publishers).

Munaiz, C. (2005), '¿Por qué somos violentos?' *Prensa Libre*, 9 October, <http://www.prensalibre.com/pl/2005/octubre/09/125167.html> accessed 9 October 2005.

Municipalidad de San Lucas Tolimán (2002a), *Plan de desarrollo comunitario 2002–2012: Casco urbano* (Sololá: Municipalidad de San Lucas Tolimán).

Municipalidad de San Lucas Tolimán (2002b), *Plan de desarrollo comunitario 2002–2012: Colonia Centro Urbano Parcelamiento Pampojilá* (Sololá: Municipalidad de San Lucas Tolimán).

Municipalidad de San Lucas Tolimán (2002c), *Plan de desarrollo comunitario 2002–2012: Colonia Santa Cruz Quixayá* (Sololá: Municipalidad de San Lucas Tolimán).

Municipalidad de San Lucas Tolimán (2002d), *Plan de desarrollo comunitario 2002–2012: Cantón San Martín* (Sololá: Municipalidad de San Lucas Tolimán).

Municipalidad de San Lucas Tolimán (2002e), *Plan de desarrollo comunitario 2002–2012: Colonia Xejuyú* (Sololá: Municipalidad de San Lucas Tolimán).

Municipalidad de San Lucas Tolimán (2002f), *Plan de desarrollo comunitario 2002–2012: Comunidad de Tierra Santa* (Sololá: Municipalidad de San Lucas Tolimán).

Municipalidad de San Lucas Tolimán (2002g), *Plan de desarrollo comunitario 2002–2012: Fincas Pampojilá, Peña Flor, Patzibir y Esmeralda* (Sololá: Municipalidad de San Lucas Tolimán).

Orobitg Canal, G. (2004), 'Photography in the field: word and image in ethnographic research', in S. Pink (ed.) *Working images: Visual research and representation in ethnography* (New York: Routledge).

Parr, H. (2004), 'Medical geography: Critical medical and health geography?' *Progress in Human Geography* 28: 245–57.

Pink, S. (2001), *Doing visual ethnography* (Thousand Oaks: SAGE Publications).

PNUD-Guatemala (2002), (*Guatemala: Desarrollo humano, mujeres y salud. Informe nacional de desarrollo humano 2002.* Guatemala: Programa de las Naciones Unidas para el Desarrollo) <http://www.pnudguatemala.org/informesdesarrollohumano/idh2002/index.html> accessed 7 February 2005.

Seijo, L. (2005), 'Crímenes provocan sicosis', *Prensa Libre*, 1 July: 8.

SNU-Guatemala (2003), *Guatemala: Una agenda para el desarrollo humano, informe de desarrollo humano* (Guatemala: Sistema de Naciones Unidas en Guatemala) <http://www.pnudguatemala.org/documentos/pdfs/informesdh/2003/Informe%20Desarrollo%20Completo.pdf> accessed 22 January 2005.

Sperling, J.M. (2006), 'The Influence of Poverty and Violence on the Therapeutic Landscapes of the Kaqchikel', (Waterloo, Ontario: University of Waterloo, Department of Geography Master's Thesis).

Tiney, J. (2000), 'Lucha por la tierra', in *Tierra y espiritualidad maya*, ed. Agencia Ecuménica Solidaridad y Comisiariato Central Misionero (Guatemala: Ak' kutan /Voces del tiempo).

Vela, M., Sequén-Mónchez, A. and Solares, H.A. (2001), *El lado oscuro de la eterna primavera* (Guatemala: FLACSO).

Ventura, T. (2000), 'Acceso-tenencia de la tierra', in *Tierra y espiritualidad maya*, ed. Agencia Ecuménica Solidaridad y Comisiariato Central Misionero (Guatemala: Ak' kutan/Voces del tiempo).

Wakefield, S. and McMullan, C. (2005), 'Healing in places of decline: (re)imagining everyday landscapes in Hamilton, Ontario', *Health and Place* 11: 299–312.

Wang, C. and Burris, M.A. (1994), 'Empowerment through photo novella: portraits of participation', *Health Education Quarterly* 21: 171–86.

Wang, C. and Burris, M.A. (1997), 'Photovoice: concept, methodology, and use for participatory needs assessment', *Health Education and Behavior* 24: 369–87.

Williams, A. (1998), 'Therapeutic landscapes in holistic medicine', *Social Science and Medicine* 46: 1193–1203.

Williams, A. (1999), 'Place identity and therapeutic landscapes: The case of home care workers in a medically underserviced area', in A. Williams (ed.) *Therapeutic*

Landscapes: The Dynamic between Place and Wellness (Lanham, MD: University Press of America).

Wilson, K. (2003), 'Therapeutic landscapes and First Nations peoples: an exploration of culture, health and place', *Health and Place* 9: 83–93.

Chapter 16

Restoration or Risk? Exploring the Place of the Common Place

Christine Milligan

Introduction

During the last decade or so, Gesler's concept of the therapeutic landscape has permeated into mainstream health geography. Even a cursory review of recent conference proceedings and geography journals bears testimony to the increased importance geographers have attached to examining the complex, but potentially beneficial, effects to health and well-being of engaging with particular types of landscapes. Significantly, we have seen a gradual shift within this body of work from the historical to the contemporary (for example, Kearns and Collins, 2000; Gesler et al., 2004) and from the individual and unique to the commonplace (for example, Williams, 2002; Milligan et al., 2004; Wakefield and McMullan, 2005). This extension of the concept to ordinary, everyday places has made way for a more nuanced understanding of the ways in which everyday landscapes can contribute to the maintenance of human well-being (for example, Andrews, 2002; Johnsen et al., 2005). Such environments are not necessarily 'wild places' untouched by human intervention (if, indeed, such places still exist in modern societies), rather they are landscapes that have been modified or specifically created. That is, they are places that have been cultivated, designed or domesticated to meet a specific social or cultural need, or that have been adapted by groups or individuals to meet their own particular requirements.

The vast majority of the literature that engages with the ideas that underpin the concept of the therapeutic landscape focus on the relationship between places and 'wellness'. Indeed, it can be argued that attaching the prefix 'therapeutic' to a geographical interpretation of landscape, whether natural or created, brings with it an assumption that certain places have inherent qualities that promote physical or mental well-being. These effects are often analysed in relation to specific social or cultural groups such as those with mental ill-health (Geores and Gesler, 1999), children (Thurber and Malinowski, 1999), Indigenous populations (Madge, 1998; Wilson, 2003), but they can also reflect highly individualised and temporal responses. For example, enjoyment of the garden when an individual is fit, healthy and able to benefit from the restorative effects of garden work may alter as the individual ages or develops a chronic illness that constrains his/her ability to maintain the garden and increases his/her reliance on others to do so (Milligan et al., 2004). Similarly, a young person's active engagement with natural woodland in childhood – as a space in which to play and stretch the imagination, may become seen as 'risky' in young adulthood as myth and media begin to influence his or her response to woodland (Milligan and Bingley, 2007).

With one or two exceptions (for example, Conradson, 2005), what has been less well teased out in the therapeutic landscapes literature is, firstly, the relational and often highly individualised dynamics of people's encounters with nature and landscape; and secondly, how these can change over time from positive to negative encounters and vice versa. It is the duality of the 'therapeutic' engagement with the natural environment that is addressed within this chapter. In doing so, the chapter firstly draws on newly emerging sub-fields of interest within geography (for example, around phobic and emotional geographies) as well as work in both environmental psychology and philosophy to consider how they might help to broaden our understanding of the complex relationship between people, places and well-being. The latter two disciplines have a long tradition of working with notions of therapeutic and restorative environments, yet the therapeutic landscapes literatures rarely engage with this work (though see Williams, 1999; 2000). Secondly, the chapter draws on a range of studies that focus on people's engagement with everyday – or 'common places' – to demonstrate how people's interrelationship with the natural environment may shift from one that can be epitomised as restorative or therapeutic to one characterised by fear or risk.

Observations from Parallel Literatures

Environmental Psychology

As indicated above, geographical narratives around therapeutic landscapes have much in common with some of the work in environmental psychology. Researchers working within this discipline have sought to examine the link between landscape and what they refer to as the 'restoration' of physical and emotional well-being for over 20 years (for example, Ulrich, 1979; Kaplan and Talbot, 1983; Cooper Marcus and Barnes, 1999; Hartig et al., 2003). How restoration ensues is seen to be connected not only to the characteristics of the individual and the demands he or she faces, but also to the links between environmental preferences and the social, physical and temporal characteristics of a potentially restorative environment (van den Berg, 2003). Attempts to prevent disease or promote health and well-being, then, are seen to require an understanding of the environmental characteristics that support or promote restoration (Hartig and Staats, 2003).

Theory in this field has been dominated by two positions, one focuses on how natural landscape can contribute to stress *reduction* (for example, Ulrich 1984), the other focuses on the *restoration* of well-being and concentration (for example, Kaplan, 2001; van den Berg, 2003). In examining variations in people's responses to natural and built landscapes, for example, environmental psychologists have demonstrated that not only do natural environments elicit stronger restorative effects than built environments, but they can also promote a greater improvement in mood and concentration (for example, Hartig et al., 1996). This highlights the importance of not only the theoretical work around restorative landscapes but also the practical aspects of it. Theoretically, this work involves identifying those qualities of person–landscape transactions that promote restoration. Practically, it also involves an engagement with policy-makers, planners, landscape architects and the like to develop, modify and maintain landscapes that can be seen to possess the physical and social characteristics that promote restoration (Kaplan et al., 1998). Both Ulrich (1984) and Kaplan's (2001) work on the psychological effects of a view, for example, point not only to the benefits of ensuring that people are provided with windowed, rather than windowless environments, but it also demonstrates the restorative potential of including a natural element in the view from that window, providing what Kaplan refers to as a 'micro-restorative experience' (2001, 508). Hartig et al. (2003) concluded that in an era of escalating health care costs and declining environmental quality, health care strategies that have a natural environment component to them may have particular value for an increasingly urbanised population.

Of course, connecting well-being and natural environments is not new – indeed it has been the focus of much of the early work on therapeutic landscapes, where Gesler (1993; 1998) and others sought to examine the link between nature and the healing process at well-known historical sites such as Epidauros, Lourdes and Bath. Nevertheless, as recent work reveals, geographers have now begun to critically engage with ideas about the design of contemporary healing spaces in health care settings (Moon et al., 2006; Gesler et al., 2004) that have, in the past, been the purview of architects and environmental psychologists. Where the emerging geographical work in this field differs is that it tends to adopt a more subjective and person-centred approach than is the case in much of the environmental psychology literature (Gesler et al., 2004).

Emotional Geographies

A second strand of literature that bears scrutiny comes from work around emotional geographies. In 1999, Hasse pointed to the importance of understanding the role of emotions in the people–environment relationship, calling on human geographers to express a greater sensibility to emotional behaviour. Anderson and Smith (2001, 8) too have called for a greater engagement with the 'emotive topographies' that act to shape person–landscape transactions. Understanding these issues, they maintain, is important if geographers are to retain their critical edge and avoid the pressures to focus on policy-oriented or economic goals. Whilst echoing the importance of a geographical engagement with emotion, Bondi et al. (2005) caution that this is a tricky endeavour, pointing to the difficulties of communicating those emotions, feelings or affect that are often at play beneath the actions of our everyday lives. Despite these problems, there is a growing body of work that pays close attention to the interrelationship between people, place and emotion (see special issues in *Gender, Place and Culture*, 2004; *Social and Cultural Geography*, 2005; and the edited volume by Davidson et al., 2005).

Work within this field demonstrates that how people experience places is inextricably linked not only to feelings and emotions *about* these places, but also emotions engendered *by* them. In their work on place, death and bereavement, for example, authors such as Brown (2003) and Morris and Thomas (2005) draw not only on the links between emotions and place, but also the therapeutic potential of such places as home, hospice and the institution in facilitating the 'good death'. Morris and Thomas note that where that death takes place is not just a matter of individual choice, it is also bound up in pre-existing relationships with people and places. Fear of crisis and speed of access to support, or having a loved one die in your own home can also mean that for some, home is no longer viewed as a safe or unproblematic space. So while the home death is often held up as the ideal amongst palliative care commentators, this often fails to recognise that 'the home' is not homogenous and therefore it cannot always be seen as an unproblematic or therapeutic environment. Indeed, it can also be a site of fear, shame, stress, struggle and emotional labour – particularly for some women and family care-givers (see Burman and Chantler, 2004 and Milligan et al., 2005a).

Other work in this field has drawn out not only the beneficial aspects of particular landscapes, but also how these can be mediated by both the self–landscape encounter (Conradson, 2005) and by the wider community's relationship with place and landscape. Parr et al.'s (2005a; 2005b) work on rural mental health in the Scottish highlands, for example, refers us to a 'tyranny of repressive normality' (2005a, 94) where, in some places, people's actions and emotions become mediated by a culture of resilience and repression rooted in historical and religious traditions. The emergent normativity is characterised by a people who actively avoid any overt emotional expression, creating huge pressures on those with mental or emotional problems not to disclose their mental state for fear of transgressing these cultural norms and finding themselves faced with community rejection. The uncovering of a rural culture underpinned by resilience and repression is starkly at odds with the, often idealised, images of rural environments that are characterised by peace and tranquillity (for example, within tourist or other promotional literatures), and as places that individuals should visit in order to experience restoration or renewal.

It seems evident, then, that a close connection exists between work around emotional geographies and that of therapeutic landscapes – at least in terms of issues related to health or well-being. A significant strand of work around therapeutic landscapes, for example, has been concerned with issues of mental and spiritual well-being, feelings of identity and belonging – in other words, people's emotional response to particular places – and how this impacts on their health. Indeed, Wilson (2003) maintains that the notion of the therapeutic landscape is important because it allows us to embrace geographical perspectives that draw on meaning and emotional responses to places. What is striking, however, is that whilst the 'emotional turn' has spawned a proliferation of work around feelings, emotion and affect – issues that are central to much of the work around therapeutic landscapes – any engagement with these two sets of interrelated sets of discourses as a means of furthering our understanding of the links between the emotional and the therapeutic encounter with places has been marginal at best.

Fear and Phobic Geographies

Clearly, the interrelationship between people and places is complex. Places that represent sites of peace, safety, calm and tranquillity for some people at some times in their lives can also prove to be sites of fear, anxiety and stress for others. Hence, to imbue any single place with qualities of restoration or renewal can fundamentally misrepresent how the interrelationship between people and places can shift over time and with changing circumstances. Nowhere is this demonstrated more visibly than in the literatures around geographies of fear and phobic geographies.

In his early geographical work, Tuan (1979) drew on the metaphor of 'landscapes of fear' as a means of facilitating study into the interrelationship between people, fear and landscape. This work stretches across the imagined landscapes of myth and fairytale to the perception of natural hazards. Tuan's focus on landscape drew together ideas around people's attitudes and values and their physical responses to, and interventions in, landscape. With the odd exception (see Jarosz, 1994), subsequent work on geographies of fear has tended to focus on the real or perceived effects of crime, terrorism, contestation, conflict and marginalisation (for example, Sparks, 1992; Lupton and Tulloch, 1999; Pain, 2000; Gold and Revill, 2003). This work seeks to elucidate how certain built, social or isolated landscapes can act to increase people's fear of crime, attack, exclusion and so on. While much of this work focuses on run-down urban neighbourhoods, a recent paper by Panelli et al. (2004) drew together ideas around safety, fear, emotion and landscape as they examined women's differing experiences of the rural in the New Zealand context. Rather than experiencing rural communities as safe, peaceful and harmonious spaces, they argued that in practical terms, women often experience such places as sites of crime and fear, hence their response to rural landscapes is often 'less certain, more troubled, and even at times contradictory' (p. 463). The confluence of work around landscapes of fear and therapeutic landscapes emerges through Frazier and Scarpaci's (1998) work on state violence and mental health. Drawing on the aftermath of the violence and human rights abuses in Chile during the later part of the twentieth century, they demonstrate how a community struggled to reclaim its landscape and mental well-being through protest and testimony following the uncovering of mass graves across the country. The scars of state violence that are left on the bodies, minds and landscapes of Chile and its people, they maintain, render internal 'connective tissues of minds, bodies, communities and places' (p. 56), such that 'ill-health and place are inextricably bound in Chile's human rights calamity' (p. 71).

Attempts to understand cultural constructions of people–landscape interactions, thus, need to be alert to the potential difference and inconsistency in people's response to the world around them.

A second small, but important, literature that draws out the interconnectedness between fear and place is that which focuses on 'phobic geographies' (Davidson, 2003). Within this genre, work has considered not only agoraphobia (a spatially mediated panic or anxiety disorder) but also bio-phobias – that is specific fear and anxiety-based disorders linked to the natural world (see Davidson and Smith, 2003; Smith and Davidson, 2006). Here, often irrational fear and disgust of natural phenomena such as birds, animals, insects, snakes, moths, dirt and so on, become manifest in disorders that are directly oppositional to the claims for restoration and renewal that underpin much of the environmental psychology and therapeutic landscapes literatures. As Davidson (2003) points out, however, phobic conditions such as agoraphobia do not consist of fear of a single entity, but rather are made up of a complex and amorphous cluster of fears that are experienced in different ways, by different people, in different socio-spatial settings. While there may be similarities running through people's experiences of these conditions, they are essentially individualised events.

Restoration or Risk?

Phobic responses to landscapes or nature represent the extreme end of a spectrum of responses to the environment. What is also evident from the discussion so far is that while natural landscapes are more likely to have restorative effects than built ones, places are rarely inherently restorative or risky. Though some commonalities and wider socio-cultural responses to landscape can often be drawn out, people can also experience them in different ways at different times and for different reasons. A person's therapeutic (or non-therapeutic) encounter with place may, thus, be a highly individualised event. To illustrate this, the remainder of this chapter draws on a range of experiential narratives drawn from work with people across various ages and cultures in the UK as a means of illustrating how and why ordinary, everyday landscapes can be interpreted and understood by individuals as both places of peace, tranquillity and restoration, and places of fear, anxiety and stress. In doing so, some of the factors that can contribute to these differential experiences are also examined.

'Wild Natures' – Wooded and Rural Landscapes

There is a deep-rooted belief in contemporary Western society that an engagement with nature and the natural landscape brings with it a range of personal, social and health benefits that are conducive to well-being (Parry-Jones, 1990). A person's experience of landscape through the senses is inseparable from the social and psychological context of that experience. Hence, an individual's relationship with 'wild nature' such as local woodland or rural landscape can be framed as either a social engagement that includes the presence of others (whether human or animal) or as a unique and individual subjective experience. Such engagements may be fleeting, offering what Kaplan (2001, 508) referred to as 'micro-restorative experiences' or sustained. Understanding how this might be differentially experienced by people from a range of ages and cultural backgrounds is important if we are to have a clearer insight into how we might create or maintain natural environments that are conducive to the promotion or maintenance of well-being.

Environmental psychology work on the restorative potential of natural environments has tended toward the use of experimental methodologies that seek to draw out some generalisation of causality and effect. That is, it is predisposed towards hypotheses that look for a binary link – natural landscapes either do, or do not, have a role to play in improving health and well-being. While this provides some interesting insights into the relationship between people and landscape, it does tend to subsume the importance of individual experience. What is missing, then, is a wider understanding of how particular engagements with nature can be therapeutic for some individuals whilst at the same time prove risky and anxiety-inducing for others.

Natural settings are seen as discrete from the built environment in which many people live in that they consist of rich and coherent ecosystems that can be both observed and explored. However, these need not be distant, hard-to-reach environments, but can also include local woodland, countryside, nearby riverside or seashore.

In considering the therapeutic potential of people–landscape transactions, four distinct restorative benefits of the natural landscape can be identified (Knopf, 1987; Herzog et al., 1997; Herzog and Barnes, 1999).

i) Nature Restores

Firstly, the natural landscape offers a restorative setting for attentional recovery from the demands placed upon people in their everyday lifeworlds. Here, natural landscapes are seen to be associated with feelings of peace, tranquillity and restoration. Studies of both older people and young adults in the north west of England (see Milligan et al., 2004; 2005b; Bingley and Milligan, 2007; Milligan and Bingley, 2007) indicate that despite substantial age differences, people from the same cultural background (in this case, white British) can experience the restorative effects of natural landscapes in similar ways. Florence (aged 73), for example, noted that for her, the countryside represented a place of renewal, where she could escape from the tensions of her everyday life. Referring to the difficult relationship that existed between herself and the adult son with whom she shared her home, she remarked:

> On Saturday, things were so fraught that I just went off out in the car and sat in the countryside for an hour enjoying the birds and views, out of his way, and remembering my late husband.

Similarly, work with young adults regarding their engagement with woodland revealed that they, too, found natural settings helpful as sites for relieving some of the tensions and stresses they experienced in their everyday lives. Ros (aged 17) remarked:

> It clears your mind … It's somewhere where you can think without people going. You can't hear anybody talking so you can think, and you're on your own and it's peaceful …. Like an hour outside, it flows through your mind and everything goes back into perspective instead of thinking it's the end of the world – which it never is!

This positive engagement with nature appears to be neither age- nor gender-specific. Both older and younger men and women referred to its restorative potential. Jack (aged 21), for example, explained that for him, woodland offered a place of sanctuary from difficult circumstances at home:

> I mean there were things at home that weren't good. But, when they weren't good it [local woodland] was a way of getting away from it.

Chawla (1992) and Chatterjee (2005) have argued that having access to restorative places that are located away from their everyday lifeworlds is important for young people as they move from childhood to young adulthood, with a subsequent decline in their social attachment to the family base. The experience of older people, however, indicates that this may be equally important at the other end of the age spectrum and that older people, too, benefit from having access to such restorative spaces.

ii) Nature provides a site for reflection and diversion

Secondly, natural landscape provides a site for reflection, where the pleasure taken from the aesthetic beauty of the natural environment offers a setting in which it is possible to think through immediate and unresolved problems. As Terry (aged 18) explained:

> sometimes you just need to get out of the house and go somewhere where you can get your head right and stuff. And the forest and the woodland areas are nice places to go and sit and collect your thoughts.

Here, the landscape is experienced passively – internalised as a scenic gaze that impacts on a person's emotional psyche in ways that are both positive and beneficial. As one 73-year-old woman also put it:

> I enjoy, you know, seeing the acorns and the different seasons, and when you see the first green shoots, the snowdrops starting, and you have the berries on the trees.

This type of response illustrates the positive impact that growth and renewal of the natural landscape can have on an individual's sense of well-being. For others, however, the natural landscape offers an opportunity to 'get out' and enjoy the peace and calm of the countryside. 'Meryl' (aged 72), for example, noted:

> I love being out in the fresh air and exercise, gentle exercise. … I think you spend enough time in the house.

Young adults, too, noted their enjoyment of nature's diversions. 'Ruth' (aged 18) commented:

> they've built a pathway right up along by the river and it's just lovely. I walk on me own and enjoy it – watching all the ducks in the river and the herons and the birds.

iii) Nature facilitates competence building

A third claim for the restorative benefits of nature is that it facilitates competence building. Here, however, commentators such as Rohde and Kendle (1994) maintain that older people respond differently to aspects of the outdoor environment than do young people – so how nature facilitates competence building may vary across different age groups. A recent report by Ward-Thompson et al. (2002) on access to woodland noted, for example, that childhood experience, and the frequency of visits to woodland in childhood, are the most important predictors of how people feel in woodland as adults and how often they are likely to visit it. Lowered anxiety in or around woodland appears to be linked to the extent of unsupervised play in wooded areas in childhood and lower levels of parental anxiety in relation to that play. Milligan and Bingley (2007) stress the importance that young people place on being able to participate in unstructured, less supervised play, when young. This is seen by young adults as a means of allowing them to develop both a sense of agency and a sense of confidence in their own assessment of what is safe and what is not. Such competence building serves young adults well in their teenage years and beyond. As one young woman (aged 21) put it:

> if you learn to be aware of the risks when you're little and think, 'that's safe – aaah maybe not I got hurt' … you learn that maybe you shouldn't climb things that are six times taller than you or whatever, 'cos there's a lot of judgement of, you know, self, sort of your own abilities.

Of course the 'risky' aspect of woodland can also be one of its appeals – providing a place in which young people can learn and test their physical agility, contributing to increased levels of confidence and independence that, in turn, provide a source of inner strength that can be drawn upon in subsequent challenging or stressful situations.

For older people, the relationship between nature and competence building can occur in different, but no less important ways. Maintaining fitness in retirement through such activities as rambling or hill walking can evoke a sense of achievement and exhilaration. As one 73-year-old woman explained:

> I think if you get to the top of a mountain … you've a wonderful feeling. You can't really explain it.

Though this, of course, depends much on a level of physical fitness that can decline with age, it is often replaced with less taxing engagements with nature – such as walks around a favourite (and usually flatter) beauty spot closer to home. For others, it may be the linking of scenic qualities of the natural environment to a new hobby such as painting or photography. As one 69-year-old man commented:

> when I retired I felt, well I haven't seen that much of the Lake District … so I just used to take off for two or three days sometimes … It was just the sheer beauty of it, you know. But also I had gone into photography, so they complemented each other. And I got quite a boost out of, you know, taking the photographs of some of the views … I can see a photograph now and I'm not just seeing a photograph, I can kind of understand what goes into getting it.

iv) Nature Carries Symbols that Affirm the Culture or Self

People's response to 'managed' or 'wild' nature and the therapeutic or non-therapeutic experience it engenders is shaped not only by individual subjective experiences, but also by the socio-cultural environment within which they are embedded. Work with older people in the north west of England, for example, illustrated how the local landscape is seen to symbolise a shared identity and heritage (in this case as uniquely English) (Milligan et al., 2004). The Lake District, a landscape of outstanding natural beauty and tranquillity in the north west, portrayed not only a place that people sought to access for both restoration and activity, but it also symbolised, for them, a quintessential 'Englishness'. This is wrapped up not only in the visually aesthetic landscape of lush woodland, vegetation and tranquil lakes, but in a specifically English literary culture associated with the area through 'local heroes' such as Beatrix Potter, Samuel Taylor Coleridge and William Wordsworth.

Tolia-Kelly (2004; 2006), however, demonstrated how such quintessentially 'English' landscapes can also be appropriated by non-white English people. Her work with British Asians reveals how memories of the wide range of non-British landscapes that they are connected with – stretching from the Indian sub-continent to East Africa – are deeply embedded in their environmental practices as they seek to make the UK landscape meaningful and inclusive for themselves. Contrary to the expectation that non-white English people would feel alienated by landscapes redolent of a white 'English' rural idyll, she found that the English pastoral landscape was a site of desire and pleasure for Asian women, reminding them of life in such places as rural India, East Africa and Bangladesh. As one Asian woman, within her study, noted:

> That's the sort of thing I enjoy. To be in the scenery, to be in peace, and you know … more natural environment (Tolia-Kelly, 2006, 346).

It is worth noting, however, that Tolia-Kelly qualifies the pleasure gained from an engagement with the English landscape by drawing attention to the way in which these Asian women carefully consider the extent to which this will involve contact with 'English' (that is, white English) people during these landscape encounters.

Others have also noted how the person–landscape transactions of migrant communities are shaped by culturally imbued memories of 'home'. Lalich (2006), for example, illustrates how the communal efforts of twentieth century migrants to Australia have contributed to the development of 'managed natures' that provided a sense of familiarity, security and belonging in an otherwise alien landscape. Thomas's (1999) work on garden landscapes and migrant communities demonstrated, however, that the person–landscape transaction amongst people from different cultural backgrounds can be experienced in very different ways. His work highlights how Vietnamese migrants in Australia, used to domestic environments characterised by open, unbounded landscape with plenty of space for social interaction, found that the cultural practice of developing housing in streets with fenced-in gardens provoked both unease and anxiety amongst this community.

By bringing together both the imagined and observed, local landscapes can be seen to give a sense of order and relatedness that operate at both a conceptual and a perceptual level in a way that creates a meaningful whole. Engagement with such landscapes can range from walking and climbing in local beauty spots, to observing, working on/with or peaceful meditation within it. So while a person may gain specific physical health benefits from an active engagement with the landscape, individuals may also derive considerable psychological benefit from their passive involvement with nature (Ulrich and Addoms, 1981).

From Positive to Negative People–Landscape Transactions

Whilst natural landscapes provide a restorative environment for some people, it cannot be assumed that this is the same for all. Though negative person–landscape transactions are less well documented (but see Burgess, 1996, Ward-Thompson et al., 2004), they nevertheless exist and are important in understanding what contributes to people's engagement with, or avoidance of, specific forms of natural landscape. The extent to which an individual experiences landscape as either conducive or deleterious to the promotion or maintenance of their well-being can be mediated by a range of differing influences. These can include declining health or physical mobility, differing cultural perceptions of what constitutes a restorative environment, childhood experiences, peer group and familial influences, and the influence of myth and media.

While the domestic garden is often seen as conducive to well-being, for example, there is an acknowledgment amongst older people that advancing age can limit their capacity to cope with it. Some older people prepare for this eventuality by adapting the domestic garden to minimise the physical activity required to maintain it. 'Ralph' (72), for example noted:

> I've got it [the garden] slabbed and it's in squares right the way round. I've done this about four years ago, when I knew this [old age] was coming on and I said, well I'm going to get it done and make it my 'half hour garden' (Milligan et al., 2004, 1787).

As an outwardly visible part of the home, the ability to keep a neat and tidy garden can be very important to older people – reflecting their ability to manage the home itself. For those older people who are unable to manage their garden, Milligan et al. (2004) point to three possible outcomes:

i) reliance on a family member or neighbour for help in the heavier gardening tasks – reducing the older person's independence (a factor seen as critical to older people's definition of health); or
ii) the older person having to move from his/her familial home to alternative accommodation such as a flat or sheltered housing; or
iii) more affluent older people may be able to afford to engage a gardener for a few hours a week, enabling them to retain their garden. However, this is a limited option for the many older people who survive on very limited incomes.

For some older people, then, an unmanageable garden can lead to a loss of independence, frustration and/or depression with negative impacts on the individual's well-being.

In thinking about younger people's engagement with natural environments, three sets of influences that affect their response to such natural environments can be identified:

i) Formative experiences and familial influences Parental anxieties resulting in clearly delimited 'safe' play zones during childhood can influence the extent to which people engage with the natural environment in later life. Young adults brought up in rural areas, for example, describe having been told by parents that playing in woodland was 'too risky' unless accompanied by an adult or an older sibling. One consequence of this is that in young adulthood wooded areas can be seen as 'scary places' that would not be accessed unless accompanied by others (Milligan and Bingley, 2007). This reflects Aitken's (1994) claims that parental anxieties can considerably dilute potentially important experiences of independence in childhood.

ii) Dirt and density For some, concerns around woodland are linked to specific types of wooded areas. Certain forest formations – particularly densely wooded areas or those with narrow paths and overhanging trees that shut out sunlight can make people feel intimidated, trapped or shut in.

As Sophie (aged 18) commented:

> I get a bit claustrophobic sometimes and I feel like I'm trapped by all these plants.

Also reflecting on this type of woodland, Ella (aged 17) noted:

> it was dark 'cause all the trees are over your head and all that … and there was like a run down house which was quite scary to walk past. It was the darkness, what's hiding in there, crawling in there.

Dense woodland can also raise concerns about visibility – in part this relates to the animal and insect life that may be 'lurking within'. Further, while muddiness in woodland is often associated with enjoyable childhood play, for some, it raises concerns around dirt, uncleanliness and potential contamination from touching fallen leaves, twigs or vegetation. In part, the density of some forms of woodland is linked to fears about being trapped if attacked. This can be particularly so for young women, as one commented:

> I might get mugged or something … and it's just such a confined space that you're in. If anything did happen when you're in a wood you wouldn't be able to really get out, 'cause all the trees are there … no-one would know you were there.

While these concerns do not extend as far as the bio-phobic experiences outlined by Davidson et al. (2005), they do nevertheless illustrate that natural landscapes cannot automatically be viewed as inherently restorative places.

iii) Myth and media A third set of factors contributing to perceived risk in accessing woodland relate to the influences of myth and media. The powerful (and usually negative) influence of the media can be seen in the often sensationalised reporting of accidents, murders and attacks in wooded areas.

As Ruth put it:

> Like you see all these things in the news like people getting buried out in the woods and stuff, and like that's quite scary.

Media sensationalism not only acts to feed fear of attack by strangers, particularly (but not solely) amongst young women, but it also feeds into existing fears about supernatural forces in woodland drawn from childhood fairytales, myth and contemporary horror films (for example, Little Red Hiding Hood, the Blair Witch Project and so on). As one young woman commented:

> You know, a lot of the ghost stories are set in woods … and you get the actual things that are in the news [and the] supernatural. I think a lot of it is supernatural which I think goes back to like the whole, you know, lady in the woods thing and Blair Witch …

Others have drawn on high profile criminal cases – such as the infamous Moors murders in the UK in the 1960s – to highlight how adverse perceptions of natural woodland can be accentuated by media reporting.

It is clear from this section, then, that while there are some common factors that affect how people experience local landscapes as inherently restorative or risky, there are also important differences that can arise as a consequence of both age-related and cultural experiences that vary across time.

Concluding Comments

This chapter has sought to draw attention to some of the wider literatures both within geography as well as elsewhere, as a means of highlighting some of the interconnections between these works and the therapeutic landscapes literatures. Environmental psychologists and those working in the fields of emotional or phobic geographies have both demonstrated how issues of the psyche, the mind and emotion can also be an integral part of understanding the complexities of people's interrelationships with places. Indeed, the environmental psychology literature has much to tell us about the risk reductive capacity and the restorative potential of natural landscapes and how this might be integrated into the design of health policy. Much of this work, however, has been situated within a quantitative tradition that relies heavily on interventionist approaches. Hence, it fails to engage with the individual, subjective experience of the person–landscape transaction that underpins much of the contemporary work around therapeutic landscapes. So, not only do health geographers need to begin to engage with the environmental psychology literature in a more sustained and critical way, but it also needs to be recognised that environmental psychologists and policy-makers can also learn much from the work of health geographers.

Secondly, in thinking about therapeutic landscapes, I have suggested that we need to distinguish not only between the *restorative* and the *curative* effects of places (Geores and Gesler, 1999), but also between *restorative* and *risk/fear-inducing* effects of the natural environment. So, for example, while visiting woodland or tranquil and scenic areas may have a (temporary) restorative effect on an individual's sense of well-being, it may not have the longer term curative effects on an individual's physical or mental well-being that places such as hospitals or retreats may have. Similarly, for some, natural landscapes such as wooded areas or even the everyday landscapes of the home or garden may induce stress, fear and anxiety rather than peace and tranquillity. So what may offer a restorative landscape for one person can prove stressful or anxiety-provoking for another. Moreover, these responses can shift and alter over time for a wide variety of reasons that may be related to age or culture, or the influence of a peer group, personal relationships and experiences or myth and media. In other words, people's responses to landscape are complex and it cannot simply be assumed that natural landscapes equate to therapeutic landscapes for all.

Acknowledgments

Some of the material within the third section of this chapter draws on joint research undertaken with Amanda Bingley and Tony Gatrell. I would like to acknowledge the work and contribution of both these people; they have been a pleasure and inspiration to work with.

References

Aitken, S.C. (1994), *Putting Children in their Place* (Washington, DC: Association of American Geographers).

Anderson, K. and Smith, S. (2001), 'Editorial: Emotional Geographies', *Transactions of the Institute of British Geographers* 26: 7–10.

Andrews, G. (2002), 'Towards a More Place-Sensitive Nursing Research: An Invitation to Medical and Health Geography', *Nursing Inquiry* 9: 221–38.

Bingley, A. and Milligan, C. (2007), '"Sandplay, Clay and Sticks": Multi-Sensory Research Methods to Explore the Long-Term Influence of Childhood Play Experience on Mental Health', *Children's Geographies* 5:3, 283–296.

Bondi, L., Davidson, J. and Smith, M. (2005), 'Introduction: Geography's "Emotional Turn"', in J. Davidson, L. Bondi and M. Smith (eds) *Emotional Geographies* (Aldershot: Ashgate) pp. 1–18.

Brown, M. (2003), 'Hospice and spatial paradoxes of terminal care', *Environment and Planning A* 35: 833–51.

Burgess, J. (1996), 'Focusing on fear: the use of focus groups in a project for the Community Forest Unit, Countryside Commission', *Area* 28: 130–35.

Burman, E. and Chantler, K. (2004), 'There's no place like home: emotional geographies of researching "race" and refuge provision in Britain', *Gender, Place and Culture* 11: 375–97.

Chatterjee, S. (2005), 'Children's friendship with place: a conceptual inquiry', *Children, Youth and Environment* 15: 1–26.

Chawla, L. (1992), 'Childhood place attachments', in A. Altman, and S. Low (eds) *Place and Attachment*, vol. 12 (New York: Plenum).

Conradson, D. (2005), 'Landscape, care and the relational self: Therapeutic encounters in rural England,' *Health and Place* 11: 337–48.

Cooper Marcus, C. and Barnes, M. (1999), *Healing Gardens: Therapeutic benefits and design recommendations* (New York: Wiley).

Davidson, J. (2003), *Phobic Geographies: the phenomenology and spatiality of identity* (Aldershot: Ashgate).

Davidson, J., Bondi, L. and Smith, M. (eds) (2005), *Emotional Geographies* (Aldershot: Ashgate).

Davidson, J. and Smith, M. (2003), 'Bio-phobias/Techno-philias: virtual reality exposure and treatment for phobias of "nature"', *Sociology of Health and Illness* 25: 644–61.

Frazier, L. and Scarpaci, J. (1998), 'Landscapes of state violence and the struggle to reclaim community' in R. Kearns and W. Gesler (eds) *Putting Health into Place: landscape, identity and well-being* (Syracuse: Syracuse University Press) pp. 53–74.

Geores, M. and Gesler, W. (1999), 'Compromised Space: Contests over the Provision of a Therapeutic Environment for People with Mental Illness', in A. Williams (ed.) *Therapeutic Landscapes: the Dynamic Between Place and Wellness* (Lanham, MD: University Press of America) pp. 99–122.

Gesler, W. (1993), 'Therapeutic landscapes: theory and a case study of Epidauros, Greece', *Environment and Planning D* 11: 171–89.

Gesler, W. (1998), 'Bath's reputation as a healing place', in R. Kearns and W. Gesler (eds) *Putting Health into Place: Landscape, Identity and Well-being* (Syracuse: Syracuse University Press) pp. 17–35.

Gesler W., Bell, M., Curtis, S., Hubbard, P. and Francis, S. (2004), 'Therapy by design: evaluating the UK hospital building program', *Health and Place* 10:117–28.

Gold, J. and Revill, G. (2003), 'Exploring landscapes of fear: marginality, spectacle and surveillance', *Capital and Class* 80: 27–50.

Hartig, T., Böök, A., Garvill, J., Olssen, T. and Gärling, T. (1996), 'Environmental influences on psychological restoration', *Scandinavian Journal of Psychology* 37: 378–93.

Hartig, T., Evans, G., Jamner, L., Davis, D. and Gärling, T. (2003), 'Tracking restoration in natural and urban field settings', *Journal of Environmental Psychology* 23: 109–23.

Hartig, T. and Staats, H. (2003), 'Guest Editors Introduction: Restorative environments', *Journal of Environmental Psychology* 23: 103–7.

Hasse, J. (1999), 'Overlooking the emotional domain in human geography', *Geographische Zeitshcrift* 87: 63–83.

Herzog, T.R. and Barnes, G.J. (1999), 'Tranquillity and Preference Revisited', *Journal of Environmental Psychology* 19: 171–81.

Herzog, T.R., Black, A.M., Fountaine, K.A. and Knotts, D.J. (1997), 'Reflection and Attentional Recovery as Distinctive Benefits of Restorative Environments', *Journal of Environmental Psychology* 17: 165–70.

Jarosz, L. (1994), 'Agents of Power, Landscapes of Fear – the vampires and heart thieves of Madagascar', *Environment and Planning D* 12 (4): 421–36.

Johnsen, S., Cloke, P. and May, J. (2005), Transitory spaces of care: serving homeless people on the street', *Health and Place* 11: 323–36.

Kaplan, R. (2001), 'The nature of the view from home: psychological benefits', *Environment and Behaviour* 33: 507–42.

Kaplan, R., Kaplan, S. and Ryan, R. (1998), *With People in Mind: design and management of everyday nature* (Covelo, CA: Island Press).

Kaplan, S. and Talbot, J. (1983), 'Psychological benefits of a wilderness experience', in I. Altman and J. Wohlwill (eds) *Behaviour and the Natural Environment* (New York: Plenum Press) pp. 163–203.

Kearns, R. and Collins, D. (2000), 'New Zealand children's health camps: therapeutic landscapes meet the contract state', *Social Science and Medicine* 51: 1047–59.

Knopf, R.C. (1987), 'Human behaviour, cognition, and affect in the natural environment', in D. Stokols and I. Altman (eds) *Handbook of Environmental Psychology* (New York: John Wiley) pp. 783–825.

Lalich, W. (2006), *Measurement and Spatial Effects of the Immigrant Created Cultural Diversity in Sydney* (Milan: Fondazione Eni Enrico Mattei).

Lupton, D. and Tulloch, J. (1999), 'Theorising fear of crime: beyond the rational/irrational opposition', *British Journal of Sociology* 50: 507–23.

Madge, C. (1998), 'Therapeutic landscapes of the Jola, The Gambia', *Health and Place* 4: 293–311.

Milligan, C. and Bingley, A., (2007). 'Therapeutic places or scary spaces? The impact of woodland on the mental well-being of young adults', *Health and Place* 13:3, 799–811.

Milligan, C., Bingley, A. and Gatrell, T. (2005a), 'Healing and Feeling: the place of emotions for older people', in J. Davidson, M. Smith and L. Bondi (eds) *Emotional Geographies* (Ashgate: Aldershot) pp. 49–62.

Milligan, C., Bingley, A. and Gatrell, T. (2005b), 'Digging deep: Using diary techniques to explore the place of health and well-being amongst older people', *Social Science and Medicine* 61: 1882–92.

Milligan, C., Gatrell, T. and Bingley, A. (2004), '"Cultivating Health": therapeutic landscapes and older people in Northern England', *Social Science and Medicine* 58: 1781–93.

Moon, G., Kearns, R. and Joseph, A. (2006), 'Selling the private asylum: therapeutic landscapes and the (re)valorization of confinement in the era of community care', *Transactions of the Institute of British Geographers* 31: 131–49.

Morris, S. and Thomas, C. (2005), 'Placing the dying body: emotional, situational and embodied factors in preferences for place of final care and death in cancer', in J. Davidson, L. Bondi and M. Smith, (eds) *Emotional Geographies* (Aldershot: Ashgate) pp. 19–32.

Pain, R. (2000), 'Place, social relations and the fear of crime: a review', *Progress in Human Geography* 24: 365–87.

Panelli, R., Little, R. and Kraack, A. (2004), 'A Community Issue: Rural women's feelings of safety and fear in New Zealand', *Gender, Place and Culture* 11: 445–67.

Parr, H., Philo, C. and Burns, N. (2005a), 'Not a display of emotions: emotional geographies in the Scottish Highlands', in J. Davidson, L. Bondi and M. Smith (eds) *Emotional Geographies* (Aldershot: Ashgate) pp. 87–102.

Parr, H., Philo, C. and Burns, N. (2005b), 'Social geographies or rural mental health: experiencing inclusions and exclusions', *Transactions of the Institute of British Geographers* 29: 401–19.

Rohde, C. and Kendle, A. (1994), *Report to English Nature – Human Well-being, Natural Landscapes and Wildlife in Urban Areas: A Review* (Bath, UK: University of Reading).

Parry-Jones, W. (1990), 'Natural landscape, psychological well-being and mental health', *Landscape Research* 15: 7–11.

Smith, M. and Davidson, J. (2006), '"It makes my skin crawl ...": The embodiment of disgust in phobias of nature', *Body and Society* 12: 43–67.

Sparks, R. (1992), *Television and the drama of crime: moral tales and the place of crime in public life* (Bucks: OU Press).

Thomas, M. (1999), *Dreams in the shadows: Vietnamese-Australian lives in transition* (Sydney: Allen and Unwin).

Thurber, C. and Malinowski, J. (1999), 'Summer Camp as a Therapeutic Landscape', in A. Williams (ed.) *Therapeutic Landscapes: the Dynamic Between Place and Wellness* (Lanham, MD: University Press of America) pp. 53–70.

Tolia-Kelly, D. (2004), 'Landscape, race and memory: biographical mapping of the routes of British Asian landscape values', *Landscape Research* 29: 277–92.

Tolia-Kelly, D. (2006), 'Mobility/stability: British Asian cultures of "landscape and Englishness"', *Environment and Planning A* 38: 341–58.

Tuan, Y-F. (1979), *Landscapes of Fear* (London: Blackwell).

Ulrich, R.S. (1984), 'View through a window may influence recovery from surgery', *Science* 224: 420–21.

Ulrich, R.S. (1979), 'Visual landscapes and psychological well-being', *Landscape Research* 4: 17–23.

Ulrich, R.S. and Addoms, D.L. (1981), 'Psychological and recreational benefits of a residential park', *Journal of Leisure Research* 13: 43–65.

Van den Berg, A.E. (2003), 'Environmental preference and restoration: (How) are they related?' *Journal of Environmental Psychology* 23: 135–46.

Wakefield, S. and McMullan, C. (2005), 'Healing in places of decline: (re)imagining everyday landscapes in Hamilton, Ontario', *Health and Place* 11: 299–312.

Ward-Thompson, C., Aspinall, P., Bell, S., Findlay, C., Wherrett, J. and Travlou, P. (2004), *Openspace and social inclusion: local woodland use in Central Scotland* (Edinburgh: Forestry Commission).

Ward-Thompson, C., Aspinall, P., Bell, S. and Findlay, C. (2002), *Openspace and social inclusion: local woodland use in Central Scotland* (Edinburgh: Edinburgh College of Art; OPENspace publications).

Williams, A. (1999), 'Place identity and therapeutic landscapes: the case of home care workers in a medically underserviced area', in Williams, A. (ed.) *Therapeutic Landscapes: The Dynamic between Place and Wellness* (Lanham: MD: University Press of America, 71–98.

Williams, A. (2002), 'Changing geographies of care: employing the concept of therapeutic landscapes as a framework in examining home space', *Social Science and Medicine* 55: 141–54.

Wilson, K. (2003), 'Therapeutic landscapes and First Nations peoples: an exploration of culture, health and place', *Health and Place* 9: 83–92.

Chapter 17

Surviving Profoundly Unhealthy Places: The Ambivalent, Fragile and Absent Therapeutic Landscapes of the Soviet Gulag

Geoffrey DeVerteuil and Gavin J. Andrews

Introduction

Since its inception (Gesler, 1992), the concept of therapeutic landscapes has emerged as an idealized framework and geographical metaphor to approach the health/place relationship and the search for physical, psychological and social well-being. Despite considerable and varied empirical attention, at least three important critiques have emerged that we feel have been insufficiently extended. The first is that places can both hurt *and* heal, having negative and positive health effects (Wilson, 2001; Baer and Gesler, 2004; Wakefield and McMullan, 2005; Wilton and DeVerteuil, 2006). While a few works have examined less than healthy places to identify their redeeming features (Wakefield and McMullan, 2005), we need to expand our knowledge of profoundly unequal and unhealthy places and times, and investigate more systematically their anti-therapeutic effects as well as potential efforts to sustain therapeutic 'havens' in them (Kearns and Gesler, 1998; Pinfold, 2000; Gastaldo et al., 2004). In extreme places such as war zones and death camps, creating and sustaining therapeutic landscapes becomes more than simply promoting well-being; it can mean the difference between life and death. Examining such profoundly unhealthy places necessarily involves linking therapeutic landscapes with survival strategies both physical and mental (Gastaldo et al., 2004). No geographical works have investigated profoundly harmful, seemingly irredeemable places that stretch the concept of therapeutic landscapes into unknown territory (but see Frazier and Scarpaci, 1998 for the legacy such places leave behind, in the case of post-Pinochet Chile).

Second, the therapeutic landscapes concept has been criticized for overlooking the importance of everyday and ordinary geographies, in favor of extraordinary, one-off events such as visiting spas (Wilson, 2003; Baer and Gesler, 2004; Andrews and Kearns, 2005). We feel this critique can be extended to encapsulate what we deem the 'extreme everyday': acute conditions experienced on an everyday basis in profoundly unhealthy places (for example, war zones, death camps and epidemics).

In other words, setting the everyday within an extraordinarily difficult context under which marginal actors must struggle mightily to navigate and survive.

Third and finally, an ongoing critique focuses on the fact that therapeutic landscapes can be as much imagined as physical, thereby precluding the need for human co-presence (see Williams, 1998; Andrews, 2004; Gastaldo et al., 2004). We wish to usefully extend the work on 'therapeutic landscapes of the mind', relating the importance of 'non-physical' (and physical) landscapes to individualized well-being and survival in decidedly harmful places.

In this chapter, we expand upon each of these critiques by using an exploratory case study of a profoundly unhealthy place in historical perspective: the Soviet Gulag during Stalin's reign (1929–53). We rely on a limited sample of four personal memoirs of survivors to see *whether* therapeutic landscapes could be usefully created and sustained within the Gulag as part of broader physical and mental survival strategies. We seek evidence of any intersections between survival strategies and therapeutic landscapes within such an unpromising environment.

Despite its limited and exploratory nature, our research extends the concept of therapeutic landscapes into largely unknown territory and challenges some of its underlying assumptions. We build on how therapeutic landscapes may be created and sustained in unpromising locales as a matter of survival, under 'extreme everyday' conditions, and through both physical and mental strategies. Our study of the Gulag also contributes to historical research in health geography, particularly in the reconstruction of historical landscapes of health, health care and disease (Craddock, 2001; Andrews and Kearns, 2005). Indeed, the twentieth century witnessed the unparalleled ability – both in terms of state authority and technological advances – to terrorize and liquidate entire peoples, thereby creating profoundly unhealthy places that belied the tremendous progress occurring simultaneously in medical science.

A historically enduring example of this inhumanity was the Soviet Gulag under Josef Stalin. The infamous Gulag (*Glavnoe Upravlenie Lagerei*, meaning Main Camp Administration) forced labor camps came to signify:

> Not only the administration of the concentration camps but also the system of Soviet slave labor itself, in all of its forms and varieties: labor camps, punishment camps, criminal and political camps, women's camps, children's camps, transit camps. Even more broadly, 'Gulag' has come to mean the Soviet repressive system itself, the set of procedures that prisoners once called the 'meat-grinder': the arrest, the interrogations, the transport in unheated cattle cars, the forced labor, the destruction of families, the years spent in exile, the early and unnecessary deaths (Applebaum, 2003, xv–xvi).

The Gulag was chosen not only for its unhealthiness, but also for its marginal status, occupying the geographical fringes of the Soviet Union as well as being literally 'off the map' within geographical scholarship (Andrews and Kearns, 2005). We move beyond mapping what is already famous to discover and articulate the neglected, marginal, and everyday histories of individuals, groups and places. The study of life in the Gulag presents an everyday social and health history of a notorious yet understudied place, rarely analyzed in terms of its negotiation, subtleties and details.

Upon detailing the unhealthy environments of the Soviet Gulag, we outline our methodological approach and analysis of the four memoirs. The small sample of memoirs, along with issues of selective memory, necessarily limits the representativeness of the analysis, as well as the ability to speak to larger issues of survival and dehumanization in twentieth-century totalitarian camps. As such, we cast the analysis and results as exploratory. Three themes materialized, emerging as a process whereby early efforts at creating and sustaining therapeutic landscapes (remembering former places, fantasies about Siberia, 'havens' in hospitals and other safe spaces) wilted under the corrosive effects of long-term life in the camps. The conclusion speaks to how the case study may inform a future convergence between health and political geography.

The Soviet Gulag as a Profoundly Unhealthy Environment

As a system of prisons, forced labor camps and places of exile, the Gulag was the most prominent and enduring symbol of the Stalinist terror machinery. Somewhat contradictorily, the Gulag was both a political tool to intimidate and eliminate 'enemies of the state' (real or imagined) as well as an economic tool for industrial and agricultural expansion, the construction of enormous public works, and the development of the Far North (Nordlander, 1998; Viola, 2001). However, with Stalin's death in 1953, it had become apparent that the system of forced labor was '… wasteful, corrupt, and, above all, unprofitable' (Applebaum, 2003, 471). On a much smaller scale, the Gulag would remain in operation until the end of the Soviet Union.

Three major waves of state terror, each with their own designated enemies of the state, fed the Stalinist Gulag: 1930–32 (forced collectivization, deportation of prosperous peasants [*kulaks*], Ukrainian famine); 1936–38 (mass purges of Communist officials, intelligentsia, military cadres, bureaucrats, and so on); and 1946–48 (mass internment of returning Russian soldiers from Europe, foreign prisoners of war). No fewer than 18 million people were estimated to have passed through the Gulag between 1929 and 1953 (Applebaum, 2003). On a scale never witnessed before, millions of people were deprived of their homes, uprooted and sent to isolated forced labor camps, never to return to the mainstream; 'excess deaths' from the Gulag alone stand at a minimum of 2.7 million (Pohl, 1997; Rosefielde, 1997). Moreover, unlike the short-lived Nazi death camps, the Gulag existed long enough to infect a whole new generation.[1] Solzhenitsyn (1998) would even compare the Gulag to a cancer metastasizing across the land.

---

1 This is not to say that the Soviet Gulag was equivalent to the Nazi death camps. First, the primary rationale for Gulag was not to exterminate rigidly defined racial or religious groups, but the flexible elimination of enemies of the state and, concomitantly, the rapid development of the Soviet economy (Parrish, 1996). Under the auspices of Article 58 of the prison code, virtually anyone could be arrested for 'counter-revolutionary' crimes. Second, we know measurably less about the Gulag than the Nazi system, for the former's machinery was never laid bare, and 'many of the same people who had committed the political crimes were still in political office' after Stalin's death (Adler, 1999, 6). The scope of the camps and the terror was carefully concealed to both Russians and Westerners. As a result, neither group has truly confronted its unpleasant legacy (Reynolds, 2002). Within Russia, there have

That the Stalinist Gulag was unhealthy, if not lethal, is hardly debatable. What is more important for our purposes is *how* various aspects of the Gulag were unhealthy. This involves examining the vast array of places that functioned as bridges from arrest and interrogation to transport, to the forced labor camps and, if the prisoner survived, on to exile. The process began with arrest, which took place in public spaces such as squares, train stations and subways, as well as private spaces such as apartments and offices. Especially during the 1936–38 period, the police were simply fulfilling (or over-fulfilling, to show their worth) quotas from above, so arrests were quite arbitrary. If interrogation followed arrest, then the shattered prisoner had to endure the absolute power of the interrogator to administer punishment and extract confessions.

Once 'processed,' prisoners were transported to the far-flung forced labor camps, a massive internal movement from cities and the countryside to the undeveloped North (Viola, 2001). Transport included *Black Marias* (specially built trucks to transport prisoners by street), red cattle cars (single-purpose train convoys with up to 25 cars), and marine transport (barges along rivers that flowed to isolated camps in the north, steamers that crossed the Sea of Okhotsk to Magadan) (Bollinger, 2003). Given the fact that trips could take months, the transportation stage represented a disorienting interstitial space. It was, literally, a journey into the abyss: freezing cold in the winter, suffocating heat in the summer, non-existent hygiene, starvation rations. Many of the deprivations stemmed from both the utter inefficiency of the Gulag as a whole, as well as the indifference of the convoy administration, who after all were transporting enemies of the state (Applebaum, 2003).

Located in the bleakest parts of the Soviet Union, forced labor camps were designed with bluntly economic and carceral, rather than therapeutic, aims. There was never enough food for the working prisoner, as supplies would have been plundered ahead of time by trusties, thieves, and administrators (Applebaum, 2003). Survival very much depended on a panoply of strategies: adapting to camp mores, making yourself noticed and useful to rise above the horde of prisoners, and acquiring light and/or indoor work (or avoiding work altogether), because those that slogged in the mines, forests or construction sites were frequently 'last-leggers' (a term used for those near death) after several months (Conquest, 1978).

Early release was unthinkable, unless it was through death, which in the camps was frighteningly anonymous. Great efforts were made to conceal death (and suicide) in camps: bodies were dumped in common graves, family members were never notified, and connections to the prisoner's former life were lost. By 1949, Stalin had instituted the high-security Special Camps, where the links to the outside world were even more tenuous, the ultimate 'island' of the Archipelago (Applebaum 2003). Most prisoners who survived were exiled upon release (Adler, 2002); for them, their new 'home' was frequently dreary communities in close proximity to the Gulag itself. After Stalin's death in 1953, many prisoners were formally released and rehabilitated, but would continue to endure stigmatization. Although profoundly unhealthy, most processed through the Gulag did survive – so how can we then

been no commissions seeking the truth, no funds compensating the injured, and victims have mostly been forgotten, with only modest memorials (Daniszewski, 2002).

understand whether their survival techniques overlapped in any way with the creation and sustenance of therapeutic landscapes?

Methods

Understanding whether prisoners created and sustained therapeutic landscapes under 'extreme everyday' conditions necessitates a broadly qualitative approach, one that incorporates first-person, narrative and autobiographical elements. There has been recent attention in historical geography to representation and the writing of historical narrative (Wishart, 1997). Approaches include travel writings (Gruffudd et al., 2000), biography (Lorimer, 2005) and oral history (Andrews et al., 2005) in the reconstruction of historical landscapes. However, in comparison to the scope and volume of historical geography as a sub-discipline, this is a relatively rare undertaking. This state of affairs might be due to the restricted historical range that narratives are able to recall, giving rise to concerns over the credibility of such 'evidence' and perhaps even because of the lack of expertise in such approaches (Andrews et al., 2005). Nevertheless, we argue that they provide a means to investigate and articulate the everyday social geographies located in the recent past, potentially paving the way towards a rich source of data.

In this chapter, we adopted the perspective of those who directly experienced (and survived) the Gulag, using memoirs. The reading of memoirs counterbalances the state's accounts of its own actions, as well as dry statistical and historical studies that ignore individual survival techniques. As autobiographical narratives, camp memoirs were essentially based in a strong urge to record what was happening to the individual and others around, so that the ephemeral nature of places, events and people under mass terror could hopefully be memorialized. This witnessing was key, since 'millions were destroyed in the camps and disappeared without leaving a grave behind them or any other material evidence that they ever existed' (Razgon, 1997, 121). Despite their usefulness, however, there are several limitations to memoirs, including their inability to formulate the 'big picture.' As Solzhenitsyn would himself admit, he had but a 'peephole into the Archipelago,' not a 'view from the tower' (1998, 7). The 'whole story' may also not be told for other reasons, such as the need to conceal self-incriminating and evil acts. Memoirs should therefore be seen as selective and partial, and what is 'real' in these biographies are the memories and perceptions of the Gulag landscapes themselves, not some positivistic notion of reality.

Out of hundreds of potential published and unpublished works and testimonials (some of which remain untranslated), we drew on material from four individuals' camp memoirs. Our sample therefore cannot be considered representative of the enormous diversity within the Gulag survivor literature, so we cast the analysis and results as preliminary and exploratory. Nevertheless, the four memoirs were chosen to broadly hint at the typical experiences of those urbane, educated individuals arrested as political prisoners under Article 58 (the primary 'legal' instrument through which political prisoners were arrested). The first two memoirs were published while the Soviet Union was still in existence: Ginzburg's *Journey into the Whirlwind* (first appearing in 1967, translated in 1967) and the companion piece *Within the Whirlwind* (first appearing in 1979, translated in 1981), as well as Shalamov's *Kolyma Tales* (translated in 1980, 1994 version used). We also included two memoirs that emerged in the more open post-Soviet period that largely confirmed the findings of the earlier works: Bardach's *Man is Wolf to Man* (1998) and Adamova-Sliozberg's 'My Journey,' part of a larger collection of edited stories on women in the Gulag – *Till My Tale is Told* (first appearing in 1989, translated and abridged in 1999). Although an indispensable source and the most recognized work on the Gulag, we have purposefully left out Solzhenitsyn's three-volume *Gulag Archipelago*. This is because the work is less of a personal memoir and far more of a reading on the broader nature of totalitarianism and human suffering as a whole.

Altogether, over 1,800 pages of text were analyzed, spanning close to 50 years in forced labor camps, prison, intermittent civilian life, and exile (see Tuchman, 1994 and Atkinson, 1998 for more information on documentary/memoir/diary analysis). Each memoir was read twice by the first author, with the first reading focused on identifying (and coding) emerging personal narratives that clearly overlapped with the concept of therapeutic landscapes. The second reading involved isolating the coded material in order to identify common themes that ran through each of the four memoirs. Each narrative of survival also revealed how therapeutic landscapes emerged as complex, unbounded and dispersed interactions between individual survival tactics (both mental and physical) and the broader physical and social environment (Gastaldo et al., 2004).

The nature of survival in totalitarian camps – whether in the Gulag (Alexandr Solzhenitsyn in particular), Nazi concentration camps (the example of Primo Levi's memoirs on his survival at the Auschwitz death camp) or beyond – has generated an extraordinarily large and complex literature. Given the limited scope of the chapter and its focus on therapeutic landscapes, we examine only three everyday needs: (1) economic/material subsistence, (2) social subsistence, and (3) psychological/cognitive adaptation. The first basic need involves a range of material requirements for physical survival, such as food, warm clothing, and material for barter, such as tobacco. As previously mentioned, these materials were always in short supply and required constant struggle and scheming. The second basic need involves social interactions and networks among the prisoners, from which they may gain material and/or emotional support and facilitate overall survival. The need for social interactions, however, was frequently tempered by cruel competition and a 'you die today, I'll die tomorrow' outlook (Solzhenitsyn, 1998). Psychological/cognitive adaptation involves the ways in which the prisoners make sense of their predicament, a process that emphasizes the individual's personality and morality. We were particularly interested in how the securing of these three basic needs overlapped – if at all – with the creation and sustaining of therapeutic landscapes.

Those who experienced the Gulag firsthand agreed that the first year in camp was critical in determining the fate of the prisoner, including overall demeanor and the strategies used to obtain material needs. Of course, what it took to survive the Gulag was very much dependent on external conditions and individual resourcefulness. A penal camp in Siberia in 1942 was substantially more difficult to survive than an agricultural camp in Western Russia in the early 1930s (Applebaum, 2003).

Some individuals were more mentally prepared for the rigors of camp than others; physically, most (if not all) arrived at the camps depleted.

Four Tales of Survival

The map below locates places in the Stalinist Gulag that figure prominently in the four memoirs.

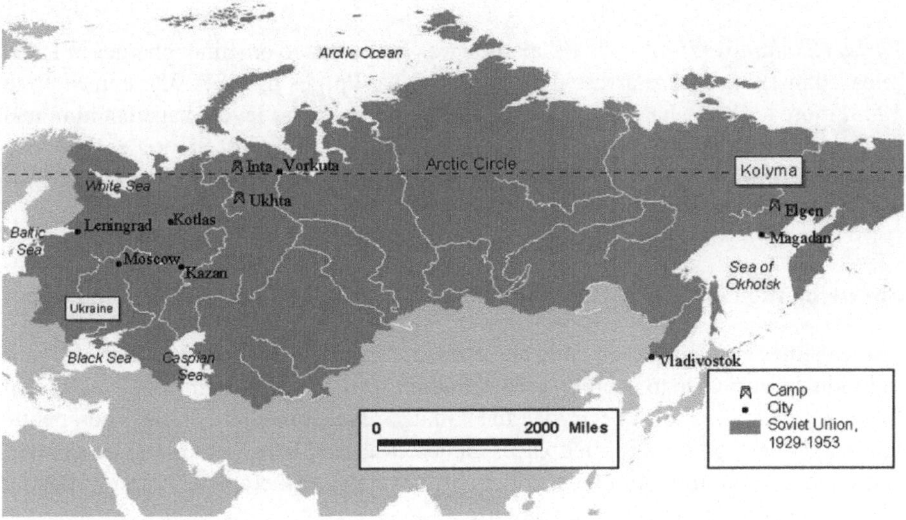

Figure 17.1 Geography of the Stalinist Gulag

Olga Adamova-Sliozberg (1902–92) In 1936, Adamova-Sliozberg lived a comfortable existence in Moscow, with two young children and a husband who was a university professor. This all changed, however, with the arrest of her husband. Although impending, her own arrest came as a shock, and she would spend the next 18 torturous years in (and out of) the Gulag system, beginning with the Lubyanka and Butyrki Prisons in Moscow, loaded onto a train north to the Solovki Prison near the White Sea, onto the Kazan Prison and then the long trip eastward to Siberia in 1939. Released in 1944, she returned to Moscow, only to be re-arrested as a 'secondtimer' in 1949 and exiled to Kazakhstan until 1954.

Janusz Bardach (1919–2002) Unlike the other authors, Bardach was a foreigner, a Polish citizen who saw his village occupied by the Soviets in 1939, and was then drafted into the Soviet military in 1940 at the age of 21. During the disastrous summer campaign of 1941 against the invading Germans, he was court-martialed for incompetency at the front, but managed to avoid execution through a family friend. He was sentenced to ten years of hard labor under Article 58. His brother Julek, now an officer in the Polish Army, secured his release in late 1945, but their entire family had been murdered by the Nazis during the occupation of Poland.

Eugenia Ginzburg (1896–1977) Ginzburg was a loyal Communist Party member and university professor in Kazan until her sudden arrest in December 1937. From Kazan, she spent two years of solitary confinement in Yaroslavl, was transported by cattle car to Vladivostock, by ship to Magadan, and then to the Elgen labor camp in Kolyma in 1939. She would spend the next 16 years in Siberia, first in camps, then in self-exile in Magadan where she arranged with great difficulty to have her last surviving son join her. She re-married, presuming that her husband had died after his arrest in 1937, and was then re-arrested in 1949, returning to the 'mainland' only in 1955.

Varlam Shalamov (1907–82) Shalamov was first arrested on minor charges in 1929, released in 1932, and re-arrested during the Great Purges in 1937. While in camp in 1943, more serious charges of 'anti-Soviet agitation' were leveled against him, and he would not be released from camp until 1951; in 1953 he left Siberia entirely and returned to Moscow. Fully rehabilitated by the state in 1956, he thereafter began writing the short stories that would comprise *Kolyma Tales*.

Surviving the Gulag: Fragile, Ambivalent and Absent Therapeutic Landscapes

Three themes emerged organically and consistently across the four memoirs. All four individuals were able to usefully create, but not truly sustain, fragile and ambivalent therapeutic landscapes to survive the Gulag. Over time, the initial therapeutic landscapes derived from former places, fantasies about Siberia as a romantic frontier, and havens in camp would fade as the corrosive and untherapeutic nature of camp life increasingly structured everyday life.

Fragile Therapeutic Landscapes of the Mind: Remembering Former Places and Fantasies about Siberia

The disorienting displacement of being transported thousands of miles, and the utter bleakness of the Gulag's infrastructure, encouraged the four authors to initially use former places as therapeutic landscapes of the mind to sustain their morale. Gastaldo et al. (2004: 171) found a similar process for immigrants, who used therapeutic thoughts of previous places to provide a sense of 'identity and. … .belonging, [and] stability in a process of displacement.' Sense of place was therefore more imagined than real, as the authors tried to mask their current realities. This was a state of 'in-betweenness,' whereby the new world of the Gulag – forced labor camps, isolators, transit prisons, and so on – was too brutal and alien to ever constitute a true 'home,' and the former world had disappeared, seemingly forever.

Both mentally and physically, Adamova-Sliozberg was totally unprepared for life in the Gulag, and for the first several years in prison she continued to fervently think of her family and home in Moscow. There she encountered long-time political prisoners who 'had gotten used to prison and were well accustomed to its awful routines. All their ties with the outside world had been severed; they had no children, no parents, and their husbands had either perished long since or, like them, were wandering from prison to prison' (Adamova-Sliozberg, 1999, 21). In her heart, she never wanted to become like those women. Similarly, Bardach would learn to 'shut out everything, to escape into the safety of my own space' (208). Pleasant memories of places in his past – '... all the places of my worry-free years, when everyone I loved was alive and happy' (380) – were mobilized to ward off feelings of depression and hopelessness. Conversely, because his village was now occupied by the Nazis, his home-based memories could also be a source of worry. What had happened to his family, his friends?

Also initially, the authors depended on fantasies of life in Siberia as therapeutic landscapes of the mind. Ginzburg was actually excited to leave the Yaroslavl prison for 'penal servitude' in Siberia – it meant 'traveling to new lands, rugged ones perhaps, but there would be fresh air, wind, and sometimes even sunshine, cold as it might be. And there would be people, hundreds of new people among whom we would live and work' (258). The imagined therapeutic effects of traveling to Siberia to work outdoors – a sort of 'promised land' of fresh air, sea breezes, stars – sustained Ginzburg during the long trip eastward and strengthened her resolve to survive. Even the hellish week-long steamer ride from Vladivostock to Magadan could not entirely erase Ginzburg's sense of hope in her new land:

> the sky began to glow with wonderful shades of mauve and lilac; my first Kolyma dawn was breaking. I felt, all of a sudden, a strange lightness and acceptance of my fate. True, this was a cruel and alien land. Neither my mother nor my sons would find the way to my grave. But all the same, it was part of mother earth – I had reached land, and need no longer fear the leaden, shark-infested waters of the Pacific Ocean (360).

For Adamova-Sliozberg, the first four years in prison cells also made her yearn for the outdoors, so her spirits lifted when she was able to see the sky during her transport to Siberia in 1939:

> The sky! If people weren't used to seeing it every hour and day of their lives, they would be astonished at the sight and would travel halfway around the world to marvel at it, as landlocked people travel to the sea. I couldn't get my fill of it, and sat for hours at the window (43).

To Bardach, the 'beautiful wilderness' of Kolyma was initially both alluring and foreboding. The stunning mountains, forests and skies excited his sense of adventure, but this landscape was counterposed with the grim infrastructure of camps and mines. These therapeutic landscapes of the mind were fragile, since they existed only in the imagination of desperate individuals caught up by forces well beyond their control.

Ambivalent Therapeutic Landscapes Surrounding Havens in Camp

Within the first year in camp, all the authors would seek physical havens to survive. While they lasted, these physical havens – literally spaces of recovery – constituted a paradise where inmates would be treated as human beings instead of units of economic production, shielded from the cold and thieves. Shalamov would occasionally feign illness (or pretend he knew carpentry) simply to obtain an easier work assignment, or just to spend a few hours indoors for warmth, knowing full well he would be exposed as a 'faker' sooner than later. Being sent from the mines to pick needles on trees – what he refers to as a 'pushover job' – allowed him to escape the dreary gold mines for a few days and be refreshed by the starkness of the Kolyma landscape. Several years later, 'a physician took interest in him and managed to have him assigned to paramedical courses' (Glad, 1994, xiv), which involved working indoors. The hospital became a haven from the cruel and indifferent camp world, a place where he could derive some everyday security.

Similarly, Bardach recognized that survival depended on not just personal strength but also finding therapeutic havens. In 1943, suffering from scurvy and night blindness, Bardach went to the field hospital and pretended to be a medical student, which then enabled him to be placed in a large central hospital. In the hospital, TB and eventually psychiatric wards, he was able to survive physically and busy himself with the small details of the patients and daily life: 'the ability to feel elated by the smallest things was the best defense I had against the spirit-killing force of the camps' (303).

However, all the authors, and many other ex-prisoners, recognized the moral ambivalence between creating and sustaining therapeutic landscapes and maintaining one's humanity. Solzhenitsyn (1998) argued that obtaining indoor jobs in hospitals, or acting as trustees and norm-setters, allowed certain inmates to enjoy life-giving, therapeutic havens not available to the rest of the inmates. That is, creating and sustaining therapeutic havens in the Gulag could simultaneously occur at the expense of the health of less fortunate inmates, as well as serving to buttress the overall system. Herein lies the possibility of considering the moral geographies of therapeutic landscapes, specifically how certain moral codes are implicated in therapeutic experiences of place. One moral aspect of therapeutic landscapes is the contradiction between how negative health impacts might co-exist in one place with therapeutic effects. Kearns and Barnett (1999) found a specific moral geography in the therapeutic landscape of Auckland's Starship hospital, in the contest between commercial interests and health care workers and patient groups over the siting of a fast food chain in the hospital's atrium.

Adaptation to Camp Life and the Fading/Absence of Therapeutic Landscapes

Over their long sentences, all of the authors would adapt to certain aspects of 'camp identity' (Solzhenitsyn, 1998; Applebaum, 2003), including prisoner psychology (fatalistic, future tense hardly ever used), language (camp slang that would not be understood on the mainland), food (the less said the better), attitude towards rehabilitation (cynical, but still had to praise the system that had enslaved them), arrival and departure (arrest and exile), appearance (severe), and, ultimately, the loss of initial therapeutic landscapes. As Shalamov emphasized, it was advantageous to forget about one's former life (and places). Grudging attachment to the anti-therapeutic landscapes of the Gulag was above all reflected by the relative dearth of fantasies about escape, given that such fantasies could have constituted important therapeutic landscapes of the mind. Beyond the fact that geography conspired to isolate the prisoners (most camps were surrounded by large expanses of tundra, snow, ice, taiga or sand), the prisoner had to ask the question – escape to what? There was no 'home' to return to anymore, as ties to the former life had largely withered through sheer distance and long periods of time. Within this context, Bardach likened his fate, and the fate of other prisoners, to '… a grain of sand floating in a sea of terror and brutality. No one would ever care if we lived or died. No one was interested in who we were, we who lived torturous lives and were buried without names of markers' (296). The only goal was to survive the present and not to worry about the future nor past, and this meant eschewing therapeutic landscapes altogether.

Once the harsh reality of Siberia was upon them, all came to fear and hate the physical landscape, with its impossibly cold winters and death-filled forests, as well as forget former places and people. Ginzburg began to despise the harshness of the landscape and its 'inhuman' inhabitants. During an arduous convoy through the Kolyma's farthest reaches, Ginzburg (1967, 395) remembered:

> ...a sense of being at the end of the earth and cut off from civilization … 'You know,' whispered the women next to me through chattering teeth … 'it wouldn't surprise me in the least if a mammoth came from behind that hill.' I agreed with her. I too felt that we were getting farther and farther away, not only from towns but from the present era, toward the neolithic age.

Over time, Ginzburg adapted her survival strategies and shifted her geographic references inward to the camps. Mentally and physically, she began adapting to the rhythms of camp life, and no longer found the memories of past places – her home, university, city – as therapeutic, as remembering them seemed like an act of self-torment. Instead, she identified with the camps, a new way of life that frightened her (1981, 78):

> By now I had come to resemble Egor the toolsetter … He pined for the central compound at Elgen more than he did for his own native village. And I caught myself pining for the dairy farm's chicken house more than for Kazan University. This gave me a fright. It seemed an outward sign of spiritual emptiness …

Adamova-Sliozberg would succumb to these same patterns of survival and 'cease to remember' (30). She would bury her thoughts of home and family, value a piece of bread over her memories, and even become a trustee in charge of setting work norms.

Like his fellow inmates, Shalamov's primary mental strategy was to narrowly focus on the bleak day-to-day existence, forgetting his former life and places – 'a human being survives on his ability to forget … we had all been permanently poisoned by the North, and we knew it' (43). The pitiless drive to survive corrupted and demoralized himself and other inmates, to the point where they were 'overwhelmed with indifference' (33). This is amply illustrated by his description of a fellow prisoner, Potashnikov (17):

> He didn't want to die here in the frost under the boots of the guards, in the barracks with the swearing, dirt, and total indifference written on every face. He bore no grudge for people's indifference, for he had long since comprehended the source of that spiritual dullness. The same frost that transformed a man's spit into ice in midair also penetrated

> the soul. If bones could freeze, then the brain could also be dulled and the soul could freeze over … Potashnikov has lost everything except the desire to survive, to endure the cold and remain alive.

Discussion and Conclusions: Therapeutic Landscapes Meet the Homicidal State

Within a context of a profoundly unhealthy place, we were able to reveal a sequence of strategies surrounding therapeutic landscapes, beginning with fragile memories of former places and fantasies about new ones, followed by ambivalent havens in camp, and ultimately fading and absence under the extreme everyday conditions of the Gulag. Given that the small sample of memoirs necessarily limits the representiveness of the analysis, as well as the ability to speak to larger issues of survival and dehumanization in twentieth-century totalitarian camps, we have cast these results as exploratory and preliminary. The admittedly limited nature of the findings do not, however, negate the important, creative and provocative findings set within spaces most geographers fear to tread.

We also realize that creating and sustaining therapeutic landscapes was a relatively minor element in the survival strategies of the prisoners, prominent only in the beginning of the process of adapting to the Gulag. Nevertheless, that these therapeutic landscapes existed at all, in such an unpromising and soul-destroying locale, testifies to the flinty strength of human survival. The Gulag experience also problematizes what is 'therapeutic' in therapeutic landscapes – that one person's haven can be at another person's expense. In so doing, we have contributed to three specific areas of knowledge. First, we have contributed to the incipient field of reconstructing health histories of marginal places (Andrews and Kearns, 2005), in this case the Stalinist Gulag. Second, we have contributed to the geographical knowledge of the Soviet Gulag. Third, we have contributed to better understanding the therapeutic landscapes of marginal groups (see, for example, Pinfold, 2000, for individuals with mental illness), the marginal group being political prisoners in the Gulag. Finally, the focus on therapeutic landscapes in the profoundly unhealthy spaces of the Gulag also suggests at least two areas of future research, both of which align health geography with concerns within political geography. First, there is need to engage and reconstruct the historical health geographies of other profoundly unhealthy places and times in the twentieth century, including Europe under the Nazis (1939–45), the Khmer Rouge in Cambodia (1975–79), and the Rwandan genocide (1994). Second, future geographical research needs to focus on the myriad issues raised by terror initiated by the state (Hewitt, 2001), spanning how state terror reshapes social geographies and place-identities. We also need to entertain the global nature of state terror, its ability to crosscut both capitalist *and* command economies, the inherent difficulties in using more direct research approaches such as ethnography and, ultimately, the crafting of a geographical theory of state terror that would address issues of unevenness and regularities, place-making and place-destruction.

Acknowledgments

We wish to acknowledge Michael Dear, Alan Emery, Kenneth Hewitt, Allison Williams, Jennifer Wolch, and the inspiration of the memoirs of the four authors. All errors are ours alone.

References

Adamova-Sliozberg, O. (1999), 'My Journey', in Vilensky (ed.) *Till My Tale is Told* (Bloomington and Indianapolis: Indiana University Press).

Adler, N. (1999), 'Life in the "big zone": The fate of returnees in the aftermath of Stalinist repression', *Europe-Asia Studies* 51: 5–19.

Adler, N. (2002), *The Gulag Survivor: Beyond the Soviet System* (London: Transactions Publishers).

Andrews, G. (2004), '(Re)thinking the Dynamics Between Healthcare and Place: Therapeutic Geographies in Treatment and Care Practices', *Area* 36: 307–18.

Andrews, G. and Kearns, R. (2005), 'Everyday Health Histories and the Making of Place: The Case of an English Coastal Town', *Social Science and Medicine* 60: 2697–713.

Andrews G.J., Kearns, R.A., Kontos, P. and Wilson, V. (2005), "Their Finest Hour": Older people, Oral histories and the Historical Geography of Social Life', *Social and Cultural Geography* 7: 153–77.

Applebaum, A. (2003), *Gulag: A History* (New York: Doubleday Books).

Atkinson, R. (1998), *The Life Story Interview (Qualitative Research Methods)* (Thousand Oaks: Sage Publications).

Baer, L. and Gesler, W. (2004), 'Reconsidering the Concept of Therapeutic Landscapes in JD Salinger's *The Catcher in the Rye*', *Area* 36: 404–13.

Bardach, J. (1998), *Man is Wolf to Man* (Berkeley: University of California Press).

Bollinger, M. (2003), *Stalin's Slave Ships: Kolyma, the Gulag Fleet, and the Role of the West*. (Westport CT: Praeger Press).

Conquest, R. (1978), *Kolyma: The Arctic Death Camps* (London: Macmillan Press).

Craddock, S. (2001), 'Engendering/Endangering: Women, Tuberculosis and the Project of Citizenship', *Journal of Historical Geography* 27: 338–54.

Daniszewski, J. (2002), 'Skeletons of History in Russian Graves', *Los Angeles Times*, 18 November.

Frazier, L. and Scarpaci, J. (1998), 'Landscapes of State Violence and the Struggle to Reclaim Community', in R. Kearns and W. Gesler (eds.) *Putting Health into Place* (Syracuse, NY: University of Syracuse Press).

Gastaldo, D., Andrews, G.J. and Khanlou, N. (2004), 'Therapeutic Landscapes of the Mind: Theorizing Some Intersections Between Health Geography, Health Promotion and Immigration Studies', *Critical Public Health* 14: 157–76.

Gesler W. (2003), *Healing Places* (London: Rowman and Littlefield).

Ginzburg, E.S. (1967), *Journey into the Whirlwind* (New York: Harcourt, Brace and World).

Ginzburg, E.S. (1981), *Within the Whirlwind* (New York: Harcourt, Brace and World).

Glad, J. (1994), *Kolyma Tales* by V.T. Shalamov (translation from Russian) (London: Penguin Books).

Gruffudd, P., Herbert, D.T. and Piccini, A. (2000), 'In Search of Wales: Travel Writing and Narratives of Difference 1918–50', *Journal of Historical Geography* 26: 589–604.

Hewitt, K. (2001), 'Between Pinochet and Kropotkin: State Terror, Human Rights and Geographers', *The Canadian Geographer* 45: 338–55.

Kearns, R. and Gesler, W. (1998), 'Introduction', in R. Kearns and W.M. Gesler (eds) *Putting Health into Place* (Syracuse, NY: University of Syracuse Press).

Kearns, R. and Barnett, J. (1999), 'Auckland's Starship Enterprise: Placing Metaphor in a Childrens' Hospital', in A. Williams (ed.) *Therapeutic Landscapes: The Dynamic Between Place and Wellness* (Lanham, MD: University Press of America).

Lorimer, H. (2005), 'Herding Memories of Humans and Animals', Personal website, Department of Geography, University of Glasgow.

Nordlander, D. (1998), 'Origins of a Gulag Capital: Magadan and Stalinist Control in the Early 1930s', *Slavic Review* 57: 791–812.

Parrish, M. (1996), *The Lesser Terror: Soviet State Security, 1939–1953* (Westport, CT: Praeger Press).

Pinfold, V. (2000), "Building up Safe Havens ... All Around the World": Users' Experiences of Living in the Community with Mental Health Problems', *Health and Place* 6: 201–12.

Pohl, J. (1997), *The Stalinist Penal System* (Jefferson NC and London: McFarland).

Razgon, L. (1997), *True Stories* (Dana Point, CA: Ardis Publishers).

Reynolds, M. (2002), 'A Soviet Legend Dies Hard', *Los Angeles Times*, 12 November.

Rosefielde, S. (1997), 'Documented Homicides and Excess Deaths: New Insights into the Scale of Killing in the USSR During the 1930s', *Communist and Post-Communist Studies* 30: 321–31.

Shalamov, V.T. (1994), *Kolymskie rasskazy* (translation by J. Glad) (New York: Penguin Books).

Solzhenitsyn, A. (1998), *The Gulag Archipelago 1918–1956: An Experiment in Literary Investigation III–IV*, Volume Two (Boulder, CO: Westview Press).

Tuchman, G. (1994), 'Historical Social Science', in N.K. Denzin and Y.S. Lincoln (eds) *Handbook of Qualitative Research* (Newbury Park: Sage).

Viola, L. (2001), 'The Other Archipelago: Kulak Deportations to the North in the 1930s', *Slavic Review* 60: 730–55.

Wakefield, S. and McMullan, C. (2005), 'Healing in Places of Decline: (Re)imagining Everyday Landscapes in Hamilton, Ontario', *Health and Place* 11: 299–312.

Williams, A. (1998), 'Therapeutic Landscapes in Holistic Medicine', *Social Science and Medicine* 46: 1193–1203.

Wilson, K. (2001), Book review of 'Therapeutic Landscapes: The Dynamic Between Place and Wellness', *Health and Place* 7: 346–8.

Wilson, K. (2003), 'Therapeutic Landscapes and First Nations People: An Exploration of Culture, Health and Place', *Health and Place* 9: 83–93.

Wilton, R. and DeVerteuil, G. (2006), 'Spaces of Sobriety/Sites of Power', *Social Science and Medicine* 63: 649–61.

Wishart, D. (1997), 'The Selectivity of Historical Representation', *Journal of Historical Geography* 23: 111–18.

PART 5
Transcending Geography: Applications in the Anthropology of Health

Introduction

Understanding Therapeutic Landscapes from the Perspectives of Medical Anthropology and Environmental Psychology

Julie Pranikoff and Setha Low

Medical anthropology has a long-standing concern with the health and well-being of individuals as understood within their cultural contexts (Foster and Anderson, 1978). Much work has focused on the complex social relationships and lay and professional networks that are activated when a sick person searches for medical care and emotional support (Kleinman, 1980). As early as the 1950s, medical anthropological research identified the problems inherent in treating people in hospitals and other medical institutions where patients are separated from their homes, families, and culturally meaningful surroundings (Taylor, 1970). With the increase of cultural diversity in the United States, questions arose about whether healing could actually take place in these medical institutions, particularly for people who do not share the beliefs and assumptions of Western biomedicine (Harwood, 1981). Interestingly, at the same time that medical anthropologists were critiquing Western hospitals and clinics, cultural anthropologists in Latin America, Southeast Asia, and Africa were writing ethnographies about the intimate relationship between landscape and body, and how the landscape plays a central role in healing practices (Bastien, 1985; Turner, 1968; Douglas, 1970).

It is somewhat surprising, therefore, that the study of therapeutic landscapes within the discipline of medical anthropology is still in its infancy, and that this collection is the first of its kind to be published. The following chapters explore four different types of therapeutic settings, ranging from mental health institutions and hospitals to rural roadsides. Because of the limited theoretical writing on this subject, each author defines 'therapeutic landscape' and then utilizes ethnographic descriptions of how these places function to support health and well-being to begin the hard work of building anthropological theory in this area.

Environmental psychologists, on the other hand, have spent the past three decades developing two theoretical approaches to understanding the healing potential of landscapes and exploring the reasons why certain environments are thought to be therapeutic. In this introduction, not only do we discuss each author's ideas of what constitutes a therapeutic landscape, we also include theoretical insights from environmental psychology and the ongoing research of Julie Pranikoff.

Theories of Therapeutic Landscapes in Environmental Psychology

One of the principal theories in environmental psychology that focuses on the therapeutic landscape was developed by Stephen and Rachel Kaplan (1989) after discovering that an Outward Bound-style program provided participants with 'a sense of well-being, of being renewed, of being restored.' Their insight that the natural environment has an impact on individuals' mentality led to their investigation of the aspects of the natural environment that support this 'restoration.' In the context of their theory, restoration occurs when people recover from 'directed attention fatigue.' Directed attention fatigue is the mental fatigue one feels after a prolonged effort concentrating on specific material. As a survival mechanism, directed attention is used to decrease confusion brought about by stressful situations and allow an individual to focus on stimuli without losing concentration.

This 'Attention Restoration Theory' posits that mental restoration occurs in environments that contain four core elements: 1) being away, 2) fascination, 3) extent, and 4) compatibility (see Kaplan and Kaplan, 1989). In these environments, individuals' minds are cleared from distracting 'cognitive leftovers' that inhibit concentration, so that attention can be redirected to things that have been on their mind, but have not yet been dealt with. Finally, there is a period of self-reflection where individuals can think about their lives and their priorities in a clearer, rational manner.

Attention restoration theory has been the basis of many studies analyzing the therapeutic components of environments. Hartig et al. (1991) compared the restorative powers of different types of vacation settings and concluded that backpackers who spent time in the wilderness benefited more in terms of attention capacity than backpackers who vacationed in an urban setting or took no vacation at all. Similar results were found when they compared the restorative quality of an outdoor walk – undergraduate students who walked in a natural setting, as opposed to those who walked in an urban setting or just relaxed in a chair, faired the best on a proofreading attention task. Furthermore, the four elements of attention restoration theory were more often found occurring in the natural setting.

Some studies have questioned the *potential* for restoration to occur in certain environments, asking participants to imagine either a hypothetical or past situation and then investigating which type of environment (as shown projected with slides) participants consider to be more restorative. Herzog et al. (1997) found that college students rated slides of natural environments higher than slides of sports/entertainment and urban settings with regard to the pictured environment's potential to support an individual's ability to focus on personal problems. Herzog et al. (2003) also found that undergraduate participants perceived natural settings to have a higher restorative potential than urban settings when these settings were seen through slides.

During the summer of 2005, Julie Pranikoff conducted a study to examine if attention restoration theory could be applied to three healing gardens in an urban rehabilitation hospital (Pranikoff, 2006). She used open-ended interviews with patients who visited the gardens to determine if they would mention the four elements of attention restoration theory when questioned about their experience there. Her findings support the idea that gardens help in the rehabilitation process. Additionally, participants repeatedly responded that the gardens' support of social relationships also contributes to the healing effects.

The second important theory supporting the idea of therapeutic landscapes is Roger Ulrich's (1983) psycho-evolutionary theory. He considers a person's reaction to the environment to be automatic – that is, 'without cognitive mediation,' and that this 'affective state then influences psychophysiological arousal, cognition, and motivation' (Hartig and Evans, 1993). Ulrich's incorporation of the physiological and affective influences of the environment contributes greatly to the research into why some environments are therapeutic. In a groundbreaking study, Ulrich (1984) examined the effects of two distinct views from inpatient hospital windows – either deciduous trees or a brick wall – on individuals recovering from gall bladder surgery. Those viewing nature spent approximately one day less in the hospital, had fewer negative nurses' notes in their charts, and needed less potent pain medication as compared to the brick wall viewers.

In a later study, Ulrich et al. (1991) explored the impact of viewing a stress-inducing movie on undergraduate students. After watching the film, the participants were then exposed to ten minutes of another video, showing an environmental setting (either natural or urban). During recovery, those exposed to natural scenes showed stress-reduction through decreased cardiovascular activity, skin conductance levels, and muscle tension. Furthermore, a measure of affective state showed that on three dimensions (positive affects, anger/aggression, and fear), the natural scenes were more restorative.

In a similar vein, Diette et al. (2003) wanted to bring the impact of the outside environment inside with murals of nature scenes and sounds of nature. These environmental interventions were used to distract patients before, during, and after undergoing a fiberoptic bronchoscopy – typically, a painful procedure. Among additional results, it was found that only 2.4 per cent of the participants in the intervention group rated their pain as 'poor or fair' (compared to 15.4 per cent of the control group).

Thus, environmental psychologists are investigating different environmental settings and what specific elements of these settings make them therapeutic places. In the following chapters, anthropologists also incorporate some of these environmental psychological theories as well as creating their own understandings of the relationship between healing and the environment.

Anthropological Contributions

In Chapter 18, Brian Hoey discusses the Grand Traverse region of the state of Michigan where, in the nineteenth century, a reformist asylum was built to 'restore' the minds of those with mental illnesses, and where developers are now renovating the buildings and recreating the space to be used by urbanites escaping from the stresses of city life, by families in search of a place in which they can live more meaningfully, by vacationers, and by others seeking retreat for various reasons. As his title suggests, Hoey employs the idea of 'purposive community' to help define a therapeutic landscape, highlighting the idea that an environment can be specifically designed to have healing effects. He compares those who created the asylum with today's 'new urbanism,' explaining that both intend to create environments that, in some manner, 'heal' inhabitants.

The Northern Michigan Asylum, founded on the principles of Thomas Story Kirkbride, was an institution built based on his belief that the physical environment and the interactions one has with the environment have the power to alter a person's mental, or moral, self. He argued that all aspects of patients' surroundings had the ability to alter their thoughts, and that a carefully manipulated environment could heal their minds. This idea is very reminiscent of attention restoration theory in which the Kaplans recognize that coming into contact with a natural environment allows for a relief of directed attention fatigue and a focus on new thoughts.

Much like Kirkbride, the current day developer, Ray Minervini, and his colleagues are working to bring Kirkbridian ideals back to the area and develop a community in which families are able to relocate and live a more balanced life. Hoey illustrates that, as in the past, Minervini and those in the 'new urbanist' movement believe their intentional community will inspire healing changes in the people who migrate there.

Similar to the reform movement and the Northern Michigan Asylum that Hoey writes of, in Chapter 19, Athena McLean discusses the Culture Change movement and institutions designed specifically to care for dementia patients. This time, however, the author concentrates on distinctions between the movement's successful application of creating 'homelike' spaces and the potential benefits to the patients being treated. McLean argues that nursing homes for dementia patients are multifaceted institutions with many functions and user groups. Explaining that the Culture Change movement aimed to 'humanize' care, she states that there is a disconnect: there *have* been improvements in the settings (for example, creating gardens within facilities) but residents are not allowed to take advantage of these positive changes (for example, having restricted access to these outdoor spaces).

In her analysis, nursing homes for dementia patients are *intersubjective* spaces, where there is a relationship between the person and their environment, objects, and most importantly, other persons. Through these intersubjective relationships, dementia patients maintain connections to their past as well as create new relationships. Furthermore, caregivers are involved in a 'moral enterprise' that allows these elders to maintain a sense of self.

To McLean, a therapeutic environment is one that has positive 'person–environment interaction' – where improvements in the physical setting are complemented by improvements in the social environment. This idea of successful person–environment interaction – or person–environment compatibility – is often spoken of in the field of environmental psychology, especially in regard to restorative and therapeutic environments. A setting that is *compatible* is one that supports the mental or physical activities that individuals wish to take part in (Kaplan and Kaplan, 1989). As McLean makes clear, a lack of connection between aesthetic changes and personal interactions undermines the therapeutic potential of these institutions and minimizes empowerment of patients and caregivers.

Like McLean's notion that an environment cannot be therapeutic without the social environment, Dawn Einwalter argues that artwork and memorials created on the roadsides of Nevada have a therapeutic impact on their creators because the process of making these pieces of work encourages the interaction between people and environment. By citing literature in public space, landscape design and horticulture therapy, Einwalter expands the notions that 'interactive and flexible' spaces understood to be therapeutic in other settings can also be therapeutic in the rural areas of Nevada.

She describes three types of art that are found in Nevada: rock writing, in which the creators use purple volcanic stones to 'write' atop the light-colored earth; rock installations by a local artist and his followers; and roadside memorials marking the sites of fatalities. A combination of the confusion over jurisdiction of the public area by the roadsides and the expense of surveying the area have allowed the artwork to remain intact. Not only does Einwalter consider the relationship between person and landscape to be therapeutic, but when considering the act of creating memorials, she points that some consider these creations to be part of a mourning process. Again, tying her work to other spaces, Einwalter illustrates the potential for flexible land use to be therapeutic in myriad types of settings, including hospital gardens.

Lastly, Peter Collins discusses the *therapeutic-aesthetic* aspects of healing environments. He compares artwork and the reactions of multiple user groups to art in two English hospitals: the James Cook University Hospital (JCUH), which is the product of a merger of two smaller facilities, and the Middlesbrough General Hospital (MGH), one of the previously independent facilities. In comparing the two hospitals, Collins explains that the JCUH took a top-down approach when choosing its artwork, meaning the management chose what works to display, even treating the setting as a gallery and commissioning artists to create specific pieces for the environment. The MGH, on the other hand, had used a bottom-up approach to its artwork, in that staff and patient spaces were decorated by the users. Although individual patients and staff had different opinions about the success of the spaces, the lack of 'participatory design' during the planning stage of the newer institution minimized its therapeutic potential. Individuals using the facility were not included when decisions about the design of the hospital were being made, thus some space is unused and other space disliked. In connection with our discussion of environmental psychology, an excerpt from an interview with Roger Ulrich is included in the chapter and highlights the importance of considering the multiple user groups of the facility and the potential for design to alter the therapeutic effect of the hospital by buffering stress.

Concluding Comments

Combining the disciplines of anthropology and environmental psychology in the study of therapeutic landscapes not only enriches the definition of such environments, it supports the notion that a broad array of environments can and should be considered in this context. This multidisciplinary perspective allows different aspects of environments to be explored when trying to understand why a specific landscape is therapeutic. Furthermore, it encourages a combination of data collection techniques – both qualitative and quantitative – to help illustrate and elucidate how individuals react to their everyday settings. It is important to make connections like this one, and to continue to investigate what gives some surroundings their therapeutic power – using this knowledge to make the healing elements as extensive and accessible as possible.

References

Bastien, J.W. (1985), 'Qollahuaya-Andean Body Concepts: A Topographical-Hydraulic Model of Physiology', *American Anthropologist* 87 (3): 595–611.

Diette, G.B., Lechtzin, N., Haponik, E., Devrotes, A. and Rubun, H.R. (2003), 'Distraction Therapy with Nature Sights and Sounds Reduces Pain During Flexible Bronchoscopy: A Complementary Approach to Routine Analgesia', *Chest* 123 (3): 941–8.

Douglas, M. (1970), *Natural Symbols: Explorations in Cosmology* (London: Barrie and Rockliff).

Foster, G.M. and Anderson, B.G. (1978), *Medical Anthropology* (New York: John Wiley & Sons).

Hartig, T. and Evans, G. (1993), 'Psychological Foundations of Nature Experience', in T. Gärling and T.G. Golledge (eds) *Behaviour and Environment: Psychological and Geographical Approaches* (Amsterdam: North-Holland) pp. 427–57.

Hartig, T., Mang, M. and Evans, G. (1991), 'Restorative Effects of Natural Environment Experiences', *Environment and Behavior* 23 (1): 3–26.

Harwood, A. (1981), *Ethnicity and Medical Care* (Cambridge, MA: Harvard University Press).

Herzog, T.R., Black, A.M., Fontaine, K.A. and Knotts, D.J. (1997), 'Reflection and Attentional Recovery as Distinctive Benefits of Restorative Environment', *Journal of Environmental Psychology* 17: 165–70.

Herzog, T.R., Maguire, C.P. and Nebel, N.B. (2003). 'Assessing the Restorative Components of Environments', *Journal of Environmental Psychology* 23 (2): 159–70.

Kaplan, R. and Kaplan, S. (1989), *The Experience of Nature: A Psychological Perspective* (Cambridge: Cambridge University Press).

Kleinman, A. (1980), *Patients and Healers in the Context of Culture: An Exploration of the Borderland between Anthropology, Medicine, and Psychiatry* (Berkeley: University of California Press).

Pranikoff, J. (2006), *Healing Gardens and the Application of Attention Restoration Theory: An Exploratory Study*, Unpublished Manuscript.

Taylor, C. (1970), *In Horizontal Orbit: Hospitals and the Cult of Efficiency* (New York: Holt, Rinehart and Winston).

Turner, V.W. (1968), *The Drums of Affliction: A Study of Religion Processes Among the Ndembu of Zambia* (Oxford: Clarenden Press).

Ulrich, R.S. (1983), 'Aesthetic and Affective Response to Natural Environment', in I. Altman and J.F. Wohlwill (eds) *Behaviour and the Natural Environment* (New York: Plenum Press) pp. 85–126.

Ulrich, R.S. (1984), 'View Through a Window May Influence Recovery From Surgery', *Science* 27 (4): 420–21.

Ulrich, R.S., Simons, R.F., Losito, B.D., Fiorito, E., Miles, M.A. and Zelson, M. (1991), 'Stress Recovery During Exposure to Natural and Urban Environments', *Journal of Environmental Psychology* 11: 201–30.

Chapter 18

Therapeutic Uses of Place in the Intentional Space of Purposive Community

Brian A. Hoey

Introduction

This largely historical chapter is positioned at the intersection of anthropology and human geography where, with sociology, psychology, architecture and public health, an integrated science of human health is forming. The role of environment in shaping quality of public health and civic life is at the center of merging academic and applied interests (see Jackson, 2003). Although public health, and especially environmental health, has long considered the *negative* effects and risks to the physical person associated with places such as industrial sites, there has been comparatively little consideration of the health-*promoting* role of place in physical and mental health (Frumkin, 2001; 2003). In terms of public policy, change is evident in the growing number of local, state and federal initiatives such as the United States Centers for Disease Control and Prevention's 'Healthy Places,' which supports '… the design and development of built environments that promote physical and mental health by encouraging healthy behaviors, quality of life, and social connectedness' (CDC, 2006). This chapter will explore therapeutic use of place to promote health and well-being while focusing on deliberate use of geographic place, architectural space, and the dynamic interplay between both employed in an intentionally created community setting.

The chapter provides an overview of the historical context for construction of the Northern Michigan Asylum in the second half of the nineteenth century, as well as its present adaptive reuse as 'neo-traditional' community. The term neo-traditional, employed by local planners, is typically used more or less synonymously with that of 'new urbanism'. New urbanism is a relatively recent community design reform movement recognized by its emphasis on bringing together mixed uses in an attempt to recapture the presumed vibrancy and stability in urban neighborhoods of an idealized American past. The Asylum's massive redevelopment takes place after nearly 20 years of uncertainty and possible demolition following its closing in 1989. The Northern Michigan Asylum is an especially interesting case as it is not only an ideal expression of nineteenth-century asylum reform but also, in its redevelopment, an expression of the ideals of new urbanism. Both movements entail not only a basic communitarian ideal but also an environmental, or spatial, determinism. The history of reformist idealism in design of the purpose-built mental asylum based on principles of 'moral treatment' provides a lens for my discussion of contemporary reimagining of the past for present purposes seen in new urbanism. Having this perspective is especially important given how new urbanism is being embraced by growing numbers of local governments and applied to numerous community planning projects around the United States and elsewhere.

While looking at therapeutic landscape, both implicitly and explicitly understood and operationalized in ideologically related projects, I extend application of the construct from consideration of the nineteenth-century asylum to the recent new urbanist movement and its design prescriptions for the post-modern 'village.' I will argue that the initial design and early operation of asylums such as Northern Michigan share a set of critical motivations, ideals, and objectives with the contemporary planning prescriptions called for by new urbanism. At the center of these movements is a commitment to using place therapeutically, that is, with curative, restorative and remedial goals, within purposively created community. The chapter suggests how the therapeutic landscape may be connected to public health and policy debates on the subject of design, community planning, and regional development in the United States. It attempts to reveal enduring resonances between major health and design reform movements of two arguably similar historical periods of great social and cultural change – the Industrial Revolution and our current post-industrial restructuring.

Setting and Fieldwork

Roughly 250 miles northwest of Detroit, Michigan's Grand Traverse region is a one-hour drive from the nearest highway. The area consists of four adjoining counties: Grand Traverse, Leelanau, Antrim, and Benzie, which have a combined population of approximately 140,000 persons according to US Census 2000 figures. Together they comprise an area extending roughly 25 miles from the region's economic and social hub at Traverse City (population 14,500). Census figures show that population increase in these counties ranged between 20 and 28 percent during the period from 1990–2000 with in-migration contributing four times the amount attributed to natural increase.

As with many rural places in the United States dependent on natural resource extraction during the nineteenth century, rapidly diminishing stocks of timber, fish and other commodities threatened the local economy. Shrewdly reading trends, business and political leaders created new purpose for the area, reframing many picturesque hills, towering sand dunes and miles of lakeshore as enchanting objects of reflection with power to promote health and well-being. The region was revisioned as a restorative sanctuary in two temporally separate but thematically related projects. The first of these is the construction of an institution for the mentally ill. Responding to local lobbying, the State of Michigan built a mental hospital in Traverse City completed in 1885. While the area was relatively remote, the competitive site selection process was not driven by a desire to distance this facility from centers of population for a common good. Rather, siting was determined by the conviction of reformist asylum planners that patients would benefit from the abundant fresh air, clean water, wholesome food, and natural splendor afforded by distance from increasingly congested and polluted population centers. Some 40 years later, the region was again marketed as refuge but to a growing middle-class eager to seek quiet retreat from the hustle and bustle of life in a rapidly urbanizing and industrializing America, by journeying to places of great beauty and real or imagined remoteness (Hoey, n.d.; cf. Aron, 1999; Löfgren, 1999; Sears, 1989).

More recently, the theme of place as refuge has again emerged in local 'place-marketing,' including promotion of Northern Michigan Asylum's redevelopment (see Kearns and Philo, 1993 on the historical marketing of place). Like many rural places once only seasonally attractive for short-term stays, the region now finds itself a destination not only for vacationers but also migrants who seek year-round retreat (see, for example, Bonner, 1997; Jobes, 2000; Murdoch and Day, 1998; Pindell, 1995). In my fieldwork over the past seven years, I have examined utopian family projects enacted through middle-class refuge-seeking behavior, where people relocate to personally meaningful geographic places as a way to intentionally refashion lives according to 'moral narratives of self' with greater emphasis on personal fulfillment, work/family balance, belonging to community, and affective attachment to place (see Hoey, 2005; 2006; cf. Bellah et al., 1996 on 'therapeutic quest'; Halfacree, 1998, 203; Murdoch and Day, 1998; cf. MacIntyre, 1984 on narrative 'quest' and the unity of a life). For these migrants, Grand Traverse is a therapeutic landscape. Here they construct personal moral geographies in which they attempt to negotiate challenges to identity precipitated by the effects of economic restructuring on work and family life.

Until recently the concept of therapeutic landscape was applied more or less exclusively to sites that facilitated recovery from illness and achieved a reputation for healing properties including spas, places of pilgrimage, and retreats. Some scholars now extend application to everyday memories and the experience of outwardly mundane but personally meaningful places (see, for example, Conradson, 2005; Gastaldo et al., 2004). These often taken-for-granted places 'have an attraction which gives [persons] a certain identifiable sense of well-being' (Jackson, 1994, 157–8) much like what I document with urban-to-rural migrants I call 'life-style migrants' (Hoey, 2005, 2006). Their story is important not only as a research context for my current project on the therapeutic use of place, but also for the fact that their numbers and common desire for a certain experience of place and longing for community creates a demand for local new urbanist projects like the Asylum's redevelopment into the Village at Grand Traverse Commons.

Introduced to the Northern Michigan Asylum by manner of proximity, initially it served simply as trope. In writing about both literal and figurative asylum, I discussed themes of refuge-*seeking*, in the behavior of urban-to-rural migrants, and refuge-*creating*, in both the 'social construction' (Low and Lawrence-Zúñiga, 2003, 20) of the symbolic landscape of the Grand Traverse region, and in individual and social responses to uncertainty and anxiety in the wake of profound cultural, economic, and structural change. By virtue of local debate on how best to redevelop the site, if at all, the Asylum emerged over time in the unfolding story as a character in its own right. This chapter is a product of that shift. It is a preliminary overview for future ethnographic research on the Asylum's redevelopment and a speculative essay on ideological and practical connections between two reform movements in purpose-built community design and use of the therapeutic landscape concept to better understand them.

Therapeutic Landscape – Therapeutic Community

While the idea that place may be restorative to human health and well-being is by no means new, the therapeutic landscape concept is a recent social science construct conceived as a way for academics to frame both real and imagined connections between place and human health (Gesler, 1992). Since its introduction, it has become an important theoretical contribution to health geography (see Williams, 1999). Health geography's emergence parallels a related turn by cultural anthropologists away from the limiting perspective of place as merely physical landscape or material context for cultural processes.[1] Rejecting the notion that place is a largely neutral setting or container within which social and cultural life unfolds, ethnographies within medical anthropology, for example, present a dynamic, relational view of human physical and mental health. This view holds that health involves complex interactions between people and their social, cultural and material environments at different points in history (see, for example, Devisch, 1993; Fadiman, 1997; Martin, 1994).

Studies in the anthropology of health such as these present us with rich conceptualizations of place akin to the study of landscape. As used by anthropologists, cultural geographers, and environmental historians, 'landscape' is understood as a cultural production, a symbolic transformation of the natural world (Cosgrove, 1983; Cronon, 1995; Jackson, 1994). Such an understanding takes into account humans, their anthropogenic environment and the manner in which this environment is conceptualized, constructed, experienced, and symbolized in different places and times (cf. Cosgrove and Daniels, 1988; Hirsch and O'Hanlon, 1995; Meinig, 1979). Similarly, landscape is used in health geography as a metaphor for a complex layering of cultural understandings, history, social structure, and built environment that converges in particular places (Kearns and Moon, 2002; cf. Williams, 1999).

Although with notable applications in nursing and palliative care (for example, see Andrews, 2002), *therapeutic landscape* has remained a largely academic concept. That of *therapeutic community*, however, derives from an almost entirely applied context. Encompassing place-based treatment and healing processes, it focuses attention specifically on important social relationships and the total milieu of social life (Jones, 1968; Kennard, 2004; Main, 1983). It emphasizes how community may be subject to design and intent to use the social collective within a particular place for healing purposes. While the therapeutic community construct is perhaps encompassed by that of therapeutic landscape, with its broad conceptualization and application, it provides a useful way of connecting my use of the terms 'community'

---

1 For consideration of the changing treatment of place in anthropology, see, for example, Lawrence and Low (1990).

and 'landscape.' The concept distinguishes therapeutic forms of community from other types, or specifically other places of institutional treatment, through its emphasis on intent and design within a social and physical setting, that is, 'the way in which [an] institution's total resources ... are *self-consciously* pooled in furthering treatment' (Jones, 1968, 85–6, emphasis added).

As legacy of the treatment approach that founded Northern Michigan Asylum, therapeutic community emphasizes the potential for healing in carefully cultivated social relationships, meaningful work, and the quality of physical environment within which individual development and social life takes shape (see Kennard, 2004). These same themes resonate with idealistic aims of new urbanist planners in developments such as the Village at Grand Traverse Commons. One of the aims of this chapter is to show how, within different historical contexts, reformist asylum planners of over a century ago and present-day new urbanists intend to *create community* for curative and restorative purposes. But what does it mean to intentionally create community?

The varied senses and many emotional layers attached to the word and idea of 'community' in popular discourse suggest that its meaning must go beyond geography or local practice alone. Within common sense understanding in the United States, we find 'an expectation of a special quality of human relationship' (Bender, 1982, 6; cf. McMillan and Chavis, 1986; Plas and Lewis, 1996). My own definition depends on the affective and experiential dimension suggested in this understanding rather than on reference to particular physical place or social assemblage. Speaking elsewhere on the topic of intentionally created community, I have argued that while the characteristics of physical space may be planned, community is something that must be experienced. Community, or perhaps 'sense of community,' may or may not be achieved through the intent of those who plan for it (Hoey, 2003). At the same time, I agree with psychologists Jeanne Plas and Susan Lewis (1996; cf. Talen, 1999) that it is possible to objectively design and create physical environments conducive to intermediate factors associated with the subjective experience of community. I use the term *intentional space* of purposive community in order to call attention to the important distinction between ideal and real – intent, imagination, and actual placed-experience.

New Urbanism

> Simply put, we wish to improve the world with design ... We believe that the physical structure of our environment can be managed and that controlling it is the key to solving numerous problems ... We believe that design can solve a host of problems and that the design of the physical environment does influence behavior (Elizabeth Plater-Zyberk, founding member of the Congress for New Urbanism, quoted in Marcuse, 2000).

> New Urbanism is at the core of a broad process of social change and institutional reorientation in which one can see a redefinition of 'community' as a practical ideal, a reworking of the way that social order and significance are registered visually in spatial arrangements and architectural form, and a reconfiguration of our ability to recognize moral order in our visual and practical experience of the built environment (Brain, 1999).

A reformist, even activist, design movement that emerged in 1980s community planning (see Duany et al., 2000; Katz, 1994), new urbanism's aim to transform all aspects of community design from new or 'greenfield' developments to urban infill and the redevelopment of existing structures or 'brownfields' such as Northern Michigan Asylum. At the core of new urbanist reform is a call to create 'healthy neighborhoods' defined by walkable scale, open spaces for public recreation, a range of housing options and businesses in 'mixed use' design, and cultivated 'sense of place' (Duany and Plater-Zyberk, 1992; Calthorpe, 1993). New urbanist doctrine asserts not only that it is possible to plan for community, but that community can be purposively created in order to have a therapeutic effect on both the individual and social level. With its communitarian ideals, new urbanism has become a key part of the debate over how to improve both civic life and public health in the United States (see, for example, Brain, 1997; 1999; Calthorpe, 1993; Krier, 1991; MacCannell, 1999; cf. Bellah et al., 1991, 1996; Etzioni, 1998; cf. Putnam et al., 2003).

New urbanism furthers assumptions essential to the reform asylum that intentional, purposeful design of the spatial order can serve as foundation for a new moral order (see Gans, 2002; Harvey, 1996; 1997; cf. Talen, 1999 on 'spatial determinism').[2] Varied expressions of purposive community-building reflect the challenges and opportunities of their respective historical periods. These range from reformist asylums to communitarian projects that include nineteenth-century intentional communities, New Deal era 'subsistence homesteads' (see Berry, 1992; Lord and Johnstone, 1942), and new urbanism's neo-traditional developments. Nineteenth- and early twentieth-century reform asylums and contemporaneous communitarian experiments were calculated reactions to widespread and transformative social, cultural and economic changes taking place in the larger society (Rothman, 1971; Zablocki, 1980). This seems true of today's new urbanism. A common axis upon which these diverse movements turn is shared intent to provide a holistic plan for community life where everyday routines performed within carefully designed spaces create more balanced, complete persons and stronger social collectives. In so doing, they self-consciously create models of harmonious living for a society seen as out of balance. Their shared dedication seems founded on a core belief that the intentional space of purposefully created community may be utilized as a therapeutic milieu and imagined haven against an increasingly heartless world (see Hawkins, 1991).

Locating the Northern Michigan Asylum in History and Place

The Northern Michigan Asylum is a product of the reformist 'moral treatment' plan that characterized a mid- to late-nineteenth century institutional approach to mental illness (Grob, 1973; see also Edginton, 1997; Hunter et al., 1986; Williams, 1999). Built at the dawn of modern psychiatric treatment, during a period of sweeping social, cultural and structural changes in late nineteenth-century America, the Northern Michigan Asylum was founded on the approach of Thomas Story Kirkbride. His plan

2 In this way, new urbanist practice perpetuates core aspects of the architecture and planning style of modernism that the movement's principal figures claim to reject (see Gans, 2002; Harvey, 1997; cf. Talen, 1999 on 'spatial determinism').

was to construct largely self-sustaining communities where the built environment and landscape became not only a sanctuary from social disorder but also healing instruments employed to restore individual balance while projecting social harmony. Kirkbride promoted a way of life designed to attain certain therapeutic goals within an asylum community, as well as to serve as an example of harmonious living for the greater society.

Seen from the vantage of contemporary medical science, treatment regimens for psychiatric disorders common at the time of Northern Michigan Asylum's opening in the 1880s, appear as based on little more than faith in the curative power of natural surroundings, carefully cultivated parklands, and well-designed architecture (see Gesler, 1992; Dear and Wolch, 1987; Hunter et al., 1986; Edginton, 1997; Williams, 1999). Given the nearly barren pharmacological cupboard of this period, limited to measures including opium derivatives such as morphine, and reliance on practices such as cupping and purging, we might conclude that this faith was born of necessity alone. In fact, the commitment of asylum reformers to the therapeutic landscape was an innovative approach and departure from earlier models which borrowed wholesale from the penitentiary. Their new approach was a reflection of important intellectual currents of the age (Edginton, 1997, 91; Sears, 1989; Tomes, 2001). During this historical period, American artists, writers, health reformers, and community planners all pointed to the ability of certain environments to exert a powerful, restorative influence on the material and non-material self (Albanese, 1990; Glacken, 1967; Macy and Bonnemaison, 2003; Whorton, 1982).

Central figures to the relatively young field of landscape architecture, notably Frederick Law Olmsted who designed New York City's Central Park and served as first Secretary General of the United States Sanitary Commission, designed many of the period's mental hospitals (see Beam, 2001; Hawkins, 1991). Writing in 1865, Olmsted asserted that taking in the scenery of carefully designed landscapes 'employs the mind without fatigue and yet exercises it, tranquilizes it and yet enlivens it' (Olmsted in Ranney, 1990, 504–5). Olmsted's teacher, the influential architect and landscape designer Andrew Jackson Downing, found inspiration through his admiration for moral treatment asylums. Downing's approach to landscape design was a blend of arguments and goals of both the emerging parkland and sanitary reform movements (Taylor, 1999). Writing mid-nineteenth century, Downing noted that 'it is somewhat curious, but not less true, that no country-seats, no parks or pleasure grounds in America are laid out with more care, adorned with more tastes, filled with more lovely flowers, shrubs and trees, than our asylums' (Downing, 1853, 269). His praise demonstrates a shared confidence with reformist asylum planners in the therapeutic power of place. Sharing their cultural critique as well, Downing pointed to rapidly changing social, cultural and economic conditions as a principal source of emotional imbalance and nervous disorder (Hunter et al., 1986, 1037). The condition most often cited as a source of mental illness was at that time the 'progress of civilization' (Hunter et al., 1986; cf. Doerner, 1981, 80).

The application of Kirkbride's model of moral treatment materialized during rapid industrialization and urbanization in the mid- to late nineteenth century at the same time as communitarian projects of the Utopian Socialist period (Sutton, 2004). The widespread shift from highly independent lifestyles to the decidedly regulated and routinized lives of wage laborers in the service of strict overseers was a wrenching change at the individual and societal levels (Gillis, 1996; Hilkey, 1997; Zunz, 1990), which brought responses ranging from the asylum reform movement to a fluorescence of intentional communities (McEwen, 1980). The actions of reformist leaders across these various social movements reflect deeply held beliefs in the power of place realized through the careful design and execution of institutions within the built environment to affect change in personal development, health and well-being and, eventually, the broader society (cf. Hayden, 1976; Sutton, 2004).

For example, the historian Nancy Tomes (1994) has noted that utopian sects such as the Shakers and the Oneida community used building plans and other spatial arrangements to both embody and advance their vision of a more perfect social collective.[3] McEwen (1980) suggests that because both voluntary and involuntary forms of intentional community are subject to control and purposeful design, they are attractive at certain times in history when other social processes appear to be in relative chaos. Assessing the historical record, McEwen has found that interest in communitarianism occurs at times when traditional total institutions, such as the asylum, are in transition.[4]

As noted by David Rothman (1971), medical superintendents of reform asylums located the roots of mental disease in the changing nature of American society and, in particular, what they perceived as the increasing fluidity, openness and loosened bonds of community life. Moral treatment was intended as an intervention into the psychological burden and crisis of personhood generated by the chaotic effects of a rapidly changing world. Asylum reform in the United States was part of a broad effort in mid-nineteenth-century Jacksonian society to encourage a new moral order to compensate for what appeared to be a steady undoing of the fabric of American society (Rothman, 1971; McEwen, 1980, 174; Sears, 1989, 92–3). Asylum reform was intended to methodically correct deficiencies found in the greater community, create stability in the day-to-day life of individuals, and balance the indiscretions of the broader society with the aim of rehabilitating victims of a disjointed social system.

---

3 As in Kirkbride's case, Lawrence Foster (1997, 254; 269) notes how the Oneida community's founder, John Noyes, was born at a time in the early nineteenth century when the United States was undergoing a 'disquieting' range of social, political and religious changes. During this time, the young republic was leaving behind elements of a relatively more cohesive colonial past and moving into a new era of 'rough-and-tumble' nineteenth-century capitalist individualism. Underlying Noyes' communal experiments, Lawrence asserts, was his concern for overcoming 'the social and religious disorder that he perceived around him.'

4 The sociologist Erving Goffman (1961, xiii) defines a total institution as 'a place of residence and work where a large number of like-situated individuals, cut off from the wider society for an appreciable period of time, together lead an enclosed, formally administered round of life.' Rose et al. (1977) refer to total institutions as 'controlled environments' to emphasize their intentionality and broaden the term's scope of application.

The Kirkbride Plan of 'Remoralization'

While recognizing the importance of practical design matters, including low ward density, adequate lighting, circulation of fresh air, and ample drainage for providing sanitary conditions and facilitating patient recovery, Kirkbride emphasized the restorative and redemptive potential of individual experience of the therapeutic landscape. A founder of the Association of Medical Superintendents of American Institutions for the Insane (now the American Psychiatric Association) over 150 years ago, Kirkbride authored the organization's first official standards for hospital design in 1854. Although pivotal to the exercise of the Kirkbride Plan, highly detailed landscape, architectural, and logistical plans basic to the care of patients were only part of his treatment model.

Kirkbride promoted a therapeutic modality focused primarily on the 'remoralization' of patients. While this may suggest patient indoctrination into prevailing middle-class mores, the category of 'the moral' had a different set of meanings in Kirkbride's time. The moral was understood to encompass the mind, emotion and character, thereby distinguishing it from material aspects of personhood. Although independent of a physical body, Kirkbride believed that this 'non-material' self could be shaped through a patient's sensory and emotional impressions. Kirkbride's goal was to engender a 'healing sense of place' in patients (cf. Gesler, 1993: 2) through attention to individual experience of place, the way in which a person would encounter their surroundings in day-to-day practice (cf. Tomes, 1994).[5]

As superintendent of the Pennsylvania Hospital for the Insane, Kirkbride wrote that within the intentional space of the asylum's purposefully created community, 'every object of interest that is placed … every tree that buds, or every flower that blooms, may contribute in its small measure to excite a new train of thought, and perhaps be the first step toward bringing back to reason, the morbid wanderings of a disordered mind' (Kirkbride, 1880 [1854], 47; cf. Steele and Hains, 2001). Moral treatment reformers assumed that the moral could be influenced through careful manipulation of spatial order. As noted by Barry Edginton (1997, 94, emphasis added; cf. Godkin, 1980 on 'rootedness' and mental health), the focus of treatment was thus 'not on the patient but the *attachment* of the patient to the healing ability of natural and social environments' (see, for example, Altman and Low, 1992 on the idea of 'place attachment').

By removing the patient from potentially harmful routines enacted in physical and social surroundings that may support such habits, remoralization in the course of moral treatment was thought to promote psychological and emotional healing.

5 This is consistent with his concern for both moral and material aspects of personhood; questions that scholars now characterize as having to do with 'embodiment.' Theories of embodiment are concerned with the lived experience of a person's own body and specifically the way in which individuals negotiate their everyday lives through their own bodies, and how they mediate, interpret, and interact with a physical and social environment (see, for example, Csordas, 1990; Scheper-Hughes, 1994; Van Wolputte, 2004). For a discussion of embodiment and disablement see, for example, Crooks and Chouinard (2006).

The regime put to work in the Kirkbride Plan thus recognized the socially and spatially constructed nature of persons. His plan acknowledged how, through the mundane and taken-for-granted routine of everyday practice, social actors are shaped through unconscious repetition and internalization (cf. Bourdieu, 1977 on 'habitus'). Kirkbride believed that the purposively constructed environment of the asylum removed afflicted persons from potentially self-destructive habits of lives unmoored and spun out of control by the sheer force of profound social and cultural changes to which he was witness. Remoralization amounted to an environmental intervention into the personal rituals of everyday life shaped by a person's social and physical setting.

Refuge, Rehabilitation, and Restoration in the Therapeutic Landscape

> The Village will not be a gated community (Minervini Group, 2004, redevelopers of Northern Michigan Asylum).

> The Village is a place where we come together to create, exchange, inspire and innovate in an environment of historic permanence and pastoral beauty. It is a 'sense of place' that encourages creative participation and expression, in all its forms, and where interaction flourishes. The Village is a renewable resource of creative experiences that traverse social, economic, cultural and artistic ideologies (Ray Minervini, quoted by ArtServe Michigan, 2007).

Although the Northern Michigan Asylum has been closed for over twenty years, the Grand Traverse region continues to provide asylum to those who seek refuge. As already noted, my research with lifestyle migrants documents how the area has become a place of refuge to the downsized and downshifting worker. Long officially sanctioned as asylum for the mentally ill, it serves today as a place of retreat for those with a variety of disabilities, including drug and alcohol addiction. Dave, a middle-aged participant in my project on urban-to-rural migration, ran a rehabilitation center located not far from the old asylum. The clinic served a range of clients, including referrals from the State of Michigan's Alcohol Treatment Services. He described for me how addicts sought treatment here rather than in hometown programs within the depressed, deindustrializing urban areas of Flint, Saginaw, and Detroit. Many clients were downsized autoworkers whose life plans were thrown into chaos when they lost self-defining jobs. They believed themselves more likely to recover if distant from familiar but personally harmful routines. In many cases, recovering addicts opted to stay after treatment rather than return home, having apparently found more than mere distance from past habit.

The region's history as host to an institution for the mentally ill and as virtual asylum for a range of people in need of convalescence from the travails of modern life has left a deep impression in the local ethos. This continues to shape the construction of its symbolic landscape. Recent practice affirms a long-standing local perception that this particular geographic place is capable of healing. One especially notable instance occurred immediately following acts of terrorism on September 11th, 2001 in the United States. A group of some 60 local health care professionals and civic leaders, calling themselves the Healing Community Response Team, spearheaded a community-wide effort to reach out to New York City firefighters. Their unique proposal was to offer a place of restorative refuge where traumatized families could come to find respite and possibly some measure of personal recovery within the therapeutic landscape.

Within this context of ongoing social construction of this particular geographic place as refuge, the old Northern Michigan Asylum is undergoing a kind of redemption through which it appears to return to the fundamental ideals of Kirkbride's moral treatment. Having become a reluctant symbol of the Welfare State's decay during the later half of the twentieth century, after years of neglect the brownfield site reemerged as a 'Renaissance Zone' established by the State of Michigan in the late 1990s to encourage investment of physical and social capital. With the blessings of hundreds of local activists who had long struggled to save the Victorian-era buildings, an early member of the Congress for New Urbanism (and a local builder) was granted permission to perform his ambitious redevelopment.[6]

Workers busily engage in giving the vast 388,000 square foot main complex an architectural rebirth over many years, with self-contained sections being worked on gradually as they can be sold. Restoration of a blighted site, burdened with the stigma imparted by contemporary views on institutionalized care, promises conversion of the old asylum into an integrated community. Residential, commercial, and recreational spaces, the realms of work and home and domains associated with the material and moral respectively, are brought together in what new urbanist developers promise will be a harmonious, even Kirkbridian, blending. Like Kirkbride, they attempt to achieve this through careful attention to the ways in which people encounter and relate to their surroundings in day-to-day practice.

The old asylum is starting over, architecturally reborn through a program of adaptive re-use and uniquely re-imagined as sanctuary in a (post)modern envisioning of the past for present purposes in an attempt to return to a 'cherished American icon … a compact, close-knit community' (Katz, 1994). A glossy brochure asks would-be community members to 'Picture a mixed-use, walkable village environment that will offer a chance to live, work, and play in the "Central Park" setting of the Grand Traverse Commons … Picture a New World approach to creating an Old World neighborhood. Picture "The Village of Grand Traverse Commons" (Minervini Group, 2001).[7]

---

6 As of late 2006, 20 residential units had been finished and sold and thirty more were being completed. These units range from 280 to 3,500 square feet. Thirty-eight businesses now operate within the renovated portions of the asylum, including a restaurant, bakery, winery, art gallery, yoga studio, and several non-profits. A Montessori junior high school also occupies part of the renovated space. An estimated 15 percent of the total building stock of the asylum property has been renovated.

7 In its reference to Central Park, the Minervini Group invokes Olmsted as the famed park's designer. At the same time, it no doubt unwittingly evokes not only his contributions as landscape architect but also his less glamorous role as first Secretary General of the United States Sanitary Commission. In this way, calling on Olmsted appropriately connects the Asylum's story to the history of zoning laws, with roots in the work of early sanitary engineers. Alongside broad health reforms in the context of asylum design, zoning laws were

Lifestyle migrants relocate to the region longing to fulfill 'family fantasies' (Aitken, 1998), ideals of family life that underlie enduring notions of 'the good' in American society (Hoey, 2005; 2006). Here in the intentional space of purposively created community, families seeking refuge and a sense of belonging gather to fulfill personal dreams of a more balanced and integrated lifestyle. In their attempt at self-fulfillment, they may help realize the communitarian vision of new urbanist planners while embracing a faith basic to moral treatment that the therapeutic landscape can affect personal change through purposeful planning of the material conditions of everyday life.

Conclusion

As noted by David Rothman (1971, 133), nineteenth-century medical superintendents believed the asylum would offer a means to 'correct within its domain the faults of the [broader] community and, through the power of example, spark a general reform movement.' Within intentional space, a therapeutic quest or ideal entails not only consideration of individual health and well-being, but also attending to the perceived ills of the broader society through promising a model of harmonious living. This is an ideal shared by proponents of new urbanism who express their belief that a carefully constructed spatial order can be used to shape the moral order.

In responding to post-9/11 anxieties in US society, however, planning prescriptions set out by new urbanists and others may lead to the excessively ordered, 'purified' communities of a sort critiqued over three decades ago by Richard Sennett (1970). Although the Minervini Group appears to anticipate an accusation in their proclamation that 'the Village will not be a gated community,' planned communities conceived in a prevailing climate of fear do risk becoming the actually or virtually enclosed places that are widely criticized for furthering unhealthy class and racial homogeneity and promoting social isolation (see, for example, Blakely and Snyder, 1997; Low 2003). Such an outcome is clearly anything but the therapeutic, moral antidote to a variety of chronic social ills that stem from rampant individualism, economic disparities, and racial difference. It would be a shame if the contribution of new urbanism amounted to little more than an aesthetic improvement over modernist forms of community planning and so-called sprawl. David Harvey (1997) asserts that new urbanism's emphasis on creating community can appear oblivious

meant to address challenges of rapid industrialization and urbanization by segregating land uses to distinct geographic areas within communities (Peterson, 1983). In the late nineteenth century, undifferentiated density in urban areas, where residential and industrial uses were seldom separated, led to a variety of risks including spread of infectious disease and contact with obnoxious and potentially dangerous industrial processes. Well intended at their establishment, according to new urbanist thought, zoning laws have long contributed to the high 'deconcentration' at the root of 'sprawl,' which now vexes many communities beset by snarled traffic, dwindling open space and diminished opportunities for public recreation (see Duany et al., 2000; Frumkin, 2003; Jackson, 2003). Because of new urbanism's interest in high density and mixed use, its developments often require exception from current zoning regulations.

to 'the darker side' of its communitarian impulse. Harvey explains that 'community' has always been a key site of social control and surveillance, akin to the managed material and moral order of the asylum as total institution. As such, a prescribed, institutionalized form of community can impede the conflict and creativity that arise from expressions of diversity in favor of a more comfortable parochialism.

This chapter joins other work at a crossroads where fields of anthropology, human geography, and public health unite around mutual consideration of the role of intentionally constructed, purposive community in human mental and physical health. Building on the discussion presented here, together with previous work with lifestyle migration, I plan to conduct research focusing on the Village at Grand Traverse Commons. I plan to examine everyday life and the divergence between an ideal and the real. I will explore the social construction of place, place marketing and, ultimately, the social production of place as therapeutic together with its consumption.

Through the lens provided by a moral treatment asylum and its redevelopment, this chapter extends application of the therapeutic landscape construct from consideration of the intentional space of community in the reform asylum to new urbanism and its design prescriptions for a neo-traditional village. Other scholars might consider my use of this construct for suggesting how a broad range of places may be examined or re-examined. These might include intentional spaces of purposive community in cases such as retreat centers and diverse enclaves defined by shared commitment to a particular lifestyle, connection to a particular geographic location, and cultivated sense of belonging.

References

Aitken, S. (1998), *Family Fantasies and Community Space* (New Brunswick, NJ: Rutgers University Press).

Albanese, C. (1990), *Nature Religion in America from the Algonkian Indians to the New Age* (Chicago: University Chicago Press).

Altman, I. and Low, S. (1992), *Place Attachment* (New York: Plenum Press).

Andrews, G. (2002), 'Towards a more place-sensitive nursing research: An invitation to medical and health geography', *Nursing Inquiry* 9 (4):221–38.

Aron, C. (1999), *Working at Play: A History of Vacations in the United States* (New York: Oxford University Press).

ArtServe Michigan (2007), 'Cultural economic learning exchange project history', [website], <http://www.artservemichigan.org/programs/index.asp>, accessed January 30, 2007.

Beam, A. (2001), *Gracefully Insane: The Rise and Fall of America's Premier Mental Hospital* (New York: Public Affairs).

Bellah, R.N., Madsen, R., Sullivan, W.M., Swidler, A. and Tipton, S.M. (1991), *The Good Society* (New York: Knopf).

Bellah, R.N., Madsen, R., Sullivan, W.M., Swidler, A. and Tipton, S.M. (1996), *Habits of the Heart: Individualism and Commitment in American Life* (Berkeley: University California Press.)

Bender, T. (1982), *Community and Social Change in America* (Baltimore: Johns Hopkins University Press).

Berry, B. (1992), *America's Utopian Experiments: Communal Havens from Long-wave Crises* (Hanover, NH: University Press of New England).

Blakely, E. and Snyder, M. (1997), *Fortress America: Gated Communities in the United States* (Washington, DC: Brookings Institution).

Bonner, K. (1997), *A Great Place to Raise Kids: Interpretation, Science and the Urban-rural Debate* (Montreal: McGill-Queen's University Press).

Bourdieu, P. (1977), *Outline of a Theory of Practice* (New York: Cambridge University Press).

Brain, D. (1997), 'From public housing to private communities: The discipline of design and the materialization of the public/private distinction in the built environment', in J. Weintraub and K. Kumar (eds).

Brain, D. (1999), 'New urbanism and the culture of technology: Rebuilding community', *Congress for New Urbanism VII: The Wealth of Cities*, Milwaukee, WI.

Brown, S. (2002), *Intentional Community: An Anthropological Perspective* (Albany: State University of New York Press).

Calthorpe, P. (1993), *The Next American Metropolis: Ecology, Community, and the American Dream* (New York: Princeton Architectural Press).

Centers for Disease Control and Prevention (2006), 'Health protection goals: Criteria and objectives', *Office of Strategy & Innovation* [website], (updated October 4, 2006) <http://www.cdc.gov/about/goals/default.htm>, accessed November 2, 2006.

Conradson, D. (2005), 'Landscape, care and the relational self: therapeutic encounters in rural England', *Health and Place* 11: 337–48.

Cosgrove, D. (1983), 'Towards a Radical Cultural Geography', *Antipode* 15: 1–11.

Cosgrove, D. and Daniels, S. (1988), *The Iconography of Landscape: Essays on the Symbolic Representation, Design, and Use of Past Environments* (New York: Cambridge University Press).

Cronon, W. (1995), *Uncommon Ground toward Reinventing Nature* (New York: W.W. Norton & Co).

Crooks, V. and Chouinard, V. (2006), 'An embodied geography of disablement: Chronically ill women's struggles for enabling places in spaces of health care and daily life', *Health and Place* 12: 345–52.

Csordas, T. (1990), 'Embodiment as a paradigm for anthropology', *Ethos* 18, 5–47.

Dear, M. and Wolch, J. (1987), *Landscapes of Despair: From Deinstitutionalization to Homelessness* (Oxford: Polity Press).

Devisch, R. (1993), *Weaving the Threads of Life: The Khita Gyn-eco-logical Healing Cult among the Yaka* (Chicago: University of Chicago Press).

Doerner, K. (1981), *Madmen and the Bourgeoisie: A Social History of Insanity and Psychiatry* (Oxford: Blackwell).

Downing, A. (1853), *Rural Essays* (New York: Leavitt & Allen).

Duany, A. and Plater-Zyberk, E. (1992), 'The second coming of the American small town', *The Wilson Quarterly* 16: 19–32.

Duany, A., Plater-Zyberk, E. and Speck, J. (2000), *Suburban Nation: The Rise of Sprawl and the Decline of the American Dream* (New York: North Point Press).

Edginton, B. (1997), 'Moral architecture: the influence of the York retreat on asylum design', *Health and Place* 3: 91–9.

Etzioni, A. (1998), *The Essential Communitarian Reader* (Lanham, MD: Rowman & Littlefield).

Fadiman, A. (1997), *The Spirit Catches You and You Fall Down: A Hmong Child, her American Doctors, and the Collision of Two Cultures* (New York: Farrar, Straus, and Giroux).

Frumkin, H. (2001), 'Beyond toxicity: the greening of environmental health', *American Journal of Preventive Medicine* 20: 234–40.

Frumkin, H. (2003), 'Healthy places: exploring the evidence', *American Journal of Public Health* 93: 1451–55.

Gans, H. J. (2002), 'The sociology of space: A use-centered view', *City and Community* 1 (4): 329–39.

Gastaldo, D., Andrews, G.J. and Khanlou, N. (2004), 'Therapeutic landscapes of the mind: theorizing some intersections between health geography, health promotion and immigration studies', *Critical Public Health* 14: 157–76.

Gesler, W.M. (1992), 'Therapeutic Landscapes: Medical Issues in Light of the New Cultural Geography', *Social Science and Medicine* 34: 735.

Gesler, W.M. (1993), 'Therapeutic landscapes: theory and a case study of Epidauros, Greece', *Environmental Planning Design D* 11: 171–89.

Gillis, J. (1996), *A World of their Own Making: Myth, Ritual, and the Quest for Family Values* (New York: Basic Books).

Glacken, C. (1967), *Traces on the Rhodian Shore: Nature and Culture in Western Thought from Ancient Times to the End of the 18th Century* (Berkeley: University of California Press).

Godkin, M. (1980), 'Identity and place: clinical applications based on notions of rootedness and uprootedness', in A. Buttimer and D. Seamon (eds) *The Human Experience of Space and Place* (New York: St. Martin's Press).

Goffman, E. (1961), *Asylums: Essays on the Social Situation of Mental Patients and other Inmates* (Garden City, NY: Anchor Books).

Grob, G. (1973), *Mental Institutions in America: Social Policy to 1875* (New York: Free Press).

Harvey, D. (1996), *Justice, Nature and the Geography of Difference* (Cambridge, MA: Blackwell Publishers).

Harvey, D. (1997), 'The new urbanism and the communitarian trap', *Harvard Design Magazine* 1 (1): 68–9.

Halfacree, K. (1998), 'Neo-tribes, migration, and the post-productivist countryside', in Boyle et al. (eds).

Hawkins, K. (1991), *The Therapeutic Landscape: Nature, Architecture, and Mind in Nineteenth-century America*, (Rochester, NY: Department of History, College of Arts and Sciences, University of Rochester).

Hayden, D. (1976), *Seven American Utopias: The Architecture of Communitarian Socialism, 1790–1975* (Cambridge, MA: MIT Press).

Hilkey, J. (1997), *Character is Capital: Success Manuals and Manhood in Gilded Age America* (Chapel Hill: University of North Carolina Press).

Hirsch, E. and O'Hanlon, M. (1995), *The Anthropology of Landscape: Perspectives on Place and Space* (New York: Oxford University Press).

Hoey, B. (2003), 'Nationalism in Indonesia: building imagined and intentional communities through transmigration', *Ethnology* 42: 109–26.

Hoey, B. (2005), 'From pi to pie: moral narratives of noneconomic migration and starting over in the postindustrial Midwest', *Journal of Contemporary Ethnography* 34: 586–624.

Hoey, B. (2006), 'Grey suit or brown Carhartt: narrative transition, relocation and reorientation in the lives of corporate refugees', *Journal of Anthropological Research* 62: 3.

Hoey, B. (n.d.), 'Narratives of travel and conversion: life-style migration as quest for refuge and potential selves in therapeutic landscapes'.

Hunter, J., Shannon, G. and Sambrook, S. (1986), 'Rings of madness: service areas of 19th century asylums in North America', *Social Science and Medicine* 23: 1033–50.

Jackson, J. (1994), *A Sense of Place, A Sense of Time* (New Haven: Yale University Press).

Jackson, R. (2003), 'The impact of the built environment on health: an emerging field', *American Journal of Public Health* 93: 1382–83.

Jobes, P. (2000), *Moving Nearer to Heaven: The Illusions and Disillusions of Migrants to Scenic Rural Places* (Westport, CT: Praeger).

Jones, M. (1968), *Beyond the Therapeutic Community: Social Learning and Social Psychiatry* (New Haven: Yale University Press).

Katz, P. (1994), *The New Urbanism: Toward an Architecture of Community* (New York: McGraw-Hill).

Kearns, G. and Philo, C. (1993), *Selling Places: The City as Cultural Capital, Past and Present* (New York: Pergamon Press).

Kearns, R. and Moon, G. (2002), 'From medical to health geography: novelty, place and theory after a decade of change', *Progress in Human Geography* 26: 605–25.

Kennard, D. (2004), 'The therapeutic community as an adaptable treatment modality across different settings', *Psychiatric Quarterly* 75: 295–307.

Kirkbride, T. (1880), *On the Construction, Organization, and General Arrangements of Hospitals for the Insane* (Philadelphia: Lippincott & Co).

Krier, L. (1991), 'Afterward', in A. Duany et al. (eds).

Löfgren, O. (1999), *On Holiday: A History of Vacationing* (Berkeley: University of California Press).

Lord, R. and Johnstone, P. (1942), *A Place on Earth: A critical Appraisal of Subsistence Homesteads* (Washington, DC: Bureau of Agricultural Economics).

Low, S. and Lawrence-Zúñiga, D. (2003), 'Locating culture', in S.M. Low and D. Lawrence-Zúñiga (eds) *The Anthropology of Space and Place: Locating Culture* (Malden, MA: Blackwell).

Low, S. (2003), *Behind the Gates: Life, Security and the Pursuit of Happiness in Fortress America* (New York: Routledge).

MacCannell, D. (1999), '"New urbanism" and its discontents', in J. Copjec and M. Sorkin (eds), *Giving Ground: The Politics of Propinquity* (New York: Verso).

Macintyre, A. (1984), *After Virtue: A Study in Moral Theory* (Notre Dame, IN: University of Notre Dame Press).

Macy, C. and Bonnemaison, S. (2003), *Architecture and Nature: Creating the American Landscape* (New York: Routledge).

Main, T. (1983), 'The concept of the therapeutic community: Variations and vicissitudes', in M. Pines (ed.) *The Evolution of Group Analysis* (Boston: Routledge & Kegan Paul).

Marcuse, P. (2000), 'The new urbanism: The dangers so far', *DISP* 140: 4–6.

Martin, E. (1994), *Flexible Bodies: Tracking Immunity in American Culture from the Days of Polio to the Age of AIDS* (Boston: Beacon Press).

McEwen, C. (1980), 'Continuities in the study of total and nontotal institutions', *Annual Review in Sociology* 6: 143–85.

McMillan, D. and Chavis, D. (1986), 'Sense of community: A definition and theory', *Journal of Community Psychology* 14 (1): 6–23.

Meinig, D. (1979), 'Symbolic landscapes', in D. Meinig and J. Jackson (eds) *The Interpretation of Ordinary Landscapes: Geographical Essays* (New York: Oxford University Press).

Minervini Group (2001), 'The village at grand traverse commons', *The Minervini Group* [website], <http://www.thevillagetc.com/site.html>, accessed March 12, 2002

Minervini Group (2004), 'The village – residential opportunities', *The Minervini Group* [website], <http://www.thevillagetc.com/residential.html>, accessed August 21, 2007.

Murdoch, J. and Day, G. (1998), 'Middle class mobility, rural communities and the politics of exclusion', in Boyle et al. (eds).

Peterson, J. (1983), 'The impact of sanitary reform upon American urban planning, 1840–1890', in Krueckeberg (ed.).

Pindell, T. (1995), *A Good Place to Live: America's Last Migration* (New York: Holt and Co.).

Plas, J. and Lewis, S. (1996), 'Environmental factors and sense of community in a planned town', *American Journal of Community Psychology* 24: 109–43.

Putnam, R., Feldstein, L. and Cohen, D.J. (2003), *Better Together: Restoring the American Community* (New York: Simon & Schuster).

Ranney, V. (1990), *The papers of Frederick Law Olmsted: Volume V, The California frontier, 1863–1865* (Baltimore, MD: Johns Hopkins University Press).

Rothman, D. (1971), *The Discovery of the Asylum: Social Order and Disorder in the New Republic* (Boston: Little, Brown).

Scheper-Hughes, N. (1994), 'Embodied knowledge: thinking with the body in critical medical anthropology', in Borofsky (ed.).

Sears, J. (1989), *Sacred Places: American Tourist Attractions in the Nineteenth Century* (Amherst, MA: University of Massachusetts Press).

Sennett, R. (1970), *The Uses of Disorder: Personal Identity and City Life* (New York: Knopf).

Steele, E. and Hains, K. (2001), *Beauty is Therapy: Memories of the Traverse City State Hospital* (Traverse City, MI: Denali).

Sutton, R. (2004), *Communal Utopias and the American Experience: Secular Communities, 1824–2000* (Westport, CT: Praeger).

Talen, E. (1999), 'Sense of community and neighborhood form: An assessment of the social doctrine of new urbanism', *Urban Studies* 36: 1361–79.

Taylor, D. (1999), 'Central Park as a model for social control: Urban parks, social class and leisure behavior in nineteenth-century America', *Journal of Leisure Research* 31 (4).

Tomes, N. (1994), *The Art of Asylum-keeping: Thomas Story Kirkbride and the Origins of American psychiatry* (Philadelphia: University of Pennsylvania Press).

Tomes, N. (2001), 'Introduction', in H. Johnson (ed.) *Angels in the Architecture: A Photographic Elegy to an American Asylum* (Detroit: Wayne State University Press).

Van Wolputte, S. (2004), 'Hang onto your self: Of bodies, embodiment, and selves', *Annual Review of Anthropology* 33: 251–69.

Whorton, J. (1982), *Crusaders for Fitness: The History of American Health Reformers* (Princeton, NJ: Princeton University Press).

Williams, A. (1999), *Therapeutic Landscapes: The Dynamic between Place and Wellness* (Lanham, MD: University Press of America).

Zablocki, B. (1980), *Alienation and Charisma: A Study of Contemporary American Communes* (New York: Free Press).

Zunz, O. (1990), *Making America Corporate, 1870–1920* (Chicago: University of Chicago Press).

Chapter 19

The Therapeutic Landscape of Dementia Care: Contours of Intersubjective Spaces for Sustaining the Person

Athena McLean

Introduction: The Historical Landscape of Dementia Care

In *Culture Against Man*, anthropologist Jules Henry (1963) decried the horrific conditions elders faced in US nursing homes, where they were warehoused to die. Neglect, poor sanitation, restraints usage, and needless deaths led by the late 1980s to federal reforms promoting improvements. Since Henry wrote his eye-opening work, the landscape of dementia care has undergone profound change, inspired by rethinking nihilistic assumptions that nothing could be done to help the afflicted. New spatial designs shedding institutional facades, and caregiving approaches centered on the person, hoped to improve life for institutionalized elders. These changes were shaped by institutional and regulatory practices within a larger social, political and economic landscape.[1]

By the 1970s, concerns had led to altered physical spaces and staff-intensive caregiving aimed at preserving residents' functioning (cf. Liebowitz, Lawton, and Waldman, 1979). When these later became seen as 'cost inefficient,' institutions redirected priorities to more efficient maintenance care tasks. Even visionary caregiving facilities gradually reverted to these standard practices (McLean, 2007, 71–7). By the 1980s, researchers voiced concerns about the neglected 'person' with dementia and environments for improving her well-being (Kitwood, 1989). Competing fiscal pressures (rising costs and reduced federal and state subsidies) and regulatory demands that institutions extensively document care, have nonetheless continued to favor the task over the person.

In response to the persistence of institutional approaches and the growing literature directing attention to the *person* (Kitwood, 1997), a 'Culture Change' (CC) movement since 1995 has focused on transforming the therapeutic landscape of

---

1 I use 'landscape' here similarly to Eric Hirsch (1995, cited in Low and Lawrence-Zúñiga, 2003, 16), as a site of tension involving both an 'idealized or imagined' background and foreground of lived everyday experience. Even more apt to the issues discussed here are Setha Low's distinction between the processes of *social production* of the structural features of the landscape versus the *social construction* of its symbolic features to those who live and frequent it (Low, 2000, 128; Low and Lawrence-Zúñiga, 2003, 20).

dementia care by deinstitutionalizing and humanizing care environments (NCCNHR, 2006). Proponents of the movement (innovative providers and staff members of long term-care, advocates, administrators, researchers, and regulators) have been encouraged by the growing embrace of new *physical* 'homelike' spaces, but have felt that the promise of this movement has still fallen short.

In this chapter, I speculate why CC has been limited. I begin by describing the various constraints dementia imposes on its sufferers. I then discuss the complexity of the nursing home as a place and how the CC movement has hoped to transform it. I suggest that CC has been realized only partially because many nursing homes have succeeded in changing inert *physical spaces*, but neglected the importance of significant *intersubjective* ones. I then draw on the concept of therapeutic landscapes and my research in dementia caregiving and ethics to discuss the contours of those intersubjective spaces. Based on this discussion, I advocate a more holistic, relational therapeutic landscape (Gesler, 2003, 6–11; Williams, 2002, 148–9) to promote well-being in dementia. This landscape would be shaped by *animating* physical spaces that evoke meaningful connections for elders and *intersubjective* spaces which sustain personhood through flexible and dynamic ethical practice. I caution, however, that this ideal cannot be realized without accompanying social, political, and economic changes within the larger *moral* landscape (cf. Liaschenko, 2001) of dementia care.

Impairments of the Sufferer of Dementia

Senile dementia is a cognitive and behavioral disorder affecting persons age 65 and older. It results from neuropathic changes in the brain, attributed mainly to Alzheimer's disease and vascular dementia. Common symptoms include cognitive, linguistic, behavioral, and mood impairments, and disorientation to time and place. Activities of daily living are affected, and some people develop problems recognizing and using objects. Symptoms, onset, rates of decline, and length of course (three to twenty years) vary among persons (Richter and Richter, 2002, 35). In very advanced dementia, incontinence, hallucinations, and inability to speak, walk, and swallow (Josephs, 2004) are common.

Short-term, and later, long-term memory are affected, accompanied by distortions in sequence of events, shattering confidence about one's past and producing a fragmented sense of self. Moving to an unfamiliar institutional setting often exacerbates confusion and fragmentation and promotes decline. Given both fragmentation and loss, many researchers believe that the most critical aspect of dementia care is enduring attention from an attentive caregiver (Kuhn, 2002, 165–6) to reinforce identity and engagement with the world (Eggers et al., 2005, 343).

Current medications only delay progression several months or reduce symptoms without delaying progression (O'Boyle, 2003). Confusion from symptoms may lead an elder to behave in disturbed ways – for example, being repetitive, displaying agitation, or 'wandering' away toward a familiar place like 'home,' to prepare her husband dinner, or the bus stop, to meet her children, forgetting that the husband is deceased and the children are now grown. These behaviors have been variously viewed as: (1) *random* medical symptoms requiring 'treatment' by restraints or psychotropic medications, or (2) the understandable product of confusion or of difficulty communicating *legitimate* needs (Sabat and Harré, 1992; Hughes, 2001). These competing perspectives differentially shape treatment approaches, care plans, and quality of life for elders.

The Nursing Home as a Complex Place

The nursing home has assumed many forms and functions, and different meanings and stakes to those it variously affects. It may assume an institutional form like a hospital unit, with a central nursing station and long corridors. It may adopt a modified layout of open space, offering greater surveillance by staff. Alternatively, it may be refashioned after homelike 'neighborhoods' rather than 'units,' or into small houses that bear no resemblance to formal institutions. Today's nursing home may also have standard institutional furnishings, be beautifully appointed, and/or be furnished with its residents' belongings.

The nursing home is a site of multiple functions. It is a place of employment and professional responsibility for various staff. It is a business that must market its commodified wares – caregiving – in appealing ways to discerning families seeking respite. It is also the place where elders come to live out their final years.

Thus the meaning of the nursing home is unevenly charged for its various stakeholders, with the stakes clearly incommensurable. For clinicians, the nursing home reinforces their professional identity. Nursing assistants (NAs), who give direct care, may find it a place of unrelenting burden (Foner, 1994) or the compelling locus for carrying out their life's work (Luster, 2006). To its stockholders, the nursing home is an amorphous source of capital, divorced from the invisible everyday experiences and sufferings (Forbes-Thompson and Gessert, 2006) of those within it. To families it is an approximation of 'home' away from 'home' – the agonizing reminder of having forsaken a parent, the painful place to witness his/her decline, or a promising refuge. Most poignantly, to its residents, the nursing home may be a strange and frightening 'non-place' (Augé, 1995; cf. Relph, 1976) through which they must pass before leaving the world (Reed-Danahay, 2001, 50). In its artificial approximation and simulation to 'home' (Low and Lawrence-Zúñiga, 2003: 32), it is also 'hyperreal'. While marketed as particularly '*homelike*,' it may in fact bear no resemblance whatsoever to *home* for its residents. In fact, the home that elders seek, when they cry out for it, may not be an external place at all, but an interior space (Frank, 2005) marking 'existential control' and social connectedness (Reed-Danahay, 2001, 60).

Limitations of Culture Change

In 1997 proponents of Culture Change convened to discuss ways to improve the quality of life of elders residing in nursing homes. They hoped to humanize caregiving environments by placing the 'person before the task' and promoting 'deep system change' (Fagan et al., 2004 [1997]).

Stimulated by the growing acceptance of the CC movement, institutional spaces increasingly are giving way to potentially therapeutic designs that help elders with dementia more effectively navigate and function in environments (Marcy-Edwards et al., 2005) bearing a semblance of 'home' (Zeisel, 2005). Clusters of smaller neighborhoods having kitchens and living rooms are replacing undifferentiated corridors. Since pretty décor, visible gardens and appealing designs attract families, industry leaders have moved to embrace them (Keane and Shoesmith, 2005). As one Culture Change proponent admitted at a recent conference, 'Culture change is imperative to staying in business.'

Limitations to the Restricted Focus on Physical Spaces

Despite the wide adoption of new architectural designs, a 'disconnect' has remained between the physical designs and social and operational contexts (Saperstein et al., 2004; Marcy-Edwards et al., 2005), limiting potential therapeutic impacts. A 'holistic understanding' for integrating design features to improve quality of life has been absent (Zeisel, 2005). For example, attractive 'homelike' kitchens used for dishing out food prepared elsewhere, and unavailable for residents' use, are only an inert visual reminder of 'home.' They cannot evoke pleasurable memories of foods cooking or of happier times working in the kitchen (Keane and Shoesmith, 2005).[2] As Saperstein et al. (2004) suggest, visual access alone has no, or very limited, therapeutic potential '... unless objects and places are actively engaged in cognitively impaired persons' lives, and frequently reinforced ...' (Saperstein et al., 2004, 325–26). Similarly, attractive gardens with restricted access may actually serve to upset elders who are prevented from entering them (cf. Zeisel, 2005, 276–7). Dementia researchers are beginning to address these complexities of environmental relations and their implications for life quality (Marcy-Edwards et al., 2005).

Many facilities nonetheless assume that new designs will automatically lead to reduced behavioral problems and to higher functioning residents (Marcy-Edwards, et al., 2005). This has prevented their expending efforts to maximize therapeutic potentials of physical spaces and to hire and train additional caregivers – a principal source of quality dementia care.

In addition, optimally utilized spaces may also cease to benefit residents if their unique design features are later ignored or abandoned. Even historic facilities like the Weiss Institute were not immune to this fate (McLean, 2007, 71–7). When it opened in 1978, after 12 years of research, planning and collaboration among clinicians, architects, and caregivers, it hoped to capitalize on person–environment interaction of those with mild dementia. Features like color-coded door plates helped residents locate their rooms, and a visible open floor plan attracted participation in activities. By 1992, the resident population had become markedly more impaired and no longer used the special design features, like the 'therapeutic' kitchen. Residents were now forbidden from entering the gazebo, originally intended as their place of refuge, because the staff appropriated it to complete paperwork (cf. Saperstein

2 This contrasts sharply with practices at Woodside Place, where the kitchen becomes a central living area, and cooking smells are used for positive stimulation (Gesler, 2003, 97).

et al., 2004). Over time resident decline, changes in funding, regulatory pressures, and institutional priorities for marketing to families' preferences for neutral colors (cf. Saperstein et al., 2004) worked against using therapeutic features as originally intended.

The Therapeutic Landscape of Dementia Care Environments

A more encompassing concept of environments that highlights their potential to promote well-being is that of 'therapeutic landscapes.' Therapeutic landscapes extend beyond physical features of the natural and built environment to their social and symbolic aspects and relational significance for the experiencing subject (Gesler, 2003, 8–9; Williams, 2002, 148). Care environments are seen not as *fixed* places, but the product of continually changing social and institutional *relations* (cf. Williams, 2002, 145). Therapeutic landscapes thus extend beyond potential therapeutic features of physical environments to the unique significance of those environments as dynamic places to the person engaging with them. The subjective experiences persons have within places imparts special meanings that can affect their subsequent states of being and ongoing constructions of self and identity (Williams, 2002, 144–7). This is particularly relevant in dementia, where the elder's identity is fragile and threatened. Therapeutic landscapes thus assist in exploring the complexity of meanings and therapeutic possibilities of dementia care environments for elders.

Intersubjective Dimensions of a Therapeutic Environment

The Relational Person: the Basis for a Therapeutic Environment

The single most significant development in dementia research, with dramatic implications for therapeutic care, is the radical reconceptualization of the person with dementia as a conscious biographical being with legitimate needs and wants. This advancement demanded radically shifting away from a hypercognitive *individualistic* view of personhood, founded on the ability to reason, and denied to those with dementia. Dementia scholars advocated instead a *relational* ethical perspective of personhood, grounded in dignity, appreciation for the elder's self-consciousness (Post, 2006, 230), vulnerability and suffering, and in enduring social acknowledgment by others (Sabat and Harré, 1992; Kitwood and Bredin, 1992, 274–6; McLean, 2007, 42–54).

In contrast to older models of dementia care that disregard the validity of the elder's subjective perspective, and act upon her as an object, relational perspectives admit the elder's agency and subjective need to act meaningfully upon the world (Frank, 2005, 177). Custodial care is thus no longer seen as an adequate therapeutic goal (Nolan et al., 2002, 200–202). Researchers are now suggesting that quality of life in dementia depends both on: (1) the pleasure the person can derive from environmental stimuli, and (2) the positive feelings, self-esteem, and the sense of belonging (Nolan et al., 2002, 200–201) that come *from meaningfully relating with others*. This may be because such relationally derived feelings, along with perceptions of having freedom and privacy, are tied to the delicate sense of being at 'home' with one's self (Frank, 2005, 184, 187–88), so fundamental to well-being (Williams, 2002, 145–46, 148) in dementia.

Impaired memory of persons and things and disorientation to time and place result in a sense of fragmentation and disruption to one's past and continuous sense of self. When there is a breakdown in the shared public space and 'life-world' because of these losses, meaning-making, built on human encounter, also becomes threatened (Hughes et al., 2006, 23, 31). By providing ongoing opportunities for the elder to meaningfully connect with her past (Post, 2006, 226) as well as with those in her present, therapeutic landscapes help to preserve a unified sense of self (Allen and Coleman, 2006, 215), meaning, and positive identity[3] and help fend off insecurity, social isolation, and possible depression (cf. Williams, 2002, 145).

What makes a place meaningful is its 'ability to foster relationship' (Chalfont, 2004, 6), the basis of all lived experience and well-being in dementia (Hughes et al., 2006, 35). Therapeutic landscapes work to maintain personhood by sparking cognitive, sensorial, and emotional experiences that recall past memories, which can be compared against experiences in other settings (Williams, 2002, 148). They also enable the construction of new meanings through ongoing intersubjective relationships. These may involve the person in multiple overlapping relations with the environmental spaces she occupies, the objects within them, and the persons with whom she engages. These consist of intersubjective relationships between the person's *internal* subjective experiential spaces and *external* physical ones, and a *third reciprocal* relational space involving elder and caregiver in the 'embodied reality of life as lived' (Bergum and Dossetor, 2005, 153). The remaining part of this chapter will examine the contours of these intersubjective spaces.

Intersubjectivity with the Environment

Since the 1960s, environments and subjectivities have come to be seen as fluid, interpenetrating processes, rather than as separate entities (Gallagher, 1993, 103; Chalfont and Rodiek, 2005, 342; Williams, 2002, 145). Researchers have recognized the increasing importance environments assume in shaping well-being as elders with dementia become more impaired (Calkins, 2004, 165). The knowledge that the right kind and amount of environmental stimuli (for example, sound, color, lighting, texture) can affect functioning and behavior of those with dementia was used in designing therapeutic physical spaces (Gesler, 2003, 86–7) to maximize both (see cf. Calkins, 2002). Such knowledge can also be used to promote well-being, which is equally, if not more, important than functioning alone (Nolan et al., 2002, 202). Well-being can be promoted by availing the elder to potential stimuli in the environment that can foster meaningful connections with other stimuli and places from her past. In fact, by enabling these connections, the most therapeutic environments stimulate

3 'Self' here refers to the internal sense of unity and agency ('the being that I am'), often expressed through the indexical 'I.' Sabat and Harré (1992) refer to this as personal identity. This self persists late into dementia (Sabat, 2006; Sabat and Collins, 1999). See McLean (2007, 49–51) for fuller discussion.

a strong sense of place and place identity or belonging (Williams, 1999, 73) – the sense of being 'at home' with oneself. Only when place identity is threatened is one aware of it (Williams, 1999, 74), as during the familiar plea heard in nursing homes, 'I want to go home.' Emotionally significant places (like home) or memories retain meaning deep into dementia (Frank, 2005, 176) and can reinforce connectivity. Inert environments in which opportunities for such connections are not pursued assume the status of non-therapeutic non-places, and stimulate a disturbing sense of placelessness or non-belonging (Williams, 1999, 75–6) and often, a 'grieving for a lost home' (Fried, 1963). In dementia, where home is associated with an increasingly faltering identity (Frank, 2005), this can be devastating. In contrast, positive environments that favorably respond to human needs can be restorative (Williams, 2002, 146–7).

Places are linked to subjective states by their ability to stimulate memories and associated moods and behaviors, often unbeknownst to the person. These memories are themselves mutually constitutive of place and person. As Casey observes, 'we are not only *in* places but *of* them' (1996, 19). Some places, like fragrant gardens, may even reawaken a forgotten need for pleasurable sensation (Chalfont and Rodiek, 2005). Chalfont and Rodiek (2005, 345–46) note that 'the body remembers because emotion evokes place.' Indeed 'as places are sensed, senses are placed' (Feld, 1996, 91). Embodied memories, such as pleasurable smells or the recognition of a soothing touch, are universal (Zeisel, 2005), primal, and last throughout dementia. Aromas have visceral relations to memories, and thus are more evocative than visual stimulation (Calkins, 2005, 71). Yet many elders are removed from these simple pleasures, *even where available resources exist.* Very few nursing homes, for example, provide opportunities for warm, soothing *unrushed* baths rendered by familiar hands, or the smell of chicken roasting (Hodder, 2004, 130–31). Embodied memories provide a vehicle for communicating long after verbal abilities are lost. After dementia had destroyed her mother's ability to talk, Ruth Hodder remained connected to her by activating sensorial memories of pleasurable cultural experiences they had previously shared, like music and tasty cheeses and sweets, not available in her 'sterile' nursing home (see Hodder, 2004: 131). However, even when the built environment provides sensory opportunities (such as visible gardens), these may be compromised by policies that prioritize safety and caregiver control over the resident's desire for freedom (Chalfont, 2004, 4) and privacy (McColgan, 2005, 430; Sachs, 2004, 3).

Intersubjectivity with Objects in One's World

Cultural beings occupy the fluid space between subjectivity and objectivity (Marcoulatos, 2003, 269–70). The reflexive self, in interpreting lived experience, crosses that space, thereby engaging in subject/object relations (Rubinstein and de Medeiros, 2005, 48). The world of the person with dementia narrows as restricted social mobility and cognition engender social isolation. When this happens, objects assume heightened significance for the person, and object relations acquire a kind of subject relations (Rubinstein and de Medeiros, 2005, 53). For example, a former folk singer, now suffering from severe speech aphasia, showed me a poster that announced a performance she had given many years earlier, proudly declaring 'me, me.' The poster allowed her to present a particular social identity that 'positioned' her (Sabat, 2006) in a way that mattered, quite apart from her identity as an Alzheimer's patient.

The capacity of the reflexive self to assign meaning to experiences and to places associated with them enables those places to absorb aspects of the self. When fragmentation of self occurs in dementia, those aspects of self contained in environments can be projected back to the person, providing a reinforcing extension of the self (cf. Williams, 2002, 144). The same thing can be said for objects. The ongoing interpretation of lived experience by the reflexive self can be seen as a kind of narration of one's life story. As one's ability to communicate one's narrative becomes verbally compromised in dementia, objects in one's world can serve as substitutes or meaningful representations of the self in the social world (Rubinstein and de Medeiros, 2005, 49, 52).

Intersubjectivity with Other Persons

The defining characteristics of a person, according to classical attributional views of personhood, are based on the ability to reason and reflect rationally on choices about the future. Once someone loses this ability, he/she ceases to be a person and to retain rights as a moral agent. This perspective ignores recent research that affect and emotion are the basis of cognition, located in an embodied mind, not a restricted brain (Thompson, 2001, 4). Further, by excluding those with dementia from the moral status of 'person,' cognitive approaches render them vulnerable to those having power over them (McLean, 2007, 43–4). Thus therapeutic possibilities of caregiving based on cognitive approaches to the person must be questioned.

In contrast, non-cognitive approaches guarantee personhood and dignity to the cognitively frail on the basis of larger humanistic, theological, relational, and contextual considerations. Personhood is built on the understanding that human experience is shared and that vulnerability and potential suffering are common to all humans. This lends an ethical dimension of *intersubjectivity* to personhood (Plourde, 2001, 140). Personhood is not based on individual attributes, but on the relational 'act of honoring the vulnerability of the other … by righting the imbalance created in that vulnerability' (Weisstub and Thomasma, 2001, 330). This relational dimension constitutes the essence of dementia 'care' and is fundamentally therapeutic. *Thus intersubjective relations between persons*[4] *constitute the most crucial of all therapeutic relations in dementia caregiving.*

Dementia Caregiving as a Moral Enterprise

Dementia scholars regard the person with dementia as a relational and emotional being with a subjectivity and dignity unavailable to mere objects (Gadow, 1985: 39). The agency of the cognitively impaired person does not disappear as the result

4 'Person' here refers to human beings, apart from other animals. Although animals are also an important source of relationship and comfort (Savishinsky, 1985) and a key element in long-term care programs, they will not be discussed in this chapter.

of impaired memory or reason. Rather, the person retains the capacity to relate and a will to communicate deep into dementia, even if it may appear distorted or absent to others (Hughes, 2001; Sabat and Harré, 1992). A dying woman with advanced dementia demonstrated this just days before her death. Although she had been immobile for weeks, when a visitor placed his hand over hers, she raised it to her cheek, shocking everyone around her (Lenshyn, 2005).

As long as the self is nurtured intersubjectively through ongoing acknowledgment by others, it can never be lost (Kitwood and Bredin, 1992). In contrast, a person who is neglected may withdraw radically, developing an inexorable interiority and detachment from the world and those within it. This is why intersubjectivity is so vital in dementia caregving. It is also why some dementia scholars consider the principal moral task of the caregiver as 'prevent(ing) the loss of the patient to the disease' with the only alternative being 'the destruction of the radically infirm' (Martin and Post, 1992, 57–8). They would argue that a further moral task is to help the elder preserve a continuing sense of self and unity by becoming familiar with his/her biography (Post, 2006: 229, 231) or 'story' – responsibilities nursing homes are neglecting (Ward, 2005, 19).

The most dedicated caregivers I have encountered indeed discovered whatever they could about elders under their care to reinforce their identities and honor their past achievements. But in dementia, the self need not depend only on memory of the past. A person's relational capacity, through semantic memory, enables the creation of new relationships in the present (Scheibe, 1989, 151–2), and through continuing intersubjective contact, new selves emerge (Ewing, 1990). In one's home, this contact occurs with those most intimately related to the elder. In the nursing home, the nursing assistant (NA) is the principal contact and key person to reinforce or deny past selves and enable new ones through respectful intersubjective give-and-take.

Examples of Intersubjective Care

One woman who had been involved in caring for her mother before she moved to a nursing home experienced an uncommon form of embodied intersubjectivity. During visits with her mother, she regularly participated in cleansing rituals where she bathed, washed her mother's hair, and helped her change into fresh clothing. The rituals pleased the mother and the physical connectedness helped the daughter better handle the loss of their verbal communication. Through their physical engagement, the daughter discovered that she could no longer distinguish her own body smells from those of her mother (Hodder, 2004, 130). This merging of their 'olfactory identities' was a product of intimacy and embodied intersubjectivity.

I have also witnessed two gifted NAs render care (McLean, 2006a, 933; McLean, 2007, 144–7). Both worked with a woman (Mrs Noble) most of the staff regarded as 'hopelessly demented' and combative. To the NAs, however, her actions made sense; they just needed to be understood (cf. MacDermid, 1995, 122). Both enjoyed Mrs. Noble and her spunkiness. They even came to visit her during days off from work.

The younger NA (Veronica) communicated with Mrs Noble by playing games like tag with her as they proceeded with bathing, dressing or feeding. These interactions revealed a playful self in the elder that her family had never before noticed. This caregiver always secured Mrs Noble's trust by confirming her okay before leaving.

The other caregiver (Elaine) was a master in human connectivity. She respected Mrs Noble's wishes, was highly attuned to her feelings and discomfort, and became familiar with her biography. She fully involved the elder in caregiving, engaging *with* her as a person, not *on* her as an inert body. Making eye-contact before talking or touching and moving gently and patiently to 'test the waters,' she gained cooperation to groom, cleanse, and toilet her without ever resorting to force or imposing her own will. Mrs Noble would assist, moving her body towards Elaine to help her change her diaper, rather than against her, as she did with the other NAs. If Mrs Noble were unwilling, Elaine would leave and return later when Mrs Noble felt ready. After learning Mrs Noble had been a fabulous cook, Elaine understood her insistence on stirring her food before eating: it was a way of controlling her meal by participating in its preparation – a way of achieving continuity and autonomy (cf. Williams 2002, 146–7). While this ritual consumed precious time, Elaine encouraged it as a meaningful connection with the elder's past. Through intersubjective engagement, the caregiver lovingly drew out the elder, and rendered life and nurturance, preserving her remaining threads of personhood. Caregiving embodied an ethics of intimacy and love and was conducted as a spiritual practice, meaningful in itself, without any expectation for improvement (cf. Gadow, 1988). Aware that she could not effect cure, she attempted to reduce vulnerability. But she did so without losing herself in the process. Through her intimate engagement with the elder, she not only gave, but in return acquired a stronger identity and sense of belonging to something greater.

Instrumental Caregiving as Potentially Destructive

In contrast to this sensitively engaged caregiving, which fosters healing and well-being, instrumental caregiving, focused strictly on the task, is at best limited in its therapeutic potential. The goal of the instrumental caregiver is to complete a prescribed set of 'care' routines, not to address the well-being of the person. The work must be done with or without the endorsement of the elder, so the mode of interaction is one of distance from, not engagement with, her. Instrumental caregivers may disrespect elders by ignoring their questions, requests, and pleas entirely (McLean, 2007, 87; cf. Forbes-Thompson and Gessert, 2006, 245) or even bruise them in the process of conducting 'care.' At best, they ignore the subjectivity of the person to get the 'job' done; at worst, they cease to see her as a 'soulful' person, 'processing' her instead as an object on an assembly line (McLean, 2007, 184) – a dehumanizing 'I–it' encounter (Buber, 1996 [1937]). Worse still, the elder may be seen as interfering with such 'processing' and medicated to be rendered 'manageable.' Despite intentions to alleviate *physical* suffering (for example, by treating bedsores), caregiving that ignores vulnerability and *emotional* suffering of persons is destructive, not therapeutic (cf. Kitwood, 1990, 181–2). Indeed, the alleviation of such suffering is rarely part of the regular therapeutic concern of nursing home care (Forbes-Thompson and Gessert, 2006, 247) or regulatory standards they follow.

Towards an Intersubjective Ethics of Dementia Caregiving

Empathy, Vulnerability, and Suffering: the Basis of an Intersubjective Ethics

In the 'empathic or caring ideal,' mutuality becomes the moral basis for caregiving (Gadow, 1985, 43). Sally Gadow has written eloquently of ethical relations in caregiving, grounded in the caregiver's sensitivity to the other's vulnerability and willingness to admit her own (1988). With no illusions about cure, a mutual vulnerability reduces the distance between them and eliminates the asymmetry of the caregiver/receiver relation. Gadow focuses on touch as the 'most compelling' form of human contact because it exposes vulnerability. When touch is empathic, rather than a requisite contact directed at performing an instrumental task, it expresses subjective concern for another's suffering and serves to alleviate subjective withdrawal into oneself (Gadow, 1985, 40–41). Such expression captures the transition from 'I–it' to 'I–thou' (Buber, 1996 [1937]).

When caring for someone in the midst of chaos and emotional suffering, it may be easier to retreat than to risk identifying with his pain, even if it can never be experienced in exactly the manner of the sufferer (Ales Bello, 2006, 11). Although we can never feel identically what the other is feeling, what is significant is the gesture of respect conveyed in trying. A lack of empathetic regard will inevitably limit one's ability to alleviate another's pain, and may even contribute to suffering (cf. Bergum and Dossetor, 2005, 106–8). On the contrary, one daughter found that she could reduce the torment of her mother's suffering by attempting to imagine her mother's world and forcing herself to penetrate her mother's subjectivity and confront her terror (MacDermid, 1995, 121). Similarly a relational ethics in caregiving encourages an intersubjective engagement that includes suffering *with* another (Bergum and Dossetor, 2005, 104) and acknowledges the vulnerability of both oneself and the other. When the caregiver admits her own vulnerability, the power differential with the elder dissolves, and the opportunity for freer intersubjective engagement and deeper mutual understanding increases. A number of examples follow.

A pediatrician who had been caring for a terminally ill infant was upset when the baby died before his devoted parents arrived, but was uncomfortable expressing her emotions to them. To her surprise, they were prepared for the event, and in an unexpected reversal, attempted to comfort her (Bergum and Dossetor, 2005, 118).

I experienced a similar situation during a visit with an elder in my study. The woman was showing me photographs while telling me her life story (McLean, 2006b). As she reached for a picture of her daughter, she became very emotional and began describing how the young woman had died. After noticing that I had tears in my eyes, however, she comforted me and then proceeded with her story.

In another example, a woman described a frustrating event late in her mother's dementia. Her mother seemed vacant and totally unresponsive as she played the piano for her – something she always enjoyed in the past. The daughter began to cry.

When she lifted her eyes from the keyboard, she saw that her mother's eyes were also filled with tears (MacDermid, 1995).

In each of these cases, the revelation of vulnerability from the seemingly stronger subject brought out strength and empathy for her suffering from weaker or more vulnerable persons – uncommon moments of 'I–thou' reciprocity and emotional mutuality. These epitomize the reciprocal ethics of Martin Buber, built on a recognition both parties acquire through mutuality.

Flexibility and Change in an Intersubjective Ethics

For those caregivers who assume a stance of mutuality, the personal risks – loss of self or control, self-exposure, selflessness (but see Bergum and Dossetor, 2005, 110–11), and suffering – are seen as worth the gains for the patient. But there are many gains for the caregiver as well. As described above, caring for Mrs Noble was 'fun' for Veronica and reinforced Elaine's spiritual convictions. Gadow talks also of the reciprocal gift of trust (1985, 42) that emerges from mutuality. Even the dying woman who had been immobile for weeks was able to express mutuality with her visitor.

It would be naïve, however, to suppose that this empathic stance and *ethical* equality and symmetry of the subjects involved imply an equivalent *structural* equality and symmetry of their involvement. In fact, it is the *asymmetry* of the relational context that imparts a heightened dignity to the more vulnerable subject (cf. Gadow, 1985, 39). Indeed for Levinas, intersubjective space is *most ethical* when reciprocity is *not* possible (Marquez, 1987, 345). This is because it is based, not on the expectation of personal gain, but on the elevation of the other – and the relational asymmetry this entails – founded on immense respect that inspires non-egoistic responsibility (Marquez, 1987, 345–46); herein lies its transcendence.

The demands and expectations of a caregiving relation, and the contours of intersubjectivity these entail, will likely change as the dementia advances, possibly necessitating changes in the ethical relationship as well. The significance of Buber's intersubjective 'I–thou' (versus the objectifying 'I–it') relation is inestimable, and of enduring relevance throughout the caregiving relation. Buber's reciprocal 'I–thou' ethics, however, is also dialogic, requiring participation by both persons. It is through this mutual participation that a higher transcendent 'thou' is achieved (Wallace, 2001, 217). However, the radical interiority that may occur late in dementia, especially with an elder whose prior caregivers were not invested in 'drawing her out,' may reduce the likelihood of (or even render impossible) a dialogic relation. In such cases, the reciprocal ethics of Martin Buber may yield or extend to the asymmetrical intersubjective ethics of Levinas, in which the rights of the radically other obligate the caregiver, without implying any reversal (Marquez, 1987, 347). This ethics is more unconditional than love and more transcendent (Marquez, 1987; 348–50). For Levinas, this intersubjective ethics constitutes the vision of God: 'I am here' (Marquez, 1987, 250), a moral enactment in relational space (Malone, 2003; 2317).

Conclusion: The Larger Moral Landscape of Dementia Care

I began my reflection on the contour of intersubjective spaces within a broader discussion about the Culture Change movement and its impact on dementia care. Despite the intention of Culture Change to humanize nursing home care, researchers remain skeptical about its success (Marcy-Edwards et al., 2005, 37) or voice concerns about inadequacies in its adoption (Zeisel, 2005; Saperstein et al., 2004). Fully analyzing the complexity of interpenetrating micro and macro forces that may account for its limited success was beyond the scope of this chapter. Rather, I speculated about how the restricted focus on inert physical spaces as the sole locus of Culture Change by some long-term care facilities may help explain its limited success. My focus has been instead with an idealized understanding of the more localized intersubjective spaces in which the therapeutic aspects of dementia caregiving can be maximized.

There is no *cure* for dementia, only the guarantee of continued impairment, vulnerability, and dependence on others. The therapeutic possibilities for dementia lie strictly in *care* and spiritual and emotional healing, arguably a higher moral charge than cure (cf. Gadow, 1988; cf. Gesler, 2003). As moral agents, caregivers serve as 'prostheses,' helping to preserve the person (Post, 2006, 229). This is an immense charge for anyone, let alone those who occupy the lowest rungs of the caregiving spectrum – NAs who are the lowest paid, most overworked, least empowered and most socially devalued occupational group in the nursing home. Overwhelmingly women of color, they share with cognitively impaired elders (now deprived of their former social wrappings) the lowest social positions of both the nursing home (Reed-Danahay, 2001, 60) and society. Yet as direct caregivers, these are the people in closest spatial and temporal proximity (Malone, 2003, 2320) to elders, with the greatest opportunities to promote their healing.

Despite the encouraging ideologies of Culture Change to empower both caregivers and receivers, the superficiality by which its message has been received in many places and the economic pressures against increasing staff size have promoted instead a 'distal nursing' (Malone, 2003) in all but the most affluent or exacting care centers. Many facilities have embraced Culture Change as a marketing tool and invested funds in improving the appearance of the facility over hiring less visible staff to engage with residents. Federal cutbacks have continued to reduce nursing home spending and federal authorities have refused to set minimal staffing requirements for NAs (McLean, 2007, 253). Thus despite aesthetic improvements, many homes have actually decreased the caregiver/receiver ratio, disrupting previous caregiving relations (Marcy-Edwards et al., 2005, 37). In addition, in order to save costs, many homes have adopted temporary irregular staff, further impeding the development of trusting relations. Even models of care promoted as Culture Change have added housekeeping demands on caregivers that threaten the proximity of the caregiver relation (Marcy-Edwards et al., 2005, 37).

The success of the intersubjective relations I have outlined above depend not only on the immediate relations with the caregiver, but also on the larger macro-relations – institutional, regulatory, legislative, political-economic – that structure and constrain them (Liaschenko, 2001, 129). Relational ethics must extend beyond the immediate caregiver relation to *a broader set of morally responsible actors and social relations.* Threats to immediate caregiver relations must be redefined as ethical problems (Liaschenko, 2001, 129). Since a therapeutic environment depends on multiple relations, everyone involved in creating policies that restrict the development of a caregiving relation, withhold access to therapeutic spaces (like gardens), or ignore the need for therapeutic opportunities to socially connect the elder to her sensory world should also be held ethically accountable.

The intersubjective ethics of Levinas reminds us that respect in the immediate 'face-to-face' caregiving relation is respect for the Face (Marquez, 1987, 345). The selfless responsibility this calls for is not something that can be readily acquired through awareness training or even principles of Culture Change. Compartmentalized educative efforts may acquaint uninitiated nursing assistants with amended caregiving priorities focused more on the person, but unless the moral import of these adjustments is appreciated, they may be resisted or ignored (Marcy-Edwards et al., 2005, 10–11). The commoditized space of the nursing home contains too many contradictory pressures to encourage sustained support for a Levinasian ethics. The very definition of sufferer as 'consumer,' and her care as a reimbursable 'service' (cf. Liaschen, 1997, 54), opposes its practice. The presumed 'therapeutic' landscape of commoditized care lends itself instead to being unauthentic and uncaring (Williams, 1998, 1198; McKnight, 1995). Only those rare caregivers motivated by a larger moral and spiritual 'calling' (Luster, 2006; Martin and Post, 1992, 67) can resist the fracturing forces that threaten intersubjectivity and its healing potential.

References

Ales Bello, A. (2006), 'The language of our living body', *Alecta Husserliana* 89: 3–14.

Allen, F.B. and Coleman, P. (2006), 'Spiritual perspectives on the person with dementia: identity and personhood', in J. Hughes, S. Louw and S. Sabat (eds) *Dementia: Mind, Meaning, and the Person* (Oxford: Oxford University Press) pp. 205–21.

Augé, M. (1995), *Nonplaces: Introduction to an Anthropology of Supermodernity*, John Howe (trans.) (London and New York: Verso).

Bergum, V. and Dossetor, J. (2005), *Relational Ethics: The Full Meaning of Respect* (Hagerstown, MD: University Publishing Group).

Buber, M. (1996 [1937]), *I and Thou*, Walter Kaufmann (trans.) (New York: Touchstone).

Calkins, M. (2002), 'Powell Lawton's Contributions to Long-Term Care Settings', *Journal of Housing for the Elderly* 17 (1/2): 67–84.

Calkins, M. (2004), 'Articulating environmental press in environments for people with dementia', *Alzheimer's Care Quarterly* 5 (2): 165–72.

Calkins, M. (2005), 'Environments for late-stage dementia', *Alzheimer's Care Quarterly* 6 (1): 71–5.

Casey, E. (1996), 'How to get from space to place in a fairly short stretch of time: phenomenological prolegomena', in S. Feld and K. Basso (eds) *Senses of Place* (Santa Fe, NM: School of American Research Press) pp. 13–52.

Chalfont, G.E. (2004), 'Current research on connection to nature: Implications for garden design in dementia care', *Therapeutic Garden Design* IV (1): 4–6.

Chalfont, G.E. and Rodiek, S. (2005), 'Building edge: an ecological approach to research and design of environments for people with dementia', *Alzheimer's Care Quarterly* 6 (4): 341–48.

Eggers, T., Norberg, A. and Ekman, S.-L. (2005), 'Counteracting Fragmentation in the Care of People with Moderate and Severe Dementia', *Clinical Nursing Research* 14 (4): 343–69.

Ewing, K. (1990), 'The illusion of wholeness: "Culture," "self," and the experience of inconsistency', *Ethos* 18 (3): 251–78.

Fagan, R.M., Williams, C.C. and Burger, S.G. (2004 [1997]), *Meeting of Pioneers in Nursing Home Culture Change, March 14–16, 1997* (Rochester, NY: Pioneer Network).

Feld, S. (1996), 'Waterfalls of Song: Notes on a western Apache landscape', in S. Feld and K. Basso (eds) *Senses of Place* (Santa Fe, NM: School of American Research Press) pp. 91–136.

Foner, N. (1994), *The Caregiving Dilemma: Work in an American Nursing Home* (Berkeley, CA: University of California Press).

Forbes-Thompson, S. and Gessert, C. (2006), 'Nursing homes and suffering: Part of the problem or part of the solution?', *The Journal of Applied Gerontology* 25 (3): 234–51.

Frank, J. (2005), 'Semiotic Use of the Word "Home" Among People with Alzheimer's disease: A plea for selfhood?' in G. Rowles and H. Chaudhury (eds), *Home and Identity in Late Life* (New York: Springer) pp. 171–96.

Fried, M. (1963), 'Grieving for a lost home', in L. Duhl (ed.), *The Urban Condition* (New York: Simon and Schuster), cited in Williams. 1999, 76.

Gadow, S. (1985), 'Nurse and patient: the caring relationship', in A. Bishop and J. Scudder (eds), *Caring, Curing, Coping* (Tuscaloosa, AL: University of Alabama Press) pp. 31–43.

Gadow, S. (1988), 'Covenant without cure. Letting go and holding on in chronic illness', in J. Watson and M. Ray (eds) *The Ethics of Care and the Ethics of Cure: Synthesis in Chronicity* (New York: National League of Nursing) pp. 5–14.

Gallagher, W. (1993), *The Power of Place: How our Surroundings Shape our Thoughts, Emotions, and Actions* (New York: Poseidon Press).

Gesler, W. (2003), *Healing Places* (Lanham, MD: Rowman & Littlefield Publishers).

Henry, J. (1963), *Culture Against Man* (New York: Vintage Books).

Hodder, R. (2004), 'Connections: My gateway to Alzheimer's world of silence', *American Journal of Alzheimer's Disease and Other Dementias* 19 (2): 129–34.

Hughes, J. (2001), 'Views of the person with dementia', *Journal of Medical Ethics* 27 (2): 86–91.

Hughes, J., Louw, S. and Sabat, S. (2006), 'Seeing whole', in J. Hughes, S. Louw and S. Sabat (eds) *Dementia: Mind, Meaning and the Person* (Oxford: Oxford University Press) pp. 1–40.

Josephs, A.S. (2004), 'Studies Emphasize New Risk Factor for Dementia and Alzheimer's', <http://heartspring.net/alzheimers_disease_stages.html> accessed 3 December 2006.

Keane, W.L. and Shoesmith, J. (2005), 'Creating the ideal person-centered program and environment for residential dementia care: 10 steps and 10 challenges toward a new culture', *Alzheimer's Care Quarterly* 6 (4): 316–24.

Kitwood, T. (1989), 'Brain, mind and dementia: With particular reference to Alzheimer's disease', *Ageing and Society* 9: 1–15.

Kitwood, T. (1990), 'The dialectics of dementia: With particular reference to Alzheimer's disease', *Ageing and Society* 10: 177–96.

Kitwood, T. (1997), *Dementia Reconsidered: The person comes first* (Buckingham and Philadelphia: Open University Press).

Kitwood, T. and Bredin, K. (1992), 'Towards a theory of dementia care: personhood and well-being', *Ageing and Society* 12: 269–87.

Kuhn, D. (2002), 'Intimacy, sexuality and residents with dementia. (Communication)', *Alzheimer's Care Quarterly* 5 (2): 165–76.

Lenshyn, J. (2005), 'Reaching the living echo: Maintaining and promoting the spiritual in persons living with Alzheimer's disease', *Alzheimer's Care Quarterly* 6 (1): 20–28.

Liaschenko, J. (2001), 'Nursing work, housekeeping issues, and the moral geography of home care', in D. Weisstub, D. Thomasma, S. Gauthier and G. Tomossy (eds) *Aging: Caring for our Elders* (Dordrecht: Kluwer Academic Publishers) pp. 124–37.

Liebowitz, B., Lawton, M.P. and Waldman A. (1979), 'Evaluation: Designing for impaired elderly people', *American Institute of Architects Journal* 68: 59–61.

Low, S. (2000), *On the Plaza: The Politics of Public Space and Culture* (Austin,, TX: University of Texas Press).

Low, S. and Lawrence-Zúñiga, D. (2003), 'Locating Culture', in S. Low and D. Lawrence-Zúñiga (eds) *The Anthropology of Space and Place* (Malden, MA and Oxford, UK: Blackwell Publishing) pp. 1–47.

Luster, P. (2006), 'Family (services of caregivers)', *Alzheimer's Care Quarterly* 7: 150.

MacDermid, E. (1995), 'Struggling to find the exit in the dark', *Journal of Case Management* 4: 120–23.

Malone, R. (2003), 'Distal nursing', *Social Science and Medicine* 56: 2317–26.

Marcoulatos, I. (2003), 'The secret life of things: rethinking social ontology', *Journal for the Theory of Social Behavior* 33 (3): 245–78.

Marcy-Edwards, D., Grant, N.K. and Slater, S. (2005), 'The exploration of self-determined behavior and activity vignettes: a descriptive exploratory study', *Alzheimer's Care Quarterly* 6 (1): 35–51.

Marquez, C. (1987), 'The curvature of intersubjective space: Sociality and responsibility in the thought of Emmanuel Levinas', in A. Tymieniecka (ed.)

Morality Within the Live and Social World (Dordrecht, Netherlands: D. Reidel Publishing) pp. 343–52.

Martin, R. and Post, S. (1992), 'Human dignity, dementia and the moral basis of care giving', in R. Binstock, S. Post and P. Whitehouse (eds) *Dementia and Aging: Ethics, Values, and Policy Choices* (Baltimore: Johns Hopkins University Press) pp. 55–68.

McColgan, G. (2005), 'A Place to Sit: resistance strategies used to create privacy and home by people with dementia', *Journal of Contemporary Ethnography* 34 (4): 410–33.

McKnight, J. (1995), *The Careless Society: Community and its Counterfeits* (New York: Basic).

McLean, A. (2006a), 'From commodity to community in nursing homes: an impossibility?', *Ageing and Society* 26: 925–37.

McLean, A. (2006b), 'Coherence without facticity in dementia: the case of Mrs. Fine', in A. Leibing and L. Cohen (eds) *Thinking about Dementia: Culture, loss and the anthropology of senility* (New Brunswick, NJ: Rutgers University Press) pp. 157–79.

McLean, A. (2007), *The Person in Dementia: A study of nursing home care in the US* (Peterborough, ON: Broadview Press).

NCCNHR (National Citizens' Coalition for Nursing Home Reform) (2006), 'Culture Change in Nursing Homes', Consumer Fact Sheet: 19, (Washington, DC: NCCNHR) <http://www.nccnhr.org/uploads/CultureChange06pdf.> accessed 14 February 2007.

Nolan, M., Ryan, T., Enderby, P. and Reid, D. (2002), 'Towards a more inclusive vision of dementia care practice and research', *Dementia* 1 (2): 193–211.

O'Boyle, R. (2003), 'Memantine Officially Approved for Use in US', *ElderCare Online* <http://www.ec-online.net/Knowledge/Articles/memantine.html> accessed November 30, 2004.

Plourde, S. (2001), 'A key term in ethics', in D. Thomasma, D. Weisstub and C. Hervé (eds) *Personhood and Health Care (*Dordrecht, Netherlands: Kluwer Academic Publishers) pp. 137–48.

Post, S. (2006), '*Respectare*: moral respect for the lives of the deeply forgetful', in J. Hughes, S. Louw and S. Sabat (eds) *Dementia: Mind, Meaning, and the Person* (Oxford: Oxford University Press) pp. 223–34.

Reed-Danahay, D. (2001), 'This is Your Home Now: Conceptualizing location and dislocation in a dementia unit', *Qualitative Research* 1: 47–63.

Relph, E. (1976), *Place and Placelessness* (London: Pion).

Richter, R. and Richter, B. (2002), *Alzheimer's Disease* (London: Mosby Press).

Rubinstein, R.L. and de Medeiros, K. (2005), 'Home, self, and identity', in G. Rowles and H. Chaudhury (eds) *Home and Identity in Late Life* (New York: Springer) pp. 47–62.

Sabat, S. (2006), 'Mind, meaning, and personhood in dementia: the effects of positioning', in J. Hughes, S. Louw and S. Sabat (eds) *Dementia: Mind, Meaning, and the Person* (Oxford: Oxford University Press) pp. 287–302.

Sabat, S. and Collins, M. (1999), 'Intact social, cognitive ability, and selfhood: A case study of Alzheimer's disease', *American Journal of Alzheimer's Disease* 14: 11–19.

Sabat, S. and Harré, R. (1992), 'The construction and deconstruction of self in Alzheimer's disease', *Ageing and Society* 12: 443–61.

Sachs, N. (2004), 'Healing gardens', *Therapeutic Garden Design* IV (1): 2–3.

Saperstein, A.R., Calkins, M.P., Van Haitsma, K. and Curyto K.J. (2004), 'Missed opportunities: the disconnect between physical design and programming and operations', *Alzheimer's Care Quarterly* 5 (4): 324–31.

Savishinsky, J. (1985), 'Pets and family relationships among nursing home residents', *Marriage and Family Review* 8: 109–34.

Scheibe, K. (1989), 'Memory, identity, history and the understanding of dementia', in L. Eugene Thomas (ed.) *Research in Adulthood and Aging: The Human Science Approach* (Albany, NY: SUNY Press) pp. 140–59.

Thompson, E. (2001), 'Empathy and Consciousness', in E. Thompson (ed.) *Between Ourselves: Second-person issues in the study of consciousness* (Thorverton, UK: Imprint Academic) pp. 1–32.

Wallace, B.A. (2001), 'Intersubjectivity in Indo-Tibetan Buddhism', in E. Thompson (ed.) *Between Ourselves: Second-person issues in the study of consciousness* (Thorverton, UK: Imprint Academic) pp. 209–30.

Ward, R. (2005), 'What is dementia care? 1. Dementia is communication', *Journal of Dementia Care* 13: 16–17.

Weisstub, D. and Thomasma, D. (2001), 'Human dignity, vulnerability, personhood', in D. Thomasma, D.Weisstub and C. Hervé (eds) *Personhood and Health Care* (Dordrecht, Netherlands: Kluwer Academic Publishers) pp. 317–32.

Williams, A. (1998), 'Therapeutic landscapes in holistic medicine', *Social Science and Medicine* 46 (9): 1193–1203.

Williams, A. (1999), 'Place identity and therapeutic landscapes', in A. Williams (ed.) *Therapeutic Landscapes: The Dynamic between Place and Wellness* (Lanham, MD: University Press of America) pp. 71–96.

Williams, A. (2002), 'Changing geographies of care: employing the concept of therapeutic landscapes as a framework in examining home space', *Social Science and Medicine* 55: 141–54.

Zeisel, J. (2005), 'Environment, neuroscience, and Alzheimer's disease', *Alzheimer's Care Quarterly* 6 (4): 273–9.

Chapter 20

Reclaiming the Therapeutic Value of Public Space through Roadside Art and Memorials in Rural Nevada

Dawn Einwalter

Roadside Art on Public Lands

Nevada's rural roadways contain a variety of informal art and memorials that are constructed by passing motorists and can subsequently be viewed by others traveling along the highway. Created by non-professionals, roadside art and memorials incorporate locally available natural materials, such as stone, which may be combined with other objects brought to the site. Three types of roadside art and memorials are explored in this chapter: *rock graffiti*, which can be found along several major highways; *stone markers and cairns* located along 'Guru Road' at the northern edge of the Black Rock Desert in northwestern Nevada; and *roadside markers*, which indicate the sites of fatal automobile accidents, and are common throughout the state.

Although roadside art and memorials can be found throughout the American west, their prevalence in Nevada is notable. Their pervasiveness is due, in part, to the fact that roadside art and memorials are created on isolated rural public lands subject to limited control by the agencies that administer them. These agencies include, among others, the Bureau of Land Management (BLM), the Nevada Department of Transportation (NDOT), and the Nevada Highway Patrol (NHP). As this chapter shows, the extent to which these rural public lands should be actively administered is the subject of considerable public discussion in Nevada, with disagreement over the degree and nature of that control. The factors of geographic size, limited administrative control, and cultural permissiveness exist in tension with one another, allowing for the presence of flexible public spaces that may be subject to personal interpretation and expression through the creation of roadside art and memorials. This chapter considers how these flexible spaces afford individuals with opportunities to derive therapeutic benefits from interaction with and use of the environment.

The Therapeutic Value of Manipulating Public Space

Kearns and Gesler define therapeutic places as those spaces '… associated with the maintenance of health and wellness' (Kearns and Gesler, 1998). While few might be inclined to see Nevada's sometimes stark roadways as capable of promoting health, by extending the concept of therapeutic places to Nevada's roadways, the conceptualization of health and place becomes 'an important consideration for public spaces' that are not normally deemed therapeutic in design and purpose (Miyake, 2002, 56). Although even the most highly programmed space may be subject to interpretation and alternative use, flexible spaces such as Nevada's rural roadsides have the capacity to confer additional therapeutic value to their users by allowing them to leave a personalized mark on the land. By creating roadside art, individuals are able to delimit and control the landscape, as well as use it as a canvas for personal expression.

Arguing that people are more satisfied with an environment that they can modify, William H. Whyte's pioneering work on urban public space, now continued by the Project for Public Spaces, influenced the way in which public spaces have been conceived and designed for more than two decades (Whyte, 1980). Influenced by such groups as the Project for Public Spaces, landscape design professionals now widely accept the idea that spaces which can be modified by the user are more likely to be successful, and such features are now design staples for urban public spaces. The Project for Public Spaces' 'Ten Principles for Creating Successful Squares' encourages flexible design, including the incorporation of '... movable chairs, tables, umbrellas, and games [that] can be used at a moment's notice' (www.pps.org, August 2006). Such amenities must have spaces where they can be used; hence, successful public spaces often provide open plazas, grassy areas, or promenades that can be used for a variety of activities.

Landscape design professionals have also concluded that the ability to manipulate and control one's environment can confer health benefits in a direct and measurable manner (Cooper Marcus and Barnes, 1999). Hospital gardens that are designed to provide 'opportunities to make choices ... and experience a sense of control' can be facilitated by the provision of interesting outdoor spaces with multiple opportunities for use and interaction (Cooper Marcus, 2002, 222). Likewise, gardening has been used as a therapeutic technique with mental health patients to help them 'gain self-esteem [and] a sense of control over their surroundings' (Kovary, 2002, 118). Horticultural therapy has also been used by therapists at the Awaji Landscape Planning and Horticulture Academy in Hyogo, Japan, to assist elderly clients who lack mobility (Miyake, 2002, 56).

The research just noted was primarily developed in urban settings, ranging from urban public spaces to semi-public institutional settings. Although Nevada's rural roadways differ significantly from these urban settings, therapeutic benefits are likely to be derived from personal manipulation of and interaction with both urban and rural environments. The difference between the two types of spaces is more likely to arise from differing forms of use and interaction. While urban environments may be designed to provide opportunities for flexible use, as encouraged by the Project for Public Spaces, in practice these environments are less amenable to manipulation than contemporary design approaches would suggest.

Many cities and private foundations and associations have carried out sincere efforts to make urban public spaces subject to interactive and flexible use. Museum exhibits that invite the viewer to interact with and manipulate them are a popular trend in museum design. The Exploratorium in San Francisco, for instance, offers hands-on science exhibits, deploying a model which has been widely emulated by other science museums. Museums that cater specifically to children, such as exploratoriums, children's parks, and discovery parks, also emphasize interactive exhibits and hands-on activities. Even ordinary playgrounds are often subject to municipal guidelines instructing playground designers to design play areas that can be used by children in novel and creative ways rather than simply selecting off-the-shelf products. Café seating ordinances, in which restaurants are allowed to extend their facilities onto a public sidewalk or plaza in order to 'activate' the space (and, unlike the above example, are not designed to cater solely to children), have also been adopted by numerous municipalities.

Despite these efforts, however, many urban public and civic spaces are so highly programmed that they offer few opportunities to truly modify the environment in ways not anticipated by the designers. Interactive exhibits and spaces have generally been designed to appeal to a defined group of target users with an associated set of appropriate uses. The resulting range of possibilities may be quite limited, with the chief appeal of many of these spaces largely confined to the visual. This is particularly representative of those spaces designed for children, which often include bright colors and varied art forms. While these spaces may be very beautiful, there is little that a child can do to explore, learn from, and interact with a brightly colored tile mosaic, for instance.

Overdesigned and programmed spaces need not be an inevitable feature of urban public spaces, however. As the earlier examples of garden therapy suggest, it is possible to design urban spaces that provide therapeutic value through their ability to accommodate manipulation and interactive use. This point is emphasized here in order to place the following discussion of rural public lands in Nevada within a broader discussion of therapeutic value, and to avoid the possibility of reifying an urban/rural dichotomy that could obscure the potential similarities of these two types of spaces. The remainder of this chapter examines the aforementioned examples of roadside art and memorials in rural Nevada, starting with rock writing, to consider some of the ways in which the therapeutic value of public spaces might occur.

Therapeutic Space – Rock Writing in the Salt Wells Basin

Rock writing provides an example of environmental manipulation that is not subject to the limits of design, but simply the availability of local materials and the constraints of the natural environment. Rock writing is found in the Salt Wells Basin, which is located approximately eight miles east of Fallon, Nevada. The Salt Wells Basin is a playa, a type of hydrologically closed basin common to Nevada's basin and range topography. The playa's high mineral content results in a barren landscape with minimal plant and animal life. The Salt Wells Basin is similarly devoid of human structures, with the exception of: a mine in the center of the playa, the highway – United States highway 50 (US 50), adjacent power lines, and the now-closed Salt Wells brothel. Puce-colored stones, washed down from the surrounding volcanic hills, litter the white, dusty landscape.

Figure 20.1 Typical rock writing in the Salt Wells Basin

Source: Author's photograph

Motorists traveling on US 50 sometimes stop along the roadway to collect these dark stones in order to create graffiti against the lighter ground of the playa. This 'rock writing' is less notable for its creativity than it is for its magnitude; the total area of continuous writing extends for more than ten miles. Messages range from the religious to the obscene, with love messages and personal names being the most common. Several large stone sculptures and cairns have also been constructed further back from the sides of the road. Too large to be read at close range, the graffiti is often placed at an angle along the sloping walls of the drainageways adjacent to the roadway for easy viewing by other passing motorists.

The contingent nature of rock writing endows it with a transient quality not unlike the work of artist Andrew Goldsworthy, who uses such materials as ice, wood, and even leaves to create naturalistic art of such an extremely ephemeral quality that in many cases his work can only be viewed through photographs. Similarly, although rock writing is constructed of large, heavy stones, the messages are continually effaced by natural erosion, as winter precipitation draining into the playa causes them to slip into the drainageway. During the unusually harsh winter of 2005, most of the previous stone writing was damaged or washed away, but people were already filling the playa with new writing by the following summer. As older rock writing decays, it is scavenged for stones for new writing, and is constantly in the process of replacement.

Rock Writing and the Limits of Administrative Control

Rock writing is similar to urban graffiti as a form of self-expression that claims and delimits a place as one's own. Rock writing differs from urban graffiti, however, in the lack of administrative control over the former. In writing about New York graffiti, sociologist Richard Sennett notes that, 'In a city that belongs to no one, people are constantly seeking to leave a trace of themselves, a record of their story' (Sennett, 1990, 205). Sennett, like many observers of the urban public realm, views urban graffiti as the result of multiple efforts to place aggressive, illicit personal claims on public space. Highlighting the illicit nature of graffiti, the San Francisco Police Department notes on its website that, 'The difference between graffiti and art is permission' (www.sfgov.org, February 2007). This author has noted a similar competition for visual space in Sacramento, where a new restaurant had to be repeatedly repainted, as taggers returned – almost daily for several weeks – to mark the walls, windows, and doors of the building. Graffiti such as this is an appropriation of the publicly accessible, visual elements of space, whether that space is privately or publicly owned. In urban areas, this appropriation of space further signifies a general lack of maintenance and security to other individuals using or viewing it. If not removed, graffiti can lead to the decay and abuse of a building or area by signifying to potential criminals a perceived disinterest on the part of property owners. Persistent graffiti can also signal a lack of safety to users and viewers of a building or area, potentially persuading people to avoid an area that is perceived as uncared for.

The danger of spiraling decay initiated by unaddressed graffiti is typically dealt with by anti-graffiti programs. The aggressive anti-graffiti program initiated by New York's public transit department to address tagging on buses was designed to display a zero tolerance attitude toward perpetrators. Buses which had been tagged were subject to immediate repainting, sometimes the same day, before returning to service (Sennett, 1990). Many cities, ranging from Melbourne to San Francisco, also solicit public assistance with graffiti control by funding extensive public outreach and clean-up programs (www.melbourne.vic.gov.au, February 2007 and www.sfgov.org, February 2007). The non-profit organization, Graffiti Hurts, specializes in programs and outreach intended to diminish the problem of urban graffiti (www.graffitihurts.org, February 2007).

Although Nevada's roadside art and memorials, like urban graffiti, is an informal visual and physical appropriation of rural public space, it lacks the overt sense of decay that urban graffiti implies. Although some motorists in Nevada reject roadside art, interpreting it as a breakdown of administrative control over the use of public space, the sense of physical danger characteristically associated with urban graffiti is also absent. The creators of rock writing in the Salt Wells Basin have typically left the area long before the viewer arrives, leaving only their traces for passing motorists to observe, interpret, or ignore.

Rock graffiti and other roadside art and memorials persist in rural Nevada for several reasons, including isolation and limited administrative control. The geographic magnitude of Nevada's rural public lands and their relative isolation from both major and minor urban centers renders them a particular challenge for the agencies charged with their management – which have limited resources to allocate to enforcement. Many rural highways are simply narrow ribbons administered by NDOT that cut through large tracts of federal land, the latter occasionally punctuated by parcels of private property. In the absence of formal surveys, precise administrative boundaries for lands along roadways are difficult to ascertain, even for agency staff. Nevada public lands often lack clear private uses and are open to all who use the roadway, creating the opportunity for the personalization of space without the act being clearly illicit or causing damage to personal property. Combined with limited governmental resources, the isolation and lack of clearly appropriate uses has led to a tolerant multi-jurisdictional administrative climate under which roadside art is not overtly treated as a problem and is permitted to 'fly under the radar' of intervention. Rock writing thus occupies a transitional zone between the licit and illicit. The extent to which this expressive space persists can be found in the tension between popular perceptions of the right to use public lands and the agencies that monitor and control this use.

To create rock writing, motorists must pull alongside the roadway, often parking partially on the blacktop and partially on the shoulder, although a few informal pullouts can be found along the roadway. The Fallon substation of the NHP is primarily responsible for enforcing roadway safety along the portion of US 50 that traverses the Salt Wells Basin (nhp.nv.gov/central/CC_Main.shtml, July 2006); it interprets its enforcement responsibilities as limited to traffic-related incidences that occur on the paved portion of the roadway, although the NHP would also respond to accidents that left the paved roadway (personal communication, Tom Ames, Fallon NHP Substation). Under this interpretation of its jurisdiction, the NHP could legitimately cite many of the vehicles stopped along the roadway as a potential hazard because the shoulder is very narrow (four to five feet wide at most) and drops steeply to the adjacent drainageways. In practice, however, the officers at the Fallon substation typically tolerate the parked cars and rock writing, and consider the rock graffiti to fall outside their jurisdiction.

NDOT, another agency that holds jurisdiction over the public roadway, is responsible for any installations within the road right-of-way, which, for purposes of regular roadside maintenance and administration, is informally identified as 'between the fence lines' or '12 feet from the paved surface' (multiple personal communications with NHP and NDOT representatives). NDOT considers most of the rock writing to be outside of this 12-foot band along the roadway. In addition, BLM land adjoins the NDOT right-of-way on both sides of the Salt Wells Basin. Since the proximity of the graffiti to the roadway varies greatly, it is likely that the graffiti is located partially in the NDOT right-of-way and partially on BLM land. A survey would be necessary to determine the exact boundary, and neither agency is likely to engage in such a costly undertaking to determine which is responsible for it without considerable incentive to do so; thus the rock writing is likely to continue to be washed into the playa and seasonally replaced in the absence of further administrative action or public complaint, or both.

With 86 per cent of the state managed by the federal government, Nevada tops even Alaska in its proportion of federal ownership. The BLM administers the largest share (67 per cent or 48 million acres), with smaller portions managed by a laundry list of other federal agencies, including the US Forest Service (USFS) and the Department of Energy. In comparison, only 11.5 per cent of the state is privately owned, with private lands largely concentrated in the two urban centers of Las Vegas and Reno/Sparks/Carson City, while the state administers a mere 0.4 per cent (http://dcnr.nv.gov).

The combined isolation of Nevada's rural roadways, the limited enforcement of non-traffic related behavior along these roadways, and the large proportion of adjacent undeveloped public lands combine to create an effective opening for the public to engage in multiple uses and interpretations of this landscape. This opening must also be understood within the 'social and cultural categories' (Williams, 1999, 72) that affect the interpretation and use of Nevada's rural highways, including the contested significance of 'public lands' and 'Nevada' as a frontier and a refuge of the illicit. In a manner similar to the roadside crosses studied by Everett, rock writing acts as as 'extra-legal' or 'outlaw' respresentation of liminal space (Everett, 2002, 1).

The control and use of Nevada's public lands have been the subject of intensive, periodic efforts by local groups and administrative bodies to gain greater control over federal lands, with events in Nye and Elko counties garnering particular media attention. Although such efforts are typically based in local initiatives and lack a statewide or regional organizational structure, similar efforts have erupted periodically in such western states as Wyoming, Montana, and Utah, and have sufficient commonality of purpose to earn the name, 'sagebrush rebellion' (Larsen, 1995). Initially appearing in the 1970s, local efforts to wrest control of federal lands continue to erupt in western states several decades later. Events in Nye and Elko counties, beginning in the mid-1990s, are typical of such outbursts of public disaffection over widespread federal control of public lands.

Nye County is the third largest county in the US, with over 11.5 million acres of federally managed land and a small population of 41,000 residents concentrated near the Las Vegas metropolitan area. Parts of the county are heavily dependent on federal grazing and mining permits to support the limited rural economy. In response to ongoing disputes with federal entities over the regulation of these permits, however, the Nye County Board of Commissioners approved a measure in 1993 giving the County management rights over USFS and BLM lands, roads, and trails within its boundaries (Harris, 1995). Although the dispute was widely reported as an attempt to 'secede' from the US, it ultimately led to a more prosaic federal court case, with predictable results.

Elko County was the scene of a similar dispute over the use of a dirt road leading to a popular trailhead in the Jarbridge Wilderness. At the suggestion of Trout Unlimited, an environmental group advocating the protection of habitat supporting wild, native trout species, the USFS closed the road after heavy rains damaged it in 1995. Local residents protested the closure, and attempted to reopen the road in 1998, assisted by an Elko County commissioner equipped with a bulldozer. With support from sympathetic residents in nearby states, local residents again tried to open the road in 2000 with picks and shovels (Mider, 2000, Christensen 1999).

Although both attempts to control federal lands were unsuccessful, the conflicts have fed a mythology of state identity and the popularly acceptable use of rural public lands. While arising primarily in sparsely populated rural counties, the concept of light administrative control of rural public lands has leaked back into Nevada's urban areas, influencing the perceptions and expectations of urban visitors to rural areas. As Whyte and Nelson Limerick note, 'Alaska is a state with an enormous amount of unpopulated land, but by the same token, it is a state where the vast majority of the people live, not outdoors in nature, but in cities. In that very urbanized state, each automobile sports a license plate with the words, "Alaska: The Last Frontier"'(Whyte and Nelson Limerick, 1994, 79). If these Alaskan drivers share anything with urban Nevadans beyond their mutual attachment to high-clearance vehicles with large cargo holds, it would be a similar demographic profile. Indeed, despite the expansiveness of its rural areas, Nevada is an intensely urban state: of the approximately 2.5 million residents, nearly 90 per cent live in the counties of Clark (Las Vegas), Washoe (Reno/Sparks), or Carson (Carson City) (www.nsbdc.org). Like Alaskans, Nevadans also share the perception of rural public lands as a 'frontier' available for public use, and urbanites bring with them the perceptions that public ownership also implies public use that can be subject to personal interpretation.

Many of the individuals who create rural roadside art and memorials are therefore likely to be urban residents passing through or visiting rural lands, often for recreational purposes. These visitors bring complex notions of land use honed by personal experience with and exposure to the rural environment (or lack thereof) as it is mediated through the conceptualizations of these spaces gained in their largely urban lives. Rural areas, with their relative lack of administrative control, thus become an extended public canvas for personal expression. The stark barrenness of local settings such as the Salt Wells Basin extends a further invitation for individuals to articulate self through messages that can serve to bound and humanize an otherwise inhumane environment. As Williams points out, this articulation of self is an interactive relationship between the person and place, in which each becomes more firmly defined (Williams, 1999). The extent to which this expression of self is exhibited can also be found in the stone markers, cairns, and idiosyncratic stone sculptures found along Guru Road, in northwestern Nevada.

Guru Road's Rock Installations

Guru Road, also known locally as Dooby's Lane (among other variants), overlooks the northern boundary of the Black Rock Desert, which, like the Salt Wells Basin, is a large playa with limited vegetation. Guru Road was originally a BLM jeep track located approximately 50 yards upslope and parallel to the highway, State Route 34 (SR 34). Dewayne 'Dooby' Williams, a resident of the nearby town of Gerlach, Nevada, began to create informal rock installations – including inscribed markers and stone sculptures – in the late 1980s, continuing until his death in 1995. Since that time, Guru Road has been maintained and expanded by volunteer associations, broadening what was initially a form of personal expression similar to rock writing into a source of group identity.

Most of the installations along Guru Road are made of native stone supplemented by found objects. The most common installations are hundreds of small stone monuments, each designed as a symbolic headstone surrounded by smaller stones to create the impression of a grave, giving Guru Lane the overall appearance of a linear cemetery. But the mood is anything but somber, as the irreverent and often witty messages carved on them poke fun at public figures and cultural topics, including homilies expressing

Figure 20.2 'Ground Zero,' an installation on Guru Road representing a nuclear blast

Source: Author's photograph

William's preoccupation with living the good life, family, community, women, and time. Several larger installations focus on a particular theme or topic. 'Ground zero,' for instance, represents the site of a nuclear blast, with stone markers signifying various distances from the blast site.

The effort by Williams to create a lasting message in stone is suggested by the striking resemblance between his work and pioneer graffiti which can still be seen in the Black Rock Desert – High Rock Canyon Emigrant Trail National Conservation Area, a federally protected conservation area managed by the BLM and located approximately 35 miles north of Guru Road (www.nv.blm.gov/, August 2006). Older graffiti found in the conservation area was created by emigrants traveling to California along the Applegate Trail in the 1840s and 50s, and includes names, dates, and in some cases, the writer's place of origin incised into the stone walls of High Rock Canyon. The incised words were sometimes painted with materials such as axle grease to highlight their visibility. This graffiti is similar to other emigrant inscriptions, such as that of the explorer William Clark, who carved his name and the date July 25, 1806, in the sandstone monolith, Pompey's Pillar, east of Billings, Montana, to denote his passage through the area.

Despite the durability of its component materials, the installations on Guru Road have nonetheless been subject to wear and decay. Stones have toppled, grass grows up through the cemetery plots, and some of the more fragile items (a glass globe, for instance), have been broken. Heavy rain in the winter of 2005 damaged a portion of the road and nearby installations. Local volunteer groups, including Friends of the Black Rock/High Rock, maintain the site and its installations, as in 2006, when '... a diverse crew pulled weeds, set buildings upright, re-painted installations, poured concrete, scrubbed spray-painted graffiti, repaired Elvis' cloak and turned over fallen carved rocks' (Ebert, 2006, 4).

Interpreting Williams' headstones more literally than their creator, the long-distance motorcycle endurance group known as the Iron Butt Association has transformed Guru Road into a living art installation by adding a new memorial that commemorates members killed in motorcycle accidents. This recent addition to Guru Road includes a large stone memorial plaque inscribed with the names of the deceased, individual stones with these names, motorcycle license plates installed on a post, and a picnic table and hammock from which visitors can view the Black Rock Desert.

Williams was issued a right-of-way permit for the art installations in the early 1990s by the Winnemucca field office of the BLM after the site was investigated for potential cultural resources under the National Environmental Policy Act (NEPA). Right-of-way permits typically confer private use rights (the nature of which is specified in the permit) on a piece of public property. The Williams permit has not been renewed since its initial approval. Although the site is now managed by Williams' son, Duwayne Arthur, agency staff have not taken action to request renewal or otherwise enforce the permit (personal communications, D. Valentine and T. Burke, 2006). Instead, the site is tolerated, and of sufficient repute both locally and even nationally (the site is mentioned on numerous websites) that attempts to remove it would likely be controversial.

The permit for Guru Road was issued at a time when the BLM was more lenient toward such public art installations, and the addition of the Iron Butt memorial can be explained by its presence within the area of the original permit. However, the Nevada offices of the BLM now strictly interpret a national administrative policy that discourages permits for public art, memorials, or similar installations by private groups or individuals. Existing installations without a permit may be subject to removal, as in the case of informal memorials on Sand Mountain, Nevada (also located in the Salt Wells Basin) which commemorate off-road vehicle riders killed in accidents while riding. These memorials were removed in 2005 when the Fallon Paiute-Shoshone Tribe lodged a formal complaint, asserting the mountain's sacred status. Memorials that were not collected by off-road vehicle enthusiasts were subsequently removed by the BLM. In another instance, a request was also denied for the installation of a statue of a Pony Express rider and horse, along with an explanatory historical monument, at one of the state's many Pony Express sites on BLM land. The BLM takes the stance that in the absence of clear guidelines that specify which types of art, memorials, and monuments may be permitted, all requests will be denied, and other installations that are brought to their attention will be removed. The BLM is most likely to address installations that are specifically brought to their attention, as in the two examples above, as they have 'higher priorities' and 'not enough law enforcement' to address all installations within their exceptionally large jurisdiction (personal communication, Tom Burk, 2006). In practice, informal art and memorials that are not subject to complaints may or may not be removed.

Nevada's Roadside Markers

Roadside markers are a particularly poignant example of the therapeutic potential of roadside art and memorials. Roadside markers, also known as *descansos* in Spanish, are memorials erected at the site of a fatal vehicular accident. Although not unique to Nevada, roadside markers are somewhat more common than in other states, due in part to the tolerant attitude adopted by NDOT (Sharp, 2005). The markers can be found in both urban and rural areas, but are most common along rural roadways where they are more unobtrusive and where space to erect and maintain them is more available than on urban roadways.

Individuals and families are responsible for installing and maintaining these markers. Typically consisting of a cross with the victim's name and their dates of birth and death, a roadside marker may also be embellished with flowers or wreaths similar to those found in a formal cemetery setting. Personal items are often left at the site; a marker adjacent to SR 34 included a carved stone plaque, a glass angel, and a stuffed dog, while another on US 50 west of Hazen included a red, white, and blue wreath and flag. The crosses are typically painted white, but a striking cross on US 395 north of Reno has been painted Dutch blue.

NDOT historically tolerated these markers and did little to regulate them, until forced to respond to a marker erected in 2003 to commemorate the site where 9-year-old Krystal Steadman was found slain along US 50 near Lake Tahoe. The marker in question was a large steel cross (estimated at seven to eight feet high) set in concrete,

Figure 20.3 A wooden roadside marker decorated with personal items (adjacent to SR 34)

Source: Author's photograph.

which was installed when a smaller three-foot wooden cross with steel reinforcing was destroyed by repeated vandalism. The larger cross triggered the threat of a lawsuit, forcing NDOT to develop a more formal policy to address roadside markers in the public right-of-way.

NDOT opened the issue to public consideration in a series of meetings in 2004, with varied and emotional reactions from participants. Families and friends who had erected markers asserted their right to commemorate the death of a loved one. Some individuals criticized the appropriation of public space for personal use, arguing that they did not want to be reminded of the misfortunes of others, or that such markers should appropriately be in cemeteries, not creating linear cemeteries along the roadways. Some people simply thought the markers looked 'junky.' For a few, the presence of religious symbols in the public right-of-way suggested an inadequate separation of church and state (*Pahrump Valley Times,* 2006).

Although Nevada Revised Statutes provides for a 'clear zone' along the highway with encroachment permits required for any structure or installation (NRS Chapter 408, Roads and Highways), NDOT developed a more specific policy to regulate roadside markers. Following the public meeting process, three alternatives were considered: ban the markers outright; allow standardized, state-sanctioned markers without personal names or dates; or maintain the status quo, allowing markers with some restrictions.

The state ultimately opted for the third alternative, to allow the markers with some restrictions, provided that the markers are appropriately sized (generally interpreted as three feet tall or less), impermanent, and placed at a safe distance from the shoulder of the road. Markers are to be removed after two years. However, despite the policy's high public profile, it remains an unwritten, informally administered policy, or as noted by the NDOT spokesman, 'we have a policy, but don't do anything about it' (personal communication, Scott MacGruder, 2006). Thus, the administrative policy, as it is currently used, allows NDOT to interpret the suitability of markers on a case-by-case basis.

One of the stipulations, that the markers be impermanent, partially addressed a recent trend toward more solidly built markers. Many roadside markers have been subject to vandalism, particularly those along heavily traveled roads and near urban areas, prompting families whose markers were vandalized to respond by installing more permanent markers constructed of steel or stone set in concrete (Powers, 2003). This transition from the personal, ephemeral quality of older, wooden markers to the semi-permanence exhibited by a few of the more recent markers had been interpreted by some individuals as a more blatant appropriation of public space.

Another stipulation, limitation of the size of the markers, has been presented as a safety issue, with NDOT contending that larger markers are a distraction for drivers. This representation of the issue, however, conflicts with the fact that NDOT not only permits, but promotes, sanctioned art along the public right-of-way, including art on sound walls and overpasses which have been incorporated for visual aesthetics – that is, they are meant to be viewed by passing motorists. By framing the issue as one of size, NDOT has been able to skirt – at least for the moment – a more intensive effort to address the controversy over whether religious symbolism (the cross) should be allowed on public land.

By allowing the installation of roadside markers, NDOT believes it is tolerating a legitimate mourning process. Some individuals feel a strong need to have a marker at the location of the accident, and return to the site to care for the marker. However, NDOT must also balance support for individual grief with agency commitment to public safety and the need to ensure that the markers do not impinge too greatly on the general appearance of public roadways. The agency therefore occasionally works with families to find an appropriate location for markers to ensure that they are placed at an adequate distance from the pavement. If a marker is decorated with a large number of personal belongings, the agency may also request that some or all be removed to avoid a cluttered appearance. If markers are not actively maintained or are otherwise inappropriate, the agency may request that the family remove them. NDOT maintenance crews will remove markers if necessary, but prefer not to because the act of 'tossing a marker in the back of the truck and then throwing it in the garbage' makes maintenance staff uncomfortable. Thus, NDOT has opted to work with local families even though it is somewhat costly in terms of agency time and resources (personal communication, Scott MacGruder, 2006).

Discussion

Rather than viewing Nevada's roadside installations as an aberration contingent on the state's isolated roadways and its unique administrative climate, it is useful to consider them within the wider context of public land use, including the urban context. As noted earlier, many communities have created or seek to create public spaces with a flexible, interactive component. Perhaps, then, the therapeutic value of Nevada's roadside installations lies not only in the contingent opportunity for self-expression that they afford, but the flexibility with which the various mediums are used. This is a feature that could potentially influence the design of interactive spaces in urban areas, including those that are specifically intended to serve a healing purpose, such as hospital gardens. Such design features might take the form of specific areas where people are encouraged to modify the environment, with institutional settings offering a defined area where raw materials are provided, their use being limited to the users' imaginations. Sponsoring agencies or institutions would have to be willing to tolerate some 'mess around the edges', resulting in detractors who do not feel served by or supportive of such contingent spaces. Although these spaces would not appeal to all users, they could be incorporated as part of a larger designed space to offer true flexibility and the therapeutic values that might be provided it.

References

Author Unnamed (2006), 'Emotional roadside memorial fight not over; protests lodged', *Pahrump Valley Times* (October 9).

Ames, T., Sergeant with Nevada Highway Patrol, Fallon Substation, personal communication, October 2006.

Burke, T., State Archaeologist, Bureau of Land Management, Nevada State Office, personal communication, October 2006.

Christensen, J. (1999), 'Nevada rebellion ends with a whimper', *High Country News* (October 25).

Cooper Marcus, C. (2002), 'Postoccupancy Evaluation and the Design of Hospital Gardens', in C. Schoemaker (ed.) *Interaction by Design: Bringing People and Plants Together for Health and Well-Being* (Iowa City: Iowa State Press) pp. 219–26.

Cooper Marcus, C. and Barnes, M. (1999), *Healing Gardens: Therapeutic Benefits and Design Recommendations* (New York: John Wiley and Sons).

Everett, Holly, (2002), *Roadside Crosses in Contemporary Memorial Culture* (Denton, Texas: University of North Texas Press).

Harris, M. (1995), 'Legal shoot-out over federal land rights', *Vegetarian Times* (July).

Kearns, R.A. and Gesler, W.M. (1998), *Putting Health into Place: Landscape, Identity, and Well-Being* (Syracuse: Syracuse University Press).

Kovary, M. (2002), 'Healing Landscapes: Design Guidelines for Mental Health Facilities', in C. Schoemaker (ed.) *Interaction by Design: Bringing People and Plants Together for Health and Well-Being* (Iowa City: Iowa State Press) pp. 115–22.

Larsen, E. (1995), 'Unrest in the west', *Time* (October 23).

McGruder, S., Public Information Officer, Nevada Department of Transportation, personal communication, October 2006.

Mider, Z. (2000), 'Kicking and Screaming in Nevada', *High Country News* (July 31).

Miyake, F.A. (2002), 'Linking People with Nature by Universal Design', in C. Schoemaker (ed.) *Interaction by Design: Bringing People and Plants Together for Health and Well-Being* (Iowa City: Iowa State Press) pp. 53–62.

Powers, L. (2003), 'Vandals target roadside memorials', *Reno-Gazette Journal* (October 16).

Schoemaker, C. (ed.) (2002), *Interaction by Design: Bringing People and Plants Together for Health and Well-Being* (Iowa City: Iowa State Press).

Sennett, R. (1990), *The Conscience of the Eye* (New York: W.W. Norton and Co.).

Sharp, D. (2005), 'Battles over roadside shrines more common', *USA Today* (July 11).

Valentine, D., Archaeologist, Nevada Bureau of Land Management, Winnemucca, Nevada Field Office, personal communication, October 2006.

Whyte, R. and Nelson Limerick, P. (1994), *The Frontier in American Culture* (Los Angeles: University of California Press).

Whyte, W.H. (1980), *The Social Life of Small Urban Spaces* (Washington, DC: The Conservation Foundation).

Williams, A. (ed.) (1999) *Therapeutic Landscapes: The Dynamic Between Place and Wellness* (Lanham MD: University Press of America).

Internet-based References

Bureau of Land Management, Nevada (August 2006) <http://www.nv.blm.gov/>

Ebert, Matthew, 'Metric' 'Guru Road, Personified', *Friends of the Black Rock/ High Rock* (Summer 2006) <http://www.blackrockfriends.org/pdf_library/ Summer2006_4HR.pdf>

'Graffiti Abatement Program', San Francisco Police Department (February 2007) <http://www.sfgov.org/site/graffiti_index.asp?id=246>

'Graffiti Hurts' (copyright 2007) Keep America Beautiful Inc. <http://www.graffitihurts.org/>

'Graffiti Management', City of Melbourne, Australia (February 2007) <http://www.melbourne.vic.gov.au/info.cfm?top=145&pg=1150>

'Land and Management Status', Nevada Department of Conservation and Natural Resources (July 2006) <http://dcnr.nv.gov/nrp01/land01.htm>

'Nevada Highway Patrol', Nevada Department of Public Safety (July 2006) <http://nhp.nv.gov/central/CC_Main.shtml>

Nevada Revised Statutes 480 (October 2006) <www.leg.state.nv.us/NRs/NRS-480.html>

'2005 Estimates by County', Nevada State Demographer (October 2006) <http://www.nsbdc.org/what/data_statistics/demographer/pubs/pop_increase/>

'Ten Principles for Creating Successful Squares', Project for Public Spaces (August 2006) <http://www.pps.org/parks_plazas_squares/info/parks_plazas_squares_articles/squares_principles>

Chapter 21

The Construction of Place as Aesthetic-Therapeutic

Peter Collins

Introduction

> People say the effect is on the mind. It is no such thing. The effect is on the body, too. Little as we know about the way in which we are affected by form, colour, by light, we do know this, that they have a physical effect. Variety of form and brilliancy of colour in the objects presented to patients is the actual means of recovery (Florence Nightingale, *Notes on Nursing,* (1888).

In this chapter it is my intention to identify and describe the *aesthetic-therapeutic*, a new genre in the established health geography field of therapeutic landscapes (Conradson, 2003, 2005; Gesler, 1992; 2005; Gesler et al., 2004; Kearns and Barnett, 1999; Ulrich, 1984; 1991; 1992; 2002; Williams, 1999). The landscape or place I am concerned with here is the acute hospital and much of what I have to say in this chapter is not particularly new. As we read in the quote from Florence Nightingale, the design of hospital space has occupied the minds of planners and clinicians alike for at least 150 years. Similarly, in relation to art installations in hospitals, leading artists were vying for commissions from the great London hospitals even during the mid eighteenth century – indeed, William Hogarth considered such commissions to be indicative of a successful career (Uglow, 1997; 322–65). However, what is new and what I wish to focus on here is the extraordinary emphasis given nowadays to the aesthetic – to a focus on 'the beautiful' – in establishing the acute hospital as a therapeutic environment.

Broadly speaking, is the hospital that has come to carry the burden of the therapeutic at least since the Middle Ages. Taken to mean the treatment of disease or more recently the holistic notion of both physical and mental well-being, the therapeutic is the primary function of the hospital. Unfortunately, public confidence in the ability of the hospital to provide for the therapeutic has been collapsing for more than a decade. It is commonplace to read, in British newspapers, accounts that claim that the hospital is increasingly likely to cause rather than cure disease. Even allowing for the recent spate of scare stories regarding the apparently unethical retention of body parts, and the reported epidemic of SARS, MRSA, C. *difficile* and related contagions, the idea of the hospital as a therapeutic place has been undermined.

At the same time, we have seen a veritable explosion in the interest in and capital spent on improving or 'enhancing' the visual appearance of public space. In the small

British city in which I live (Durham), one can hardly move in the city centre for the burgeoning flower displays, relaid cobblestones and newly erected *ye olde worlde* street signs. Local authorities in the north east of England have commissioned a series of major public works of art in recent years, perhaps the most famous – or infamous – of all being Anthony Gormley's 'Angel of the North' in Gateshead, a huge iron sculpture of an angel commissioned by the local authority, which prompted a lengthy public debate regarding its value, in both aesthetic and economic terms. Public art is just one component of a widespread tendency to aestheticize urban public space. Furthermore, there has been a proliferation of agencies seeking to facilitate the art/public interface – including national bodies such as ACE, Arts for Health, Groundwork UK, National Network for Arts and Health, Architects *for* Health and many others.[1]

Municipalities have, during the past 20 years, competed for grants both from the central government and from the EU that can be used for 'environmental improvement' of one kind or another. It is this process that the German philosopher

---

1 ACE (http://www.artscouncil.org.uk/index.php), the Arts Council England, is the national development agency for the arts in England, distributing public money from government and the National Lottery.

Arts for Health is a national voluntary centre based in Manchester, UK, established to provide practical help, information and advice to all who are concerned with using the arts as a complementary part of health care.

CABE (http://cabe.org.uk/default.aspx?contentitemid=1), the Commission for Architecture and the Built Environment, stands for an improvement in people's quality of life through good design. Its aim is to influence and inspire those who make decisions about our built environment, so that they choose good design. 'PROJECT' ('engaging artists in the built environment').

PASW (www.publicartonline.org.uk/project), Public Art South West, which, amongst other things, seeds funding a health care public art programme, especially where a new building or major refurbishment is planned.

LIME (www.limeart.org), also based in Manchester, has over twenty years' experience of using the arts in health care contexts and has a team of in-house artists working in a variety of media.

Groundwork UK (http://www.artandregeneration.com/links/index.html): 'This site is designed to provide inspiration, ideas and practical assistance both to Groundwork Trust staff who want to use the arts within regeneration projects and to artists who want to learn more about the way that Groundwork engages with the arts.'

'PROJECT' ('engaging artists in the built environment').

National Network for Arts and Health (http://www.nnah.co.uk/membership/forums/viewtopic.php?t=348) is primarily an electronic network for those working in the areas of arts and health, including government departments and government agencies, NHS Trusts, local authorities, artists, arts and medical students, community arts organizations, and charities with a special interest in health, women, children, the elderly and other specialized areas where arts in health activities form part of the services they deliver.

Architects *for* Health (http://www.architectsforhealth.com/) is a non-profit organization whose aim is to promote and campaign for better health care environments through bringing together individuals and organizations who share an interest in excellence in the planning and design of health care facilities.

Wolfgang Welsch and others have called 'aestheticization' (Welsch, 1998). Welsch goes on to argue that the aesthetecization of everything results in a form of social 'anaesthetization', that is, the aestheticizing of everything leads inevitably to a featureless blandness in which the principle of the aesthetic is negated. Neil Leach, the architect and critical theorist, makes a similar point in relation to architecture in particular (Leach, 1999). We will not pursue this argument here.

As public spaces, hospitals have not been excluded from this policy during recent years (NHS Estates, 2002; 2003; 2004; 2005; DH, 2005).[2] In the case of hospitals, it has been a process sanctioned and largely funded by central government. The NHS Estates website 'Improving the Patient Experience: The Healing Environment', includes a question aimed, one assumes, at hospital managers, under the heading 'Improving your healing environment':

> Is your ward looking a bit tired? Does your day room need redecorating? What could you do with that long, dingy corridor?

The same site states categorically:

> Good environments make us feel better, and feeling better is the key to getting better.

The King's Fund, an independent charitable foundation working for better health, especially in London, has the remit of 'improving the healing environment' and reports on its initiative 'Enhancing the Healing Environment' thus:

> Enhancing the Healing Environment, which began in 2001, is a training and grants programme to encourage and enable NHS trusts to make tangible improvements to patient spaces.
>
> Following its success in London, NHS Estates funding has allowed the programme to be rolled out nationally in November 2003 and a further mental health wave launched in November 2004.
>
> Fifty hospitals in London have already taken part in the programme, involving over 500 NHS staff and service users.
>
> Under the November extension, each of the 23 strategic health authorities outside the capital has nominated a trust to undertake a renovation project. The schemes, expected to be complete by early 2005, range from refurbishing corridors, waiting areas and wards to the creation of gardens, artworks and quiet spaces. A further wave of 23 mental health projects has now begun.
>
> Each participating trust has formed a nurse-led multi-disciplinary team (including service users) to plan and manage their project. Trust teams receive a grant of £35,000 and attend courses to develop leadership skills and give them the practical knowledge needed to make their project a success. The courses explore ways in which health care environments

---

2 Throughout this chapter I abbreviate 'National Health Service' as 'NHS' and 'Department of Health, as 'DH'.

> can be improved by the use of colour, light, art and design, and include a session at Tate Modern (www.kingsfund.org).

Significantly, then (both for academics and the general public), the therapeutic and the aesthetic have come together in recent years in significant and complex ways in the hospital itself. Nowadays, British hospitals are conceived increasingly as aesthetic as well as therapeutic places – indeed the aesthetic and therapeutic are increasingly presented as equal partners, the one partially defining and even constituting the other. In this chapter, I consider the trajectory of this 'aesthetic-therapeutic' turn and describe some of the intricate ways in which the therapeutic becomes the aesthetic and vice versa – a cultural dynamic which, in the hospitals studied, is hotly contested.

Two Hospitals

My account draws on interdisciplinary research carried out in 2002–2004 in two hospitals in the north east of England, both a part of the NHS South Tees Acute Hospital Trust: the Middlesbrough General Hospital (MGH) and The James Cook University Hospital (The JCUH). The MGH was closed after serving the local community for over a century in 2003, in order to establish a single site facility involving the extensive redevelopment of the South Cleveland General Hospital (SCGH) – first opened in 1980. The SCGH was renamed The JCUH, and has become a large hospital with just over 1,000 beds, 6,000 employees, occupying an area of 145,000 square metres. In 1997, the government confirmed that the single site scheme would progress as one of the first phase schemes under the Private Finance Initiative (PFI), that is, the facility was built and is owned by a large private consortium, which rents it to the local NHS Trust. The contract for the 30-year partnership was signed in August 1999. At the end of 2000, a £15 million bid to expand the Trust's heart unit was successful, taking the total value of building work on site to £155 million. The increasing reliance by the government on PFIs to fund capital projects is extremely controversial and poses interesting problems of ownership and control which have a bearing on the hospital environment, further discussion of which would take me beyond the remit of this chapter.

The South Tees NHS Trust (www.southtees.nhs.uk) website includes a section entitled *Making Healing an Art Form,* thus *foregrounding* a major preoccupation of The JCUH project from the outset. As written in the final report (South Tees NHS Trust, 2006,17):

> Development of The JCUH has paid special attention to building design, therapeutic colour schemes, materials, lighting, space, and acoustics. Design features and colour schemes are intended to individualise departments within the hospital and to help create a sense of intimacy within the whole. In addition, £250,000 from the building budget was ring-fenced for the purpose of commissioning artwork for the hospital. The Trust set up a 'Healing Arts' Committee to seek further funding for art works and also to fund artists' residencies to create works appropriate to this hospital environment.

Given the scale and complexity of the building, South Tees NHS Trust have, from the outset, attempted to unify the whole (itself justified in aesthetic-therapeutic terms) by theming the hospital around the life and times of the British explorer Captain James Cook (1728–79), who was born locally. Figures 21.1 and 21.2 provide representative examples of the theme. As can be seen from these images, the provision of *a therapeutic environment* at The JCUH is bound up with the installation of artworks and design which depends on a top-down 'public art' approach – as opposed, for example, to a 'bottom-up' in-house approach, drawing on the work of artists-in-residence and patients. The works draw on international or global norms of art and design; some are, or aspire to be, 'major' works of art. And as one hospital manager averred, 'Beauty makes us better'.

Figure 21.1 South Entrance of The James Cook University Hospital (note globe and capstan)

Methodology[3]

The data on which this chapter is based derives from an NHS Estates commissioned research project (Macnaughton et al., 2004) carried out from 2002 to 2004 by a team including two anthropologists (Simon Coleman and Peter Collins), two architects (Peter Kellett and Geoffrey Purves) and two experts in the field of the arts and health (Jane Macnaughton and Mike White). We also employed a full-time researcher (Anu Suokas).

3 In this section I draw on Macnaughton et al. (2004, 20–22).

Figure 21.2 Portrait of Captain James Cook (North Entrance, The JCUH)

The project had two primary objectives: first, to investigate the briefing and design processes to assess how the visions for 'patient-centred care' were carried through into the design of the new hospital (The JCUH). In relation to the first objective we examined documents prepared to guide the briefing process, and carried out a number of interviews with key participants in the design and planning process.

Our second objective was to investigate the effect of the new JCUH environment on patients' and visitors' experience of care, the experience of staff in giving care, and on hospital users' satisfaction and sense of well-being. In relation to this second objective (which is most relevant to this chapter), we developed the following research questions:

1. What is the impact of the new hospital (The JCUH) environment on patients' and visitors' experience of care, staff's experience of giving care, and user satisfaction and sense of well-being compared to the old (MGH) environment?
2. Does the new building design take into account the needs of, and interactions among, its users better than the old design?
3. What is the user response to the artwork placed or integrated within the new hospital building?

We used a number of data collection methods: a questionnaire survey of all hospital users, including patients, staff and visitors on a number of selected wards at the MGH and The JCUH; interviews with staff and patients in each study area at both sites; interviews with commissioned artists; a photographic survey of commissioned artworks, other artworks and aspects of hospital architecture and design; a brief questionnaire survey carried out in situ in the mall, at The JCUH; and direct observation of passers by in public areas.[4]

The JCUH: Public Perceptions

During the course of the research project, we asked patients, staff and visitors for their views on the art and design of The JCUH, and what is most striking about these comments is the extraordinary range of opinion – from the very positive to the extremely negative. This is hardly surprising when one considers the perceived nature of aesthetic judgement – that it is a matter of taste. For example, opinions regarding the renaming and theming of the new hospital differed markedly. Some thought the new name appropriate, others did not; some lauded the thematic approach, others did not. Here is an orthopaedic patient (OP) talking about the renaming of the hospital from 'South Cleveland to, 'The James Cook University Hospital':

Int: … what do you think of the name of the hospital?
OP: I don't mind it being called St James.
Int: The James Cook, but apparently people have started calling it St James's or Jimmy's like in Leeds, some people have confused it with that and gone to the wrong place, it must be 90 miles.

The patient actually gets the name wrong in the first instance and the interviewer corrects him, explaining that many people seem to confuse The James Cook with

4 We received ethics approval from the local Research Ethics Committee of the NHS South Tees Acute Hospital Trust.

St James' Hospital (situated in Leeds). He goes on to cast doubt on the value of the name change:

> OP: I can't work that one out, I would sooner it was called the James Cook than the Tees Valley … James Cook, I don't know where it came from but leave it as that.
> Int: Well I didn't know but the explorer James Cook was born in Middlesbrough.
> OP: Why call the hospital, you should call a dock after it?
> Int: I think he was the famous person they could think of.
> OP: Well you could call it the Juninho hospital.

As far as the name is concerned, the patient would prefer a more contemporary reference – to Juninho, for example (a Brazilian football player with Middlesbrough Football Club who was at that time something of a local hero). The interviewer goes on to explain one reason for the name change (actually offered by a member of staff in an earlier interview):

> Int: It is basically South Cleveland but bigger, but apparently the reason they did not call it, I was asking why do you call it South Cleveland, she said they did not want to put the noses out of people at Middlesbrough General. Do you get it, because people from the General may feel it was not their hospital …
> OP: So they did not want to upset the people from the General for working for the South Cleveland, but they are prepared to upset the people from the South Cleveland to work somewhere else.

This exchange is useful in that it raises several issues regarding the renaming of the hospital and its theme. Despite the numerous references to James Cook inscribed in the fabric of the place, some patients we interviewed had little idea why the hospital had been renamed or why it had been themed around the historical character of James Cook. The interviewer explains in this last exchange that the reason for the name change had to do with incorporating the staff from both the MGH and the SCGH into the new single site, at which the patient pointed out that the name change might then alienate staff already working at SCGH!

We were surprised to find that not only do the major artworks sometimes remain invisible, but that many interviewees – including staff, patients and visitors, failed to recognize the theme:

> Actually I haven't actually thought that there was a lot of James Cook, talking of murals it would have been nice to see the history of James Cook. If this is his namesake hospital there is not much of him around to say it is him, I think it is missing … a mural of him would have been, you know, this is, do many people even now know who James Cook is and you are coming to the James Cook hospital?

One non-clinical member of staff commented favourably on the perception of some that The JCUH isn't like a hospital at all, and provided this anecdote:

> … there's a lady who works in records, healthcare records, and she's an assistant to an Arkela (leader) in a local scouts group and they came as part of their proficiency badge for arts and they asked, rather than go into a gallery, they asked if they could come into

> the hospital. So I took them round and did a guided tour and we actually gave them some fact-sheets and some photographs. And they really enjoyed it, they said it didn't feel so stuffy as a museum or a gallery, one of them said.

Here the hospital is presented not only as an art gallery, but a *superior* art gallery. Others agreed that the place had 'the feel' of an art gallery (or an airport, or a shopping mall) but thought that this was a bad thing.

Members of staff we interviewed were often so annoyed by what they saw as the poor design of their departments that when asked about the aesthetics of the place, they were largely dismissive. For example, one nurse consultant complained bitterly about the poor staff facilities in her department. Asked what she would say to planners designing a hospital,

> … a very strong, a very, very strong recommendation has to be if you're designing hospitals, please design the rooms with windows in. You cannot imagine the actual negativity that that itself has generated for the staff working here.

After listening to a series of criticisms of the work environment, I asked her about the aesthetics of the hospital:

> Oh, yeah sure. The art's very nice but when you, because you're being negative about being at work, what one does is come into work, do what one needs to do and then leave. There's no – I couldn't imagine myself ever being happy here.

A consultant neurosurgeon responded similarly. Staff often criticized the lack of facilities they felt were necessary for carrying out their job: badly designed workspaces, cramped rooms, and – especially – the absence of windows. To some extent, the art and design is being perceived negatively here – in terms of its opportunity cost: the consultant's room is small and cramped because the Atrium is vast and airy. Similarly, a number of interviewees – both staff and patients – offered rhetorical comments such as 'how many sets of surgical instruments/autoclaves/defibrillators/portascans would *that* [the globe] pay for?' and 'I wouldn't mind [referring to the cost of the 'sails' in the Atrium] but we're short of operating instruments …'. Staff, in particular, seemed suspicious of the way in which the commissioned artwork was paid for – often simply failing to believe the interviewer's comment that this funding was set aside, under the terms of the PFI contract, for this purpose and could not have been spent on clinical equipment.

Unlike those questioned at The JCUH, it was noticeable that people interviewed at the old MGH returned constantly, during interview, to the metaphor of 'home', of the domestic and the private, the familiar, the intimate and the small-scale (Tuan, 1974; 1977). Staff who were transferred from the MGH after its closure, to The JCUH often made explicit this tension which has its foundation in the local/global dichotomy. Unlike The JCUH, art and design at the MGH, was 'bottom-up'. Staff and patients contributed substantially to the form and content of that *therapeutic environment* – with an emphasis on the domestic, the cherished. There is a tension, then, between the perceptions of what constitutes *a therapeutic environment* held by users on the one hand (for example, staff and patients), and planners, policy makers and managers ('management' for short) on the other.[5] Having said that, others much preferred The JCUH to the old MGH. One cancer patient told us:

> I must admit that my memories of Middlesbrough General, a few years ago, when my mother was in, I mean it was obviously on its last legs and it was dreadful! I mean it is so much better to have everything here.

There are a number of significant outcomes arising from this tension. First of all, some of the spaces at The JCUH are, as a result of their aesthetics, blatantly ambiguous, in the sense that they seem to 'belong' to no-one (see Relph, 1976 on 'inauthentic landscapes' and Augé, 1995 on 'non-places'). Let us consider, for instance, the Atrium, a large space linking the South Entrance with the Mall. From the point of view of 'management', the space is generally identified aesthetically and should therefore be considered therapeutic. However, despite the apparent informality of the seating, this is a space which remains largely unoccupied. At one point members of staff who came here to eat their lunch were moved on by security guards; the reason given was that they were 'out of place' in their uniforms – implying the possibility of infection? This reminds one of the work of the noted British anthropologist, Mary Douglas, and specifically her concept of dirt as 'matter out of place'. Douglas (1966, 2) writes: 'Dirt is essentially disorder … it exists in the eye of the beholder … In chasing dirt, in papering, decorating, tidying, we are not governed by anxiety to escape disease, but are positively re-ordering our environment, making it conform to an idea'. The implication is that the aesthetic-therapeutic environment is an *orderly*, a *clean*, environment.

Patients avoid the Atrium because of its scale, finding it 'too impersonal', 'a bit scary', 'daunting'. Visitors can occasionally be found looking at the large tapestries but their stay is generally fleeting: they express a feeling of unease in lingering. Perhaps the severest criticism of the Atrium was made by a senior consultant surgeon who compared his own small, windowless consulting room with what he calls the 'vast space' of the Atrium. He was alluding here to the opportunity cost of art and design: his room was small *because* the Atrium is vast. In this context it is worth recalling a comment made by Roger Ulrich during a recent interview (Resources For OR Design and Construction, 2002):

> Also, is it really necessary to have an absence of windows? We require windows for patients. What about the staff? Could we provide a view out that offers a respite, perhaps on a garden or a park? Would this help to buffer stress?[6]

On the other hand, others told us that the Atrium is 'beautiful', breathtaking', 'simply wonderful'. It is unhelpful to generalize from such comments except to say that

5 Ironically, during the last ten years or so, there has been an extraordinary amount of research (especially in the US and UK) relating to the likely advantages of treating the sick in their own homes instead of in acute hospitals, thus completing the domestication of the therapeutic environment. The potential advantages are couched both in clinical and economic terms.

6 See also Mroczek et al. (2005) for further consideration of the impact of the working environment on clinical staff.

Figure 21.3 Atrium and 'sails' (The JCUH)

people varied considerably in their assessment of the Atrium. This is equally true of most aspects of The JCUH.

However, the Atrium is the best example of those more obviously public spaces whose aesthetic suggests, to users, the airport, the hotel and shopping mall. Clearly, there is an element of over-design of these spaces. They are not usable spaces and users tend to proceed along and through them without pausing to admire or even take note of the works of art along the way. As I said earlier, the art tends not to be local or vernacular but formal and international – and because of that 'high-brow' or middle class and possibly alienating – another issue that I have no time to elaborate here (though I should note the irony in the fact that the NHS Trust made a point of commissioning work from 'local' artists). The many paintings displayed throughout the hospital have drawn different responses. On the whole, the more abstract the painting the less they are appreciated (see also Ulrich and Lundén, 1990). Although planners, artists and designers assume that the aesthetic-therapeutic are conjoined, this clearly is not the case. Hospital users display ambivalence towards the aesthetic-therapeutic and it is not clear how this ambivalence is to be managed.

One of the objectives in installing the large globe just opposite the South Entrance was so that people could use it as a waymarker (Figures 21.2 and 21.4). Despite her doubts about its cost, this patient indicates that the globe succeeds in this at least:

> Patient: The globe, I actually like the feature and all around it. I think it was an awful lot of money to spend on a feature like that, but it does cause a lot of interest and it's

more of a direction point. If you get to the globe, you know you're at the South Entrance and you know you can work your way.

Int: A landmark.

Patient: It is a landmark. Even now you know, people say, you know, which ward, and you say, well if you head for the globe …

And the same research participant goes on to suggest that the globe may function symbolically:

Int: So what do you think the point of the globe is? Why that globe?

Patient: Hmm … there is a reason because I know I read it and I just can't remember. To me, it represents world class because I think it's an absolute first class hospital.

I include this extract in order to note that art and design can and sometimes does have a function (or a number of functions) other than that relating to the purely aesthetic: the art may (or may not) 'work' in a number of ways. Others we interviewed had not noticed the globe even though they had walked right past it.

The comments of those patients, visitors and staff we spoke to at The JCUH were markedly different from those collected at the older MGH. Here, we were told repeatedly that although the place was 'old', decrepit' and 'had seen better days', it was a 'homely' place. The MGH was a far smaller hospital than the new JCUH; it had little sense of unity apart from that given it by its relatively small scale and its longevity – its local identity – situated as it was in the heart of the town and not in its rather well-heeled suburbs where we find The JCUH. The 'art' and 'design' of the MGH was thought of as having developed 'from the ground up'. Wards had, through one means or another, 'scratched together the money' to paint and decorate their rooms, to buy small (usually vernacular) artwork, to build screens on which 'thank you' cards could be displayed. Given that the MGH had none of the restrictions associated with PFI facilities (permission has to be sought from 'management' before any items are attached to walls, for instance), staff found many different ways of personalizing, domesticating or 'making special' their space and allowing patients to do the same.

It is interesting that by far and away the most popular pictures displayed at The JCUH were two murals depicting the life and times of two demolished hospitals, including the MGH. In these collages, designed and produced cheaply by the hospital's public relations officer, we have an encapsulation and, ultimately, celebration of the local. The same member of staff, when asked to comment on the 'artwork', quickly focused in on these murals:

> The murals have gone down hugely well, tremendously well. We actually were very anxious to get, not just staff but former members of staff, former patients, and older people, possibly people who have a lot of memories of the hospital, to get involved with those murals and we were very conscious as well that we didn't just want to close the doors on Middlesbrough General and North Riding, switch the light off and just say, right that's it. So we spent a lot of time doing, we did a video, a typical 'day in the working life of', we did a book, working with Teesside University on the history of the hospitals, and we did the murals …

Similarly, a cancer patient, in commenting more generally on artwork around the place, told us:

> I like the paintings of the old hospitals and I'm sure lots of the older generation, Middlesbrough people, will like to see the old, because the old hospital in particular, was loved … great affection for it. Very much a community spirit around there.

This emphasis on 'local colour', 'local traditions' and 'heritage' is much in evidence in other hospitals in the north east: the most acceptable art relates most often to the locale and its history.

I have tried to show how, in 'over-designing' a hospital, architects, planners and hospital administrators can, in focusing on aestheticizing the environment, alienate those who are the chief users of the facility. Although it is 'homeliness' which seems to most impress hospital users (both staff and patients), ambient social trends (such as aestheticization and globalization) mitigate against that genre. Furthermore, the fine art displayed in the hospital is often seen only in terms of its perceived opportunity cost. Finally, there is the feeling that the artwork at The JCUH is merely a convenient veneer beneath which more fundamental weaknesses and deficits are concealed. A visitor commented:

> … I mean, I think it looks lovely. But in terms of the political side of the Trust, I think it, I don't know, I suppose it belies lack of infrastructure underneath, 'no knickers', I think.

To spell this out, the reference to 'no knickers' is a shorthand for the adage 'all furs and no knickers', suggesting that while on the surface things look good, beneath the glamorous exterior lies an inadequately clothed interior. As one consultant surgeon commented, having been asked his opinion of the hospital's art and design:

> I haven't looked. It gives a very nice environment to work in, or nice environment to walk through, until you get to your place of work.

His previous remarks amounted to a scathing indictment of the planning, design and equipping of his department. Such perceptions cause tensions within the facility (especially between clinical and non-clinical staff) which will not easily be resolved.

Discussion: Interrogating the Relation *Aesthetic-Therapeutic*

In this chapter I focus on an extraordinarily complex juxtaposition of words and images, of ideologies and environments and am still left with the question of how might we characterize the relation between these two terms, 'aesthetic' (concerning beauty) and 'therapeutic' (concerning wellness)? If Welsch and others are correct, the aestheticization of the built environment has become an integral part of the ambience of public space in the West. Given that the accepted primary function of the acute hospital is to make people well, it follows that the hospital should exhibit the qualities of the therapeutic *and* the aesthetic; I have shown that this is indeed the case.

Figure 21.4 Collage (corridor off the Mall, The JCUH)

As long ago as 1888, Florence Nightingale was stating categorically the close relationship that exists between the aesthetic and the therapeutic. The acute hospital should be clean and airy but must also include a 'Variety of form and brilliancy of colour in the objects presented to patients' which is, she states, 'the actual means of recovery' (1888, 109). The implication is that aesthetic qualities must be included in any evaluation of the acute hospital as a therapeutic environment. This perspective has come to be accepted as self-evidently true. No NHS hospital is constructed nowadays without careful attention being paid to the aesthetics of the place. However, we cannot assume that the increasingly close relationship that the aesthetic and the therapeutic share in the acute hospital results in a positive and uncomplicated mix of 'perfect environment', or that the marriage of the two is purely a matter of design. Our research identifies a variety of factors influencing the strength of the aesthetic-therapeutic.

Given the evident complexity of the relationship between the two terms, it would be a mistake to characterize one as cause and the other effect. So how *can* we represent their relationship? An approximate characterization might draw on Weber's idea of 'elective affinity'. Weber draws, in his use of the term, on Goethe's novel, *Elective Affinities* (1809). In the novel, Goethe himself borrows from the field of chemistry, and applies it to the field of social relationships. Just as with the acids and alkalis spoken of by the characters, words and images – even when seemingly contradictory – turn out to have an extraordinary affinity towards one another. But then, as one of the characters observes, such affinities are necessarily problematic, fulfilling their promise only in their generation of separations.

In his brilliant essay in comparative analysis, *The Protestant Ethic and the Spirit of Capitalism* (1958/1920), Weber finds, in Goethe's observations, a useful means of conceptualizing the synchronic interplay of elements from otherwise divergent socio-cultural systems. Weber argues persuasively that there was an *elective affinity* between the ethical norms of Puritanism and emerging capitalist practices in seventeenth-century England. He identifies the generative confluence of specific economic, political and religious practices which comprise a culture complex, eventuating in what he calls the *spirit* of capitalism. In similar terms I am proposing that in recent hospital design, an elective affinity conjoins the aesthetic and the therapeutic. At the structural level, the long-standing appreciation of the beneficial role of a beautiful surrounding in curing the sick has become the defining feature of the privatization of public space. In terms of the individual's experience of the space, evaluating the environment can no longer be couched entirely in quantitative terms. While the clinical quality of the aesthetic-therapeutic environment is relatively easy for *experts* to evaluate using transparent quantitative measures, the aesthetic quality cannot be evaluated in the same way. Judging the therapeutic quality is perceived to be an objective exercise with the possibility of a single outcome; the aesthetic is perceived to be subjective, however, and amenable to an infinite range of interpretations. In the acute hospital it is the therapeutic and the aesthetic that are contingent one upon the other, and it is the closeness of their relationship which, at least partially, accounts for the contentiousness of the place.

Our research at The JCUH clearly indicates the tensions that arise from the aesthetic-therapeutic, which deepen as some people perceive the prioritizing of the aesthetic against their better judgement as negating (and not enhancing) the therapeutic quality of the place. The staff, patients and visitors that we spoke to varied considerably both in their view of the programme as a whole and of individual artworks and design features. Our findings should provide a note of caution to those who might be tempted to make sweeping statements concerning the therapeutic nature of art and design in the acute hospital. Our research indicates that the likely outcome of 'designing for health' is a complex array of tensions giving rise to protracted negotiations between the various agents involved: artists, planners, managers, clinicians, non-clinical hospital staff, patients and visitors. We must recognize the fact that the aspirations of these many parties may overlap but rarely coincide.

The British government recently ruled that 1 per cent of the budget for any 'new build' hospital should be allocated for the provision of art. Even before this legislation, NHS Trusts have tended in recent years to devote more attention and funding to the aesthetics of hospitals. We found this to be true in our comparison of the traditional MGH and the new JCUH. I would argue that this trend is a part of a wider aestheticization of public space – already described by Wolfgang Welsch (1998) and others. At the same time, the British press has revealed with increasing exactitude the shortcomings of British acute hospitals. Stories of infections contracted by those in hospital are rife. The government itself, in setting ever higher standards of care, ensures that doubt continues to grow in the minds of the public regarding the quality of care. The NHS is about to embark on an extremely ambitious target – by

December 2008, the health service will have just 18 weeks to refer, diagnose and treat a patient. Against this, The JCUH itself was recently graded 'weak' in relation to its 'use of resources' (Healthcare Commission, 2006a; 2006b). This process can only be properly understood in the light of what appears to be a global trend, that being the privatization of public space. Institutions such as museums and universities, that were once seen as non-commercial, are now judged increasingly in terms of their economic performance. The health sector in the UK reflects and, of course, contributes to this trend. With the increasing role of the private sector in hospital construction and management, hospital space is increasingly becoming, and represented, as privatized space. We found that the trappings of private and commercial spaces, hardly evident in the older MGH, were almost shockingly manifest at The JCUH. The stated objective of this process might be the same as that defined by Florence Nightingale: to transform therapeutic spaces ('ill' places) into aesthetic spaces ('inviting' places). What *is* new, however, is the idiom in which this metamorphosis is taking place.

The key question is whether or not the aestheticization of the acute hospital has taken hold due to the fact that acute hospitals are currently almost doomed to failure, given the many and varied demands placed upon them by government and public alike. The increasingly stringent accounting and evaluation procedures to which hospitals are currently subjected, along with the ever critical eye of the media, seem to emphasize the limitations of the acute hospital. In this climate, it does not seem quite so peculiar that growing attention should now be placed on those aspects of the hospital environment that can be most easily and most readily improved – that being art and design. The problem is that whatever its benefits, art and design in the acute hospital will remain a contentious issue, partly because the benefits of such programmes are so difficult to prove, but also because of the increasingly highly publicized shortfall in clinical staffing and equipment in NHS acute hospitals. Finally, one thing that becomes perfectly clear in talking to people in acute hospitals, despite the most careful planning, is that the construction and re-construction of place and space continues unabated after the architects, the planners and the engineers have left – and *that* is a process which will interest geographers and anthropologists alike.

References

Augé, M. (1995), *Non-Places: Introduction to an Anthropology of Supermodernity* (London: Verso).

Conradson, D. (2003), 'Geographies of Care: Spaces, Practices, Experiences', *Social and Cultural Geography* 4: 251–4.

Conradson, D. (2005), 'Landscape, Care and the Relational Self: Therapeutic Encounters in Rural England', *Health and Place* 11: 337–48.

DH (2005), 'What is a healing environment?' <http://www.dh.gov.uk/PolicyAndGuidance/OrganisationPolicy/HealthcareEnvironment/fs/en> accessed 25 November 2006.

Douglas, M. (1966), *Purity and Danger: An Analysis of Concepts of Pollution and Taboo* (London: Routledge).

Gesler, W.M. (1992), 'Therapeutic Landscapes: Medical Issues in the Light of the New Cultural Geography', *Social Science and Medicine* 34: 735–46.

Gesler, W. (2005), 'Therapeutic Landscapes: an Evolving Theme (Editorial)', *Health and Place* 11: 295–7.

Gesler, W.M., Bell, M., Curtis, S., Hubbard, P. and Francis, S. (2004), 'Therapy by Design: Evaluating the UK Hospital Building Programme', *Health and Place* 10: 117–28.

Goethe, J.W. von (1971), *Elective Affinities* (*Die Wahlverwandtschaften*) R. Hollingdale (trans.) First published 1809 (Harmondsworth, UK: Penguin).

Healthcare Commission (2006a), *Results of the Annual Healthcheck 2005–06*, <http://annualhealthcheckratings.healthcarecommission.org.uk/_format/_documents/Full%20analysis.pdf> accessed 25 November 2006.

Healthcare Commission (2006b), *South Tees Hospital NHS Trust*, <http://annualhealthcheckratings.healthcarecommission.org.uk/annualhealthcheckratings/searchourfindings.cfm?cit_id=21023&widCall1=customWidgets.content_view_1&search_string=> accessed 25 November 2006.

Kearns, R.A. and Barnett, J.R. (1999), 'To Boldly Go? Place, Metaphor and Marketing of Auckland's Starship Hospital', *Environment and Planning D* 17: 201–26.

The King's Fund, <http://www.kingsfund.org.uk/> accessed 25 November 2006.

Leach, N. (1999), *The Anaesthetics of Architecture* (Cambridge, MA: MIT Press).

Macnaughton, J., Coleman, S.M., Collins, P.J., Kellett, P., Purves, G. and Suokas, A. (2004), *Designing for Health: Architecture, Art and Design at the New James Cook University Hospital, Middlesbrough*, DH Estates Project B(01)13.

Mroczek, J., Mikitarian, G., Viera, E. and Rotarius, T. (2005), 'Hospital Design and Staff Perceptions: An Exploratory Analysis', *Health Care Management* 24: 233–44.

NHS Estates (2002), *The Art of Good Health: Using Visual Arts in Healthcare*, Project 0113224982.

NHS Estates (2003), *The Architectural Healthcare Environment and its Effect on Patient Health Outcomes – A Report on NHS Estates*. Project 011322480X.

NHS Estates (2004), *Exploring the Patient Environment – an NHS Estates workshop*, Project 0113224885.

NHS Estates (2005), *Wayfinding: Effective Wayfinding and Signing Systems Guidance For Healthcare Facilities* (2nd edn), 0113226985. NHS Estates <http://patientexperience.nhsestates.gov.uk/healing_environment/he_content/home/home.asp> accessed 25 November 2006.

Nightingale, F. (1888), *Notes on Nursing: What It Is and What It Is Not* (Philadelphia: Lippincott).

Relph, E. (1976), *Place and Placelessness* (London: Pion).

Resources For OR Design and Construction, Interview with Roger Ulrich (Reprinted from *OR Manager Newsletter*, March 2002) <http://ordesignandconstruction.com/he/healresearch.htm> accessed 30 November 2006.

Rodman, M.C. (2003), 'Empowering Place: Multilocality and Multivocality', in S.M. Low, and D. Lawrence-Zúñiga (eds) *The Anthropology of Space and Place: Locating Culture* (Oxford: Blackwell).

South Tees NHS Trust, <http://www.southtees.nhs.uk/default.asp?page=HealArt> accessed 25 November 2006.

South Tees NHS Trust (2006), *Annual Report 2005/2006.*

Tuan, Y. (1974), *Topophilia: a Study of Environmental Perception, Attitudes, and Values* (Englewood Cliffs, NJ: Prentice-Hall).

Tuan, Y. (1977), *Space and Place: The Perspective of Experience* (Minneapolis: University of Minnesota Press).

Uglow, J. (1997), *Hogarth* (London: Faber and Faber).

Ulrich, R.S. (1984), 'View Through a Window May Influence Recovery From Surgery', *Science* 224: 420–21.

Ulrich, R.S. (1991), 'Effects of Health Facility Interior Design on Wellness: Theory and Scientific Research', *Journal of Health Care Design* 3: 97–109.

Ulrich, R.S. (1992), 'How Design Impacts Wellness', *Healthcare Forum Journal* 20: 20–25.

Ulrich, R.S. (2002), 'What Does Research Show About Healing Environments?' (Interview) <http://ordesignandconstruction.com/he/healresearch.htm> accessed 25 November 2006.

Ulrich, R.S. and Lundén, O. (1990), 'Effects of Nature and Abstract Pictures on Patients Recovering From Heart Surgery', *Proceedings of the International Congress on Behavioral Medicine, Uppsala, Sweden.*

Weber, M. (1958), *The Protestant Ethic and the Spirit of Capitalism*, Talcott Parsons (trans.) First published 1920 (New York: Scribner's).

Welsch, W. (1998), *Undoing Aesthetics* (London: Sage).

Williams, A. (ed.) (1999), *Therapeutic Landscapes: The Dynamic Between Wellness and Place* (Lanham, MD: University Press of America).

Index

(References to illustrations are in **bold**)

Given comments such as these, what should be punctuated is that, contrary to heterosexual stereotypes of bathhouses and indiscriminate 'knocking shops', the clientele of bathhouses can be choosy just like any individual; feeling sexual one day and not the next.

Interaction by Design

Mirroring recent therapeutic landscape research into more conventional settings (Andrews, 2004; Gesler et al., 2004), another line of inquiry focused on the role of design features in the overall place experience. The design and layout features described above in our initial description were very important in facilitating sexual activity and in creating an appropriate setting. Take, for example, the fact that whilst that particular bathhouse was maze-like, others were more regular and therefore more straightforward to navigate. Clients held firm opinions on which types of layout they preferred to navigate and why:

> Someone will walk away and you can't find them, but you have to try and find them. You get to go and look around. But other places are dead end. You see the same thing, and most people stay in their rooms and do nothing. And it's too clean.

Sensory modifications were also perceived to be very important to the overall experience. Whilst one bathhouse purposefully had reduced lighting, another was far more brightly lit. Clients possessed very definite views and preferences for these elements. One specifically commented on the anonymity, ambience and excitement that darkness provided:

> Well, honestly, if you go to the other bathhouse where it's bright and it's, well, for me, I don't like it because it looks like a hospital … It's too bright there. Like, most of the people who come here, they don't want you to recognize their face, they just want to have sex. So that's what a bathhouse is. But, you have to know that mostly at night time they are really busy and they are really drunk or they're really high on drugs and they hate light. So they want to make it a little bit darker, and you know … and darkness is more like adventure.

In addition, music was another purposeful environmental modification, which helped create an appropriate atmosphere (see Andrews and Kearns, 2007). Typically this soundscape was composed of fast synthesized dance music (often techno), though other types of music were used on occasion. A client explained:

> When you have all day at work you want to listen to relaxing music like some piano or classical music to relax you. It's the same thing here, you want to have sex you want to hear something more, what's the word? Music helps a lot to get into the mood. You can't have sex if it's so quiet – too quiet. Especially in the bathhouse.

At another level, the specifically designed theme rooms where favored more or less by clients, for personal reasons. A client commented on a steam room:

> It's fun. You see so many people there – so many hot guys. There's this whole like group sex thing. It's more *exciting*.